OXFORD STATISTICAL SCIENCE SERIES

SERIES EDITORS

OXFORD STATISTICAL SCIENCE SERIES

A.C. Atkinson: *Plots, transformations, and regression*
M. Stone: *Coordinate-free multivariable statistics*
W.J. Krzanowski: *Principles of multivariate analysis: a user's perspective*
M. Aitkin, D. Anderson, B. Francis, and J. Hinde: *Statistical modelling in GLIM*
Peter J. Diggle: *Time series: a biostatistical introduction*
Howell Tong: *Non-linear time series: a dynamical system approach*
V.P. Godambe: *Estimating functions*
A.C. Atkinson and A.N. Donev: *Optimum experimental designs*
U.N. Bhat and I.V. Basawa: *Queuing and related models*
J.K. Lindsey: *Models for repeated measurements*
N.T. Longford: *Random coefficient models*
P.J. Brown: *Measurement, regression, and calibration*
Peter J. Diggle, Kung-Yee Liang, and Scott L. Zeger: *Analysis of longitudinal data*
J.I. Ansell and M.J. Phillips: *Practical methods for reliability data analysis*
J.K. Lindsey: *Modelling frequency and count data*
J.L. Jensen: *Saddlepoint approximations*
Steffen L. Lauritzen: *Graphical models*
A. W. Bowman and A. Azzalini: *Applied smoothing methods for data analysis*
J. K. Lindsey: *Models for repeated measurements* Second edition
Michael Evans and Tim Swartz: *Approximating integrals via Monte Carlo and deterministic methods*

Approximating Integrals via Monte Carlo and Deterministic Methods

MICHAEL EVANS

Department of Statistics
University of Toronto

and

TIM SWARTZ

Department of Mathematics and Statistics
Simon Fraser University

OXFORD
UNIVERSITY PRESS

OXFORD
UNIVERSITY PRESS

Great Clarendon Street, Oxford OX2 6DP

Oxford University Press is a department of the University of Oxford.
It furthers the University's objective of excellence in research, scholarship,
and education by publishing worldwide in

Oxford New York

Athens Auckland Bangkok Bogotá Buenos Aires Calcutta
Cape Town Chennai Dar es Salaam Delhi Florence Hong Kong Istanbul
Karachi Kuala Lumpur Madrid Melbourne Mexico City Mumbai
Nairobi Paris São Paulo Singapore Taipei Tokyo Toronto Warsaw

with associated companies in Berlin Ibadan

Oxford is a registered trade mark of Oxford University Press
in the UK and in certain other countries

Published in the United States
by Oxford University Press Inc., New York

First published 2000

A catalogue record for this book is available from the British Library

Library of Congress Cataloging in Publication Data

Evans, Michael (Michael John)
Approximating integrals via Monte Carlo and deterministic methods/
Michael Evans and Tim Swartz.
p. cm.—(Oxford statistical science series; 20)
Includes bibliographical references and index.
ISBN 0-19-850278-8
1. Calculus, Integral. 2. Approximation theory. 3. Monte Carlo method. I. Swartz, T.
II. Title. III. Series.
QA311 .E92 2000 515′.43—dc21 99-052843

Typeset by Newgen Imaging Systems (P) Ltd., Chennai, India
Printed in Great Britain
on acid-free paper by
Biddles Ltd, Guildford & King's Lynn

Preface

This text is concerned with the numerical approximation of integrals. Various techniques have been developed for this problem. We provide an in-depth survey of the main methods that are used by practitioners. The theory as well as the practice of approximating integrals is discussed. While the book is not exhaustive in its coverage, we have tried to be relatively complete. Basic theoretical facts are derived for each of the methodologies discussed. A reasonably complete development is provided so that the reader is not forced too often to consult other references for basic understanding. It is our opinion that understanding each of the methodologies at a reasonably deep level will help the reader develop intuition as to when the various methods are useful.

We feel that this book is unique in that it covers approximate integration methods that are not usually combined in a single text. This is helpful, not only because we can learn about each of the methods in one place with a reasonably common notation, etc., but because there are similarities amongst the methods. Rather than seeing integration methods as competitive or exclusionary the theme of this text is that the methods are often complementary. For example, it is our view that multiple quadrature methods will never replace the need for Monte Carlo but a purely Monte Carlo approach that fails to recognize and take advantage of the efficiency improvements available with multiple quadrature is not appropriate either. We have also included a reasonably extensive discussion of asymptotic methods. These are commonly used in many fields and, for some problems, they are the method of choice. Again these are not unrelated to other approximate integration methods. For example, a suitable Monte Carlo algorithm may be based in part on the asymptotic behavior of an integrand when considered as a member of a family of integrands indexed by some parameter.

The book varies in the mathematical background required of a reader. Certainly there is a need for a basic understanding of real analysis, complex analysis and some algebra. On the other hand we have tried to keep the mathematical requirements as light as possible while not sacrificing a reasonable degree of rigor. Some background in probability and statistics is essential but not at a particularly high level. Also some prior exposure to elementary numerical analysis would be helpful. The book is aimed at physical and mathematical scientists who face challenging integration problems. Such problems also arise in other fields such as economics and mathematical finance and the coverage of topics is suitable for the kinds of integration problems encountered there. Overall a graduate student, or senior undergraduate student, with a reasonable mathematical background should be able to benefit from reading this text. The book can serve as a text in a course devoted to the numerical approximation of integrals or as part of a course concerned with scientific computing.

To implement the methods, the reader should have some familiarity with a programming language such as C or Fortran. There are other higher order languages and packages that can sometimes be used but it has been our experience that, while some of these offer benefits such as shorter programming time, they do so at the price of reduced flexibility. In particular, it is important that a practitioner have available a full range of numerical software as exemplified by the NAG (1996) or IMSL (1994) packages and this virtually entails using C or FORTRAN. Still, we acknowledge the downside associated with using these languages; namely increased programming and debugging time, and can only hope that some day a more productive platform will be developed.

Each chapter concludes with a set of exercises. For the most part these represent points that were mentioned in the text and were left to the reader to verify. Some exercises are designed to reinforce some relevant concept. In general, we have not included specific integrals to be evaluated feeling it is better if students bring these from their relevant application area.

A web page concerning the text will be maintained concerning updates to the text. A link to this can be found at

http://www.utstat.utoronto.ca/mikevans/

and communication with the authors is encouraged.

Both authors thank Elizabeth Johnston and Mahua Nandi of Oxford University Press for their encouragement and patience. In addition, author Evans thanks Rosemary and Heather and author Swartz thanks Kim, Philippa and Fiona. Both authors were partially supported by grants from the Natural Sciences and Engineering Research Council of Canada and are grateful for this support.

Contents

1
Introduction

1.1 The problem

Many applications require the evaluation of integrals

$$\int_A f(x)\,dx \tag{1.1}$$

where A is a subset of R^k and $f : A \longrightarrow R$. Rarely can an exact answer be easily obtained, even when $k = 1$, so instead an approximation is sought.

Let us be precise at this point with respect to what is meant by (1.1). When f is bounded on A and $A = \prod_{i=1}^{k}[a_i, b_i]$ with $a_i, b_i \in R^1$; i.e. A is a bounded rectangle, we mean the *Riemann integral* of f over A. This is defined as the limit of a sequence of sums

$$\sum_{i=1}^{N} f(x_i)\mathrm{vol}(R_i)$$

where A is partitioned into subrectangles $\{R_1, \ldots, R_N\}$ and $x_i \in R_i$. Then

$$\int_A f(x)\,dx = \lim_{\max\{\mathrm{diam}(R_1),\ldots,\mathrm{diam}(R_N)\}\to 0} \sum_{i=1}^{N} f(x_i)\mathrm{vol}(R_i)$$

where $\mathrm{diam}(B) = \sup\{\|x - y\| \,|\, x, y \in B\}$; i.e. the integral is defined to be the number $\int_A f(x)\,dx$ if for every $\epsilon > 0$ there exists $\delta > 0$ such that whenever a partition satisfies

$$\max\{\mathrm{diam}(R_1), \ldots, \mathrm{diam}(R_N)\} < \delta$$

then

$$\left| \int_A f(x)\,dx - \sum_{i=1}^{N} f(x_i)\mathrm{vol}(R_i) \right| < \epsilon.$$

It can be shown that (1.1) exists if and only if f is continuous on A except perhaps on a subset having volume measure 0. In particular, when f is continuous on A, then (1.1) exists. It can be shown that when (1.1) exists then

$$\int_A f(x)\,dx = \int_{a_k}^{b_k} \cdots \int_{a_1}^{b_1} f(x_1, \ldots, x_k)\,dx_1 \cdots dx_k;$$

i.e. the integral can be evaluated by iterating the standard one-dimensional Riemann integral, and further, this can be done in any order. Note that the Riemann integral is finite whenever it exists.

When f is bounded on A, and A is a bounded subset of R^k contained in a rectangle $R = \prod_{i=1}^k [a_i, b_i]$, then (1.1) is defined as the Riemann integral

$$\int_R f(x) I_A(x)\,dx,$$

when it exists, where

$$I_A(x) = \begin{cases} 1 & \text{if } x \in A \\ 0 & \text{if } x \notin A \end{cases}$$

is the *indicator function* for the set A. This definition can be shown to be independent of the rectangle R.

When f is unbounded on A or A is unbounded, then (1.1) is defined as an *improper Riemann integral*. In this case, (1.1) is defined as

$$\lim_{N\to\infty} \int_{A_N} f(x)\,dx \tag{1.2}$$

when the limits exists and is the same for any increasing sequence $\{A_N\}$ of bounded subsets of A such that f is bounded on each A_N, $\int_{A_N} f(x)\,dx$ exists and any subset of A on which f is bounded is contained in some A_N. Note that we allow for the possibility that (1.1) is infinite. Whenever (1.1) exists and is finite we say that f is *integrable* over A. Determining the integrability of an integrand is an important practical problem and we discuss this in Chapter 2. It is worth noting that whenever the integrand f is nonnegative and for some sequence $\{A_N\}$ satisfying the conditions specified above, the limit (1.2) exists, then it can be shown that this is the value of the improper Riemann integral. Also, if f and g have the same infinite discontinuity points and $|f| \le g$ where $\int_A g(x)\,dx$ exists and is finite, then (1.1) exists and is finite. Note that we can often apply this result where $g = |f|$; i.e. if we can show that

$$\lim_{N\to\infty} \int_{A_N} |f(x)|\,dx$$

exists and is finite for some appropriate sequence, then we have established the existence and finiteness of (1.1).

There is much more that can be said about the mathematical theory of the Riemann integral, and in fact many details that we have omitted, but this is not our interest here. Rather, as mentioned we are concerned with obtaining a numerical approximation to (1.1). Actually, the mathematical theory outlined above is all that we will require. See Williamson, Crowell and Trotter (1968) for more discussion of the Riemann and improper Riemann integrals and in particular for proofs of the statements made here.

Sometimes when approximating integrals, high accuracy is required, while in other situations, only a relatively low level is needed. Actually, in many applications a low level of accuracy is not only acceptable but preferable as it may be spurious to report otherwise. For example, in statistical applications, the integrand f often depends upon sampled data and therefore the error in the approximation need not be much smaller than the error in the true value of the integral due to sampling variation in the data.

On the other hand, integral expressions that are used in diverse contexts (e.g. the gamma function) should be evaluated efficiently and with a high degree of accuracy. This type of problem tends to be much more demanding but perhaps doesn't arise as frequently.

These comments concerning accuracy indicate that different algorithms are suitable in different contexts. It does not make sense to reject a class of algorithms simply because they give low level accuracy or because they are computationally expensive. For example, Monte Carlo methods are sometimes criticized as not providing sufficient accuracy but in many problems they are entirely suitable and even preferable. By contrast, although Monte Carlo methods can be applied to most integration problems, this does not mean that they are always suitable. In particular, when high accuracy is required, they are not appropriate. In any problem, it is therefore important that the practitioner determine the level of accuracy required.

We refer to the problem of evaluating (1.1) as the *integration problem*. Accordingly, we define what is meant by a "solution" to a particular problem. For us, an algorithm provides a solution if it can return an approximation together with an exact error bound and can reduce the error bound to a prescribed level of accuracy by increasing the amount of computation. Of course, this must be achievable within acceptable computation times. Some may argue that the requirement of an exact error bound is too stringent but if we are to have confidence in an approximation this seems like a necessity. Most of the approximation methods discussed in the text do not meet this requirement as weaker criteria for assessing error are typically employed. These weaker error criteria are subject to error and the fact that we can be substantially misled must be acknowledged because this does happen. Although weaker error criteria are not ideal, we gain confidence in a result when the answer is obtained using several approximations.

We do not provide a solution to the general integration problem in this text and in fact it is fair to say that a solution to the general problem does not exist. Rather, as indicated, we compromise on the conditions for a solution in a particular problem and use weaker error criteria that are designed to give the practitioner some confidence in the accuracy of the approximation. The methods of approximation are characterized to a degree by their approach to error analysis. We discuss this at greater length in Section 2.6.

There continues to be a steady development of more powerful and useful methods of integral approximation. Much of this is in conjunction with the rapid development of computer hardware as more computationally intensive methods become more feasible. Modern integration methods make full use of the raw computational power available to attempt to solve previously intractable problems.

1.2 The dimensional effect

One of the characteristics of the integration problem that contributes to its difficulty is the so-called *dimensional effect*, or the *curse of dimensionality*. The dimensional effect is the phenomenon whereby virtually all integration methods deteriorate rapidly in performance as the dimension k rises. In fact, for some methods, their use is inconceivable beyond a fairly low dimension because of an exponential increase in the number of operations necessary to fully implement the approximation. Sometimes it is claimed that Monte Carlo methods do not suffer from the dimensional effect, but strictly speaking, this is not true as it becomes increasingly difficult to determine the relevant characteristics of the integrand f and the domain A as the dimension rises. The curse of dimensionality is so central to the integration problem that we begin our discussion of it right away via a simple example.

Example 1.1 *The volume of a sphere*
Consider the problem of evaluating the volume of the unit sphere $S^{k-1} = \{x | x_1^2 + \cdots + x_k^2 \leq 1\}$ in R^k. This can be represented as the integral

$$\int_{S^{k-1}} dx_1 \cdots dx_k \tag{1.3}$$

which is well known to equal

$$\frac{\pi^{k/2}}{\Gamma\left(\frac{k}{2}+1\right)}.$$

Still, considering various methods for evaluating (1.3) illustrates the impact of the dimensional effect. Noting that the unit sphere is contained in the rectangle $[-1, 1]^k$, we can write (1.3) as

$$\int_{[-1,1]^k} I_{S^{k-1}}(x_1, \ldots, x_k)\, dx_1 \cdots dx_k.$$

A number of approaches to evaluating such an integral involve uniformly spreading points over $[-1, 1]^k$ and then averaging the integrand over these points. This could in fact be a weighted average. One method is to simply subdivide each axis into

m equal length subintervals and then compute the approximation

$$\frac{1}{m^k}\sum_{i_1=1}^{m}\cdots\sum_{i_k=1}^{m} I_{S^{k-1}}\left(-1+2\frac{i_1-1}{m-1},\ldots,-1+2\frac{i_k-1}{m-1}\right). \tag{1.4}$$

It certainly seems reasonable that we should treat each axis symmetrically and that the number of subdivisions along an axis should not decrease as we increase dimension. We observe then that this approximation involves an exponential increase in the number of function evaluations since m^k evaluations are required. For example, when $m = 10$ and $k = 10$, then 10 billion function evaluations are required and this is infeasible with current technology. In fact, it is very likely that $m = 10$ is too small to give a reasonable approximation.

This naive approach to uniformly distributing the points over $[-1, 1]^k$ is clearly infeasible even for relatively modest dimensions. More sophisticated methods have therefore been developed. One of these is to randomly place a point in $[-1, 1]^k$ in such a way that the point falls in regions of equal volume with equal probability; i.e. the point is uniformly distributed in $[-1, 1]^k$. We then independently generate N uniformly distributed points in $[-1, 1]^k$, say $X_1, \ldots, X_N$, and compute the average

$$\frac{1}{N}\sum_{i=1}^{N} I_{S^{k-1}}(X_{i1}, \ldots, X_{ik}).$$

In this case, we are in control of the number of function evaluations N. Observe the following, however. The volume of $[-1, 1]^k$ is 2^k, and therefore the ratio of the volume of the unit sphere to the rectangle is given by

$$\frac{\pi^{k/2}}{\Gamma\left(\frac{k}{2}+1\right)2^k}. \tag{1.5}$$

Using Stirling's approximation to the gamma function; namely

$$\Gamma(x) \sim \sqrt{2\pi}e^{-x}x^{x-1/2}$$

as $x \to \infty$, we have that

$$\begin{aligned}
\lim_{k\to\infty}\frac{\pi^{k/2}}{\Gamma\left(\frac{k}{2}+1\right)2^k} &= \lim_{k\to\infty}\frac{\pi^{k/2}}{\sqrt{2\pi}e^{-k/2-1}\left(\frac{k}{2}+1\right)^{k/2+1/2}2^k} \\
&= \lim_{k\to\infty}\left[\frac{e}{\sqrt{2\pi}\sqrt{\frac{k}{2}}\left(1+\frac{2}{k}\right)^{k/2+1/2}}\right]\left(\frac{e\pi}{2k}\right)^{k/2} \\
&= 0.
\end{aligned}$$

Table 1.1. Values of the proportion (1.5) as a function of the dimension k.

k	Proportion
1	1.00×10^{0}
2	7.85×10^{-1}
3	5.24×10^{-1}
4	3.08×10^{-1}
5	1.64×10^{-1}
10	2.49×10^{-3}
20	2.46×10^{-7}
50	1.54×10^{-27}
100	1.87×10^{-69}
1000	2.87×10^{-1186}

Therefore the volume of the unit sphere is a vanishingly small proportion of the volume of the rectangle $[-1, 1]^k$ as the dimension k rises. Table 1.1 gives some values of (1.5). By dimension $k = 5$ the unit sphere is occupying less than 20% of the volume of $[-1, 1]^k$ and by $k = 10$, it is an almost immaterial proportion of this set. Contrast this with the situation when $k = 1, 2$ and 3.

The implication of this for the integration technique based on uniform sampling is that $I_{S^{k-1}} = 0$ for most samples. In fact, in high dimensions, it is conceivable that we may entirely miss the region where $I_{S^{k-1}}$ is non-zero with a random placement of points; note that when $k = 50$ the probability that $I_{S^{k-1}}$ is nonzero is 1.54×10^{-27}. Therefore, even a random placement of points fails unless properties of the integrand are taken into account. From a certain point of view one may argue that 0 is not a bad approximation to (1.5). On the other hand, suppose that we want to calculate the expectation of a function with respect to the uniform density on the sphere. This problem can be expressed as a ratio of integrals where the ratio need not be anywhere near 0. In this case, a poor approximation of the denominator (e.g. 0) adversely affects the approximation of the ratio.

It is fair to say that taking into account the characteristics of the integrand is overwhelmingly more important than the specific choice of integration technique. Unless we take advantage of the structure of the integrand, we are almost certainly doomed to failure as the dimension rises. This example exhibits that the dimensional effect is more than just the exponential growth in the number of operations required for the implementation of a method.

In a sense, the dimensional effect is "the" problem in approximating integrals. Whereas for the most part we feel comfortable approximating integrals in low dimensions like $k = 1, 2$ and 3, we begin to lose our geometric intuition as the dimension increases. Example 1.1 illustrates that higher dimensions have much more "space" than perhaps our intuition permits. The effects of an integrand concentrating in a small region, or having a skewed shape or some other feature that would perhaps be only slightly significant for the success of an integration method in low dimensions can have enormous implications for such success in high-dimensional problems. In general these effects are very difficult to assess.

1.3 Examples

The approximation of integrals is a common problem in the physical sciences that results from the application of certain mathematical theories. It is worth pointing out that while we are concerned with approximating integrals over subsets of Euclidean spaces with respect to Lebesgue measure, some of the methods are also applicable to the approximation of sums. This is not surprising given that a sum can be represented as an integral with respect to counting measure. Interestingly, it is with the Markov chain methods discussed in Chapter 7 where this similarity is most apparent. In any case, successful application of the mathematical theory in question typically requires good approximation techniques.

The necessity of evaluating integrals is endemic to the applications of probability and statistics because basic quantities such as the probability of an event and the expected value of a random variable are expressed as integrals. Physics and its applications are also a healthy source of intractable integrals. For example, applications of statistical mechanics often involve the evaluation of very high dimensional integrals and sums. In financial mathematics, the application of various pricing strategies involves the evaluation of integrals. In fact, it is fair to say that intractable integrals arise in almost any field where mathematics is applied.

We discuss two specific classes of problems and refer to these throughout the book.

Example 1.2 *Expectations with respect to the multivariate normal distribution* The multivariate normal distribution is central to the fields of probability and statistics. It also plays a key role in constructing approximations to integrals. Accordingly, it is important to be familiar with this probability distribution. Reading this example and doing Exercises 7–15 at the end of this chapter should provide sufficient background knowledge. Interestingly, the multivariate normal not only plays a key role in constructing various integration strategies, but it is also a source of many intractable integrals arising as expectations with respect to this distribution. This example also introduces some key concepts and terminology from the fields of probability and statistics that are used throughout the text. As stated in the Preface, we are assuming that the reader has at least some background in probability and statistics; Casella and Berger (1990) is recommended as a good reference source.

A random vector $X \in R^k$ follows the *standard multivariate normal* distribution if it has density given by

$$
\begin{aligned}
f(x) &= f(x_1, \ldots, x_k) \\
&= (2\pi)^{-k/2} \exp\left(-\tfrac{1}{2}x'x\right) \\
&= (2\pi)^{-k/2} \exp\left(-\tfrac{1}{2}\sum_{i=1}^{k} x_i^2\right) \\
&= \prod_{i=1}^{k} (2\pi)^{-1/2} \exp\left(-\tfrac{1}{2}x_i^2\right).
\end{aligned}
$$

Note that f is the product of the density functions for k independent standard normal densities on R^1; i.e. X_i is distributed with density given by $(2\pi)^{-1/2}\exp\left(-\frac{1}{2}x_i^2\right)$. So given (Exercise 7) that

$$\int_{-\infty}^{\infty} (2\pi)^{-1/2} \exp\left(-\tfrac{1}{2}x^2\right) dx = 1$$

we see immediately that

$$\int_{R^k} f(x)\, dx = 1.$$

Since $f \geq 0$ this establishes that f is indeed a probability density function; i.e. the probability that the random vector X falls in the set A is given by

$$P(A) = \int_A f(x)\, dx_1 \cdots dx_k$$

which is in general the relevant formula for computing this probability for an arbitrary density f on R^k.

The *expectation* of a function $h : R^k \to R$ with respect to a probability density f on R^k is given by

$$E[h(X)] = \int_{R^k} h(x) f(x)\, dx$$

provided that

$$\int_{R^k} |h(x)| f(x)\, dx < \infty$$

and otherwise the expectation does not exist. The value $E[h(X)]$ is also called the *mean* of the random variable $h(X)$. Throughout the remainder of this example we assume that all discussed expectations exist.

Now consider the univariate case. The mean $E[Z]$ of a random variable Z is a measure of the location of the distribution of Z. Note that the special expectation

$$\mathrm{Var}[Z] = E[(Z - E[Z])^2] = E[Z^2] - (E[Z])^2$$

is called the *variance* of the random variable Z and is a measure of the spread of the distribution about its mean $E[Z]$. The smaller the variance, the more tightly concentrated is the distribution of Z about its mean. The quantity

$$\mathrm{Sd}[Z] = \sqrt{\mathrm{Var}[Z]}$$

is known as the *standard deviation* of Z and when $E[Z] \neq 0$

$$\mathrm{CV}[Z] = \frac{\mathrm{Sd}[Z]}{E[Z]}$$

is called the *coefficient of variation* of Z. The absolute coefficient of variation $|\mathrm{CV}[Z]|$ is perhaps the most relevant measure of the amount of variability in a random variable Z as it is independent of scale changes; i.e. if $Y = cZ$ for $c \neq 0$ and $E[Z] \neq 0$, then $|\mathrm{CV}[Y]| = |\mathrm{CV}[Z]|$.

The *covariance* of two random variables (Y, Z) is a measure of how the variables vary together and is defined by

$$\mathrm{Cov}[Y, Z] = E[(Y - E[Y])(Z - E[Z])].$$

Note that $\mathrm{Cov}[Y, Z] = \mathrm{Cov}[Z, Y]$ and $\mathrm{Cov}[Y, Y] = \mathrm{Var}[Y]$. A location-scale invariant measure (for positive rescalings) of the extent to which an affine relationship exists between two random variables is given by the *correlation coefficient*

$$\mathrm{Corr}[Y, Z] = \frac{\mathrm{Cov}[Y, Z]}{\mathrm{Sd}[\mathrm{Y}]\mathrm{Sd}[Z]}.$$

From the Cauchy–Schwarz inequality (think of $E[YZ]$ as an inner product defined on the space of all random variables satisfying $E[Y^2] < \infty$) we have that

$$-1 \leq \mathrm{Corr}[Y, Z] \leq 1$$

and $\mathrm{Corr}[Y, Z] = 1$ if and only if $P(Y = a + bZ) = 1$ for constants a, b with $b > 0$. Similarly, $\mathrm{Corr}[Y, Z] = -1$ if and only if $P(Y = a + bZ) = 1$ for constants a, b with $b < 0$. When Y and Z are independent; i.e. their joint density factors as the product of their marginal densities, then $\mathrm{Corr}[Y, Z] = 0$. However, the converse of this statement is not true in general.

For the standard multivariate normal we have that (Exercise 8) $\mathrm{Corr}[X_i, X_j] = \delta_{ij}$ where $\delta_{ij} = 1$ when $i = j$ and $\delta_{ij} = 0$ when $i \neq j$. Notice, however, that $\mathrm{CV}[X_i]$ is not defined in this case.

For a general random vector $X \in R^k$, the vector

$$E[X] = \begin{pmatrix} E[X_1] \\ E[X_2] \\ \vdots \\ E[X_k] \end{pmatrix}$$

is called the *mean vector*, and the matrix

$$\text{Var}\,[X] = \begin{pmatrix} \text{Cov}[X_1, X_1] & \text{Cov}[X_1, X_2] & \cdots & \text{Cov}[X_1, X_k] \\ \vdots & & & \vdots \\ \text{Cov}[X_k, X_1] & \text{Cov}[X_k, X_2] & \cdots & \text{Cov}[X_k, X_k] \end{pmatrix}$$

is called the *variance matrix* or the *variance–covariance matrix*. Thus we have extended the expectation and variance operators to appropriate spaces of random vectors. We can show (Exercise 13) that for $a \in R^k$ and $C \in R^{k\times k}$,

$$E[a + CX] = a + CE[X]$$

and

$$\text{Var}[a + CX] = C\text{Var}[X]C^t.$$

Therefore, the zero vector $0 \in R^k$ is the mean vector of the standard multivariate normal distribution and the variance matrix is given by the identity matrix $I \in R^{k\times k}$. Hereafter, when the random vector $X \in R^k$ is distributed according to the standard multivariate normal distribution, we write $X \sim N_k(0, I)$.

The general multivariate normal distribution is obtained by transforming a standard multivariate normal vector X by an affine transformation

$$Y = \mu + CX$$

where $\mu \in R^k$ and $C \in R^{k\times k}$ is invertible. Then, using the multivariate change of variable formula, the expectation of $h(Y) = h(\mu + CX)$ is given by

$$\begin{aligned} E[h(Y)] &= E[h(\mu + CX)] \\ &= \int_{R^k} h(\mu + Cx)(2\pi)^{-k/2} \exp\left\{-\tfrac{1}{2}x^t x\right\} dx \\ &= \int_{R^k} h(y)(2\pi)^{-k/2} |\det C|^{-1} \exp\left\{-\tfrac{1}{2}(y-\mu)^t (CC^t)^{-1}(y-\mu)\right\} dy \\ &= \int_{R^k} h(y)(2\pi)^{-k/2} |\det \Sigma|^{-1/2} \exp\left\{-\tfrac{1}{2}(y-\mu)^t \Sigma^{-1}(y-\mu)\right\} dy \end{aligned}$$

where $\Sigma = CC^t$. Therefore, the density of $Y \in R^k$ is given by

$$(2\pi)^{-k/2} |\det \Sigma|^{-1/2} \exp\left\{-\tfrac{1}{2}(y-\mu)^t \Sigma^{-1}(y-\mu)\right\}. \tag{1.6}$$

Using the properties of the expectation and variance operators, we have that $E[Y] = \mu$ and $\text{Var}[Y] = CIC^t = CC^t = \Sigma$, and we write $Y \sim N_k(\mu, \Sigma)$.

Note that Σ is a positive definite matrix and there are many matrices (square roots of Σ) that give rise to it. In fact, $\Sigma = CC^t = (CQ)(CQ)^t$ for any orthogonal matrix $Q \in R^{k\times k}$.

As mentioned earlier, there exist integrals based on the multivariate normal distribution that cannot be evaluated in closed form and which are important for applications. For example, suppose that $X \sim N_k(\mu, \Sigma)$ and h is the indicator function for the rectangle $R = \prod_{i=1}^k [a_i, b_i]$, where $a_i, b_i \in R^1 \cup \{-\infty, \infty\}$. Then

$$\begin{aligned} E[h(X)] &= \int_{R^k} h(x) f(x)\, dx \\ &= \int_{a_k}^{b_k} \cdots \int_{a_1}^{b_1} (2\pi)^{-k/2} |\det \Sigma|^{-1/2} \\ &\quad \times \exp\left\{-\tfrac{1}{2}(x-\mu)^t \Sigma^{-1}(x-\mu)\right\} dx_1 \cdots dx_k \end{aligned}$$

and, except for particular choices of the limits, this cannot be evaluated in closed form. When all the $a_i = -\infty$, this integral is the *distribution function* of the $N_k(\mu, \Sigma)$ distribution evaluated at $(b_1, \ldots, b_k)$. When $(a_1, \ldots, a_k)^t = \mu$ and $(b_1, \ldots, b_k) = (\infty, \ldots, \infty)$, this integral is called an *orthant probability*. Note that when $k = 1$, $\mu = 0$ and $\Sigma = \sigma^2 = 1$, then $X \sim N(0, 1)$ and the distribution function at x is

$$\Phi(x) = \int_{-\infty}^{x} \phi(z)\, dz$$

where

$$\phi(z) = (2\pi)^{-1/2} \exp\left\{-\tfrac{1}{2}z^2\right\}.$$

Even in this simple case, we cannot evaluate Φ without approximating the integral. In some fields, it is common to consider the *error function*

$$\begin{aligned} \mathrm{Erf}(x) &= \sqrt{2}[2\Phi(x) - 1] \\ &= 2\sqrt{2} \int_0^x \phi(z)\, dz \end{aligned}$$

defined for $x \geq 0$. Therefore $\mathrm{Erf}(x)$ is $\sqrt{2}$ times the probability that $X \sim N(0, 1)$ lies in the interval $[-x, x]$.

These integrals represent fairly challenging problems for any integration approach partly because of the lack of smoothness introduced into the problem by the indicator function. In general, however, provided that the region where f is dominant is roughly the same region where h is nonzero, then we will see that there are often good algorithms for evaluating $E[h]$.

Example 1.3 *Bayesian statistics*
One of the most fertile grounds for producing intractable integrals is the application of Bayesian inference. In fact, a criticism that use to be made of this methodology was that it could not be applied in general because of the intractability of many of the resultant integrals. With the development of better integration techniques over the past several years, this concern has dissipated. We describe some of the basics of this theory, and in particular, we describe how the integrals arise. The references Flournoy and Tsutakawa (1991) and Evans and Swartz (1995a) contain discussions of approaches to approximating integrals in this context.

The basic model for a statistical problem arises as follows. We observe data $x \in \mathcal{X}$ which is the realization of a random quantity $X \in \mathcal{X}$. What makes this a statistical problem is that we don't know the probability measure P on $\mathcal{X}$ that describes the variation in X. In fact, our purpose is to make an *inference* about the true P based on the observation x and some assumptions concerning P. These assumptions state that the true P lies in some class of probability measures on $\mathcal{X}$ denoted by

$$\{P_\theta | \theta \in \Omega\}$$

where θ is a variable indexing the class. This class is called the *statistical model*, θ is called the *parameter* of the model and Ω is called the *parameter space*. It is common to restrict the class of possible probability measures so that efficient inferences are made. In most problems, the parameter space is a "nice" subset of R^k and the probability measure P_θ has a density f_θ on $\mathcal{X}$ with respect to some support measure υ on $\mathcal{X}$. A Bayesian model then adds to these ingredients a *prior* probability measure Π which has a density π with respect to Lebesgue measure on Ω (we restrict our attention to this situation). Bayesian inference then proceeds by looking at the *posterior distribution* of θ which is the conditional distribution of θ given $X = x$. Since the joint distribution of (θ, X) has density

$$f_\theta(x)\pi(\theta)$$

(f_θ is the conditional density of X given θ), it follows that the posterior distribution of $\theta \in \Omega$ has density

$$\frac{f_\theta(x)\pi(\theta)}{\int_\Omega f_\theta(x)\pi(\theta)\, d\theta}.$$

Notice that the integral in the denominator

$$\int_\Omega f_\theta(x)\pi(\theta)\, d\theta \tag{1.7}$$

is the marginal density of X evaluated at x. The inverse of (1.7) is referred to as the *normalizing constant* and the posterior density is known provided that we can evaluate this quantity. Inference is then conducted by computing posterior expectations

$$\frac{\int_\Omega m(\theta) f_\theta(x)\pi(\theta)\, d\theta}{\int_\Omega f_\theta(x)\pi(\theta)\, d\theta} \tag{1.8}$$

for various functions $m : \Omega \to R^1$. Note that these posterior expectations are ratios of integrals.

Even simple choices of f_θ and π can result in intractable integrals. Moreover, it is rarely a good idea to make these choices solely to obtain an integral that is easy to evaluate. Realistic Bayesian modeling almost always results in intractable integrals. Clearly, good approximation methods are essential for the successful application of Bayesian inference.

Note that when π is a probability density function, then (1.7) is finite with probability 1 with respect to the distribution of X since

$$\int_{\mathcal{X}} \left\{ \int_\Omega f_\theta(x)\pi(\theta)\, d\theta \right\} \nu\,(dx) = \int_{\mathcal{X}\times\Omega} f_\theta(x)\pi(\theta)\, \nu(dx)\, d\theta = 1.$$

On the other hand, it is common amongst some practitioners of Bayesian inference to use priors π that are nonnegative but where $\int_\Omega \pi(\theta)\, d\theta = \infty$. These are called *improper priors* and are used because some believe that they are an appropriate characterization of ignorance. When improper priors are used, we have no guarantee that (1.7) is finite, and without this, the posterior distribution is undefined. Therefore, we must always check that (1.7) is finite when using improper priors. This may seem obvious, but it is rarely straightforward to verify. Moreover, as we will see in Chapter 7, there are methods of approximating (1.8) that do not require the approximation of (1.7). If (1.7) is not finite, then these methods are invalid. Of course, in general there is no guarantee that

$$\int_\Omega m(\theta) f_\theta(x)\pi(\theta)\, d\theta$$

is finite so this must also be checked, even when using proper priors.

We make one final comment which concerns *nonidentifiability* in Bayesian integration problems. This arises when the parameter θ does not index; i.e. $P_{\theta_1} = P_{\theta_2}$ for certain distinct values $\theta_1, \theta_2 \in \Omega$. This situation can create special problems in integration and we should be aware of nonidentifiability when it occurs. To a certain extent, nonidentifiability is avoided in Bayesian problems whenever $\pi(\theta_1) \neq \pi(\theta_2)$ but computational problems may still arise particularly when the amount of data x is large.

1.4 Outline of the book and additional relevant references

This book is a substantial expansion of the review paper by Evans and Swartz (1995a). We amplify on that material by providing more background and more discussion of the various methodologies.

In Chapter 2, we introduce some basic concepts that are relevant to all of the various integration methods. In particular, we introduce the important topic of error analysis. In Chapter 3, we discuss the most useful algorithms for generating variates from probability distributions. The material of this chapter underlies all of the Monte Carlo methods of integration and is essential for application of the approach. In Chapter 4, we begin the discussion of the integration techniques with a presentation on asymptotic approximations. This methodology can often be implemented without the use of a computer, and historically an extensive amount of development has gone into these methods. In particular, the Laplace method is seen as being of central importance. In Chapter 5, we develop multiple quadrature techniques which are the direct analogues of Riemann sums. In Chapter 6, we discuss the Monte Carlo method known as importance sampling together with the various variance reduction techniques associated with it. In Chapter 7, we present Markov chain Monte Carlo methods. At the end of each chapter we provide a set of exercises that test understanding of the material and add elements to the discussion that are relevant but not essential.

Approximately half of our discussion on integration methods is on Monte Carlo methods and half is on nonrandom methods. This seems like an appropriate ratio to us but others may disagree. If you are typically concerned with very high-dimensional integrals, then perhaps only the Monte Carlo methods are of value to you, and especially the Markov chain methods. Alternatively, if you need highly accurate approximations to relatively low-dimensional integrals, then Chapters 4 and 5 are the most relevant. However, we encourage the reading of all of the chapters as the methods are by no means independent of each other. Often, insights gained in one context can be transferred to another.

There are many other books on approximate integration. We mention many of these at relevant places in the text but there are a few that can be singled out for special mention. For many of the variate generation techniques presented in Chapter 3, see Devroye (1986) and Gentle (1998). For the asymptotic approximations discussed in Chapter 4, the book by Wong (1989) is an excellent and comprehensive source. Davis and Rabinowitz (1984) is an important work primarily for the quadrature methods of Chapter 5. This material has been substantially updated in Krommer and Ueberhuber (1998). For the Monte Carlo methods discussed in Chapters 6 and 7, the book by Hammersley and Handscomb (1964) is still relevant but see also Gilks, Richardson and Spiegelhalter (1996), Fishman (1996) and Robert and Casella (1999). For general background on numerical methods, the texts by Dahlquist and Björck (1974), Kennedy and Gentle (1980), Thisted (1988) and Gautschi (1997) are recommended.

We feel that one of the main virtues of this book is the fact that all the methods are discussed in one text. In our opinion, this permits easier learning as the style and notation of our discussion does not vary much from one topic to another. Further, from a practical point of view, it is our opinion that all of the methods that we discuss should be part of the practitioner's toolkit. Another feature of the text is that we have tried to present a complete development with very few gaps.

Therefore, the text serves not only as a reference on integration techniques but also as a self-contained document that facilitates learning.

1.5 Exercises

(1) Evaluate the integral $\int_0^1 x\,dx$ directly from the definition of the Riemann integral. Hint: since the integrand is continuous we know that the integral exists.

(2) Evaluate the integral $\int_1^\infty (1/x^2)\,dx$ directly from the definition of the improper Riemann integral.

(3) Show that $\int_{-1}^1 (1/x^3)\,dx$ is not improperly Riemann integrable. What is the value of $\lim_{\epsilon\to 0}\int_{[-1,-\epsilon]\cup[\epsilon,1]}(1/x^3)\,dx$?

(4) Given

$$f(x) = \begin{cases} \dfrac{(-1)^i}{i} & \text{if } i \le x < i+1,\, i \in N_0 \\ 0 & \text{if } x < 0, \end{cases}$$

determine whether or not $\int_0^\infty f(x)\,dx$ exists.

(5) Evaluate (1.4) and compare it to the exact value of the volume of the sphere for various choices of k and m.

(6) What is the variance of the Monte Carlo estimator of the volume of the sphere discussed in Section 1.2 based on N function evaluations?

(7) Establish that

$$\int_{-\infty}^{\infty} (2\pi)^{-1/2} \exp\left(-\tfrac{1}{2}z^2\right) dz = 1.$$

Hint: consider the integral

$$\int_{-\infty}^{\infty}\int_{-\infty}^{\infty} \exp\left\{-\tfrac{1}{2}(z_1^2+z_2^2)\right\} dz_1\,dz_2$$

and make the change of variable $(z_1, z_2) \to (r, \theta)$ where

$$z_1 = r\cos\theta \qquad z_2 = r\sin\theta.$$

(8) Prove that when $X \sim N_k(0, I)$ then $E[X_i] = 0$ and $\mathrm{Cov}[X_i, X_j] = \delta_{ij}$. Note that you should also justify the existence of the relevant integrals.

(9) Prove that the variance of a random variable is invariant under location changes. How does it behave with respect to scale changes?

(10) Prove that the absolute coefficient of variation is invariant under scale changes. How does it behave with respect to location changes?

(11) Establish that the density of $Y = \mu + CX$ is given by (1.6) when $X \sim N_k(0, I)$, $\mu \in R^k$ and $C \in R^{k\times k}$ is invertible.

(12) Prove that $X \sim N_k(0, I)$ is rotationally symmetric; i.e. $Y = QX \sim N_k(0, I)$ for every orthogonal $Q \in R^{k \times k}$.

(13) Prove that the expectation operator E has the property $E[a + CX] = CE[X]$ for the random vector $X \in R^k$ and the variance operator Var has the property $\text{Var}[a + CX] = C\text{Var}[X]C^t$.

(14) Show that if X_1 and X_2 are statistically independent random variables with densities f_1 and f_2 respectively then

$$E[h_1(X_1)h_2(X_2)] = E[h_1(X_1)]E[h_2(X_2)]$$

for any functions h_1 and h_2 provided that these expectations exist.

(15) Suppose that $(X_1, X_2) \sim N_2(\mu, \Sigma)$. Show that X_1 and X_2 are statistically independent if and only if $\text{Cov}[X_1, X_2] = 0$.

(16) Suppose that $X = (X_1, \ldots, X_n)^t \sim N_n(\mu 1, I)$ where $1 \in R^k$ is the vector of 1's. Further suppose that μ is unknown and has prior distribution $N(\mu_0, \sigma_0)$. Determine the form of the posterior density of μ.

2
Some basic concepts

2.1 Introduction

Before we start on a discussion of the various integration techniques there are some topics of a fairly general nature that should be considered. For example, simple exact expressions can sometimes be obtained for integrals that eliminate the need for more elaborate approximation techniques. While this situation is relatively rare, one should always be aware of the possibility. We consider this briefly in Section 2.2. Also, even though we discuss various approximations in ideal mathematical circumstances, the implementations of such approximations involve the computer and this results in further approximations. Therefore, in Section 2.3, we discuss floating point arithmetic and some elements of error analysis as they pertain to integration problems. For a more general discussion of error in numerical problems, see Dahlquist and Björck (1974), Kennedy and Gentle (1980), Thisted (1988), and Gautschi (1997). In addition, the success of most integration problems is highly dependent on the parameterization of the integral. By a *reparameterization* of an integral we mean the transformation of the integrand to a new variable of integration. Sometimes, a good reparameterization can spell the difference between success and failure or result in a substantial savings in computational time. We discuss reparameterization in Section 2.4. Finally, in Section 2.5, we mention some of the available software packages and the integration capabilities that they contain.

One further topic could be included in this chapter, as it underlies all Monte Carlo techniques; namely, algorithms for sampling from distributions. Because of its importance, however, we have devoted Chapter 3 entirely to this topic.

2.2 Exact or symbolic integration

In some important cases

$$\int_A f(x)\,dx \tag{2.1}$$

can be evaluated in "closed" form. Typically, this means that (2.1) can be expressed as some algebraic expression involving functions that are simple to evaluate although it is possible to be more precise with the definition. From a certain point of view, one could say that the integration problem is solved in such a case. Therefore it would seem apparent that an exact integration method is to be preferred if available. However, this is not an accurate characterization of the situation. For example, if we want to evaluate

$$\int_0^1 (x^2 + 2x + 1)^{20} \, dx \tag{2.2}$$

then an exact expression can be worked out for the antiderivative of the integrand and for which we substitute the limits of integration to calculate (2.2). To five significant figures, using MAPLE, this approach gives the value 5.3635×10^{10} for (2.2) while a simple approximation technique (Simpson's rule based on 100 intervals—see Chapter 5) gives the same answer. In this case, the approximate method is probably preferred as it avoids complex expressions; namely the antiderivative is a polynomial of degree 41 with some very large coefficients. It is questionable what value the exact methods are adding in such a situation.

Still, we would like to avoid truly unnecessary approximations when it is simpler to evaluate the integral exactly. How do we recognize such contexts? While experience plays a significant role, there is software available to perform symbolic integration in certain situations. In particular, the MAPLE software package, see Char *et al.* (1992), and the MATHEMATICA software package, see Wolfram (1996), contain symbolic integration routines. While these are useful in certain contexts, they can in no way be regarded as a solution to one-dimensional integration problems. This is because symbolic methods only work in limited situations and because they can give unreliable results in contexts where we believe they should be appropriate; e.g. see Krommer and Ueberhuber (1998) for examples of computing integrals of rational functions with integer coefficients where several software packages fail to give correct answers. Also, these algorithms are generally not helpful in dealing with multidimensional problems. Overall, our conclusion is that while symbolic methods have value and hold promise for the future, even in one dimension, they do not replace the need for numerical approximations. For an introduction to these methods, and a discussion of the difficulties inherent in their implementation, see Krommer and Ueberhuber (1998). For a more detailed discussion, see Bronstein (1997).

2.3 Integrability

It may seem obvious but perhaps the first thing that we should check is whether (2.1) is finite. Of course, if f is continuous on a compact set A then indeed the integral is finite, but in general the finiteness of (2.1) is not always easy to verify. Basically, the finiteness of (2.1) depends on the behavior of f near its singularities, and if A is unbounded, the finiteness of (2.1) also depends on the tail behavior of f. Typically, a lack of integrability will reveal itself in instability of the approximations as we demand greater accuracy; e.g. as we increase the sample size in Monte Carlo algorithms. While theoretically the instability must occur, in practice, it may take a substantial amount of computation before it becomes apparent. Therefore, if it is not feasible to verify the integrability of the integrand, great caution should be observed.

We take it as given that we are only considering situations where (2.1) is finite. Let $x_0 \in \bar{A}$; i.e. x_0 is a point in A or is a limit point of A. Then x_0 is a *singularity* of f if

$$\lim_{x \to x_0} f(x) = \pm\infty$$

or some derivative of f doesn't exist at x_0 where $f(x_0)$ is defined. To examine the infinite singularity x_0 of f, we look for a function g known to be integrable over a deleted neighborhood $B_r(x_0)\setminus\{x_0\} = \{x : 0 < \|x - x_0\| < r\}$, for some $r > 0$, and such that $|f| \le |g|$ on this set. To examine the tail behavior of f, we look for a compact set $B \subset A$ and a function g known to be integrable on $B^c \cap A$ and such that $|f| \le |g|$ on this set. In the case where we can find such functions, we have verified the integrability of f and can proceed to approximate its integral over A.

There are no general methods for obtaining the dominating functions g. We note, however, that the envelope methods discussed in Sections 3.6 and 5.3 can sometimes be helpful. Part of this involves finding an invertible transformation T such that $T \circ f$ is concave (or convex) on a set. In such a case we get an upper envelope to f on this set which is of the form $g = T^{-1} \circ h$ where h is piecewise linear. For certain choices of T it is easy to determine the integrability of g which can resolve the question of the integrability of f. Possible choices for T include the logarithm transformation $T(f) = \ln(f)$ and the power transformations $T(f) = f^p$.

We note that when we cannot determine the integrability of f, it is tempting to go ahead and apply an approximation anyway. For example, this has occurred in a number of applications of Bayesian inference, where amongst some practitioners, it is common to assume an improper prior; i.e. the prior does not have a finite integral. In fact, the integrability of the posteriors that arise in such a context can depend subtly on sample size. Examples have arisen where integrals assumed to be finite were subsequently shown to be infinite.

2.4 Reparameterization

It can often happen that the representation (2.1) is not the best for the ease and accuracy of an approximation. The basic tool in modifying an integral approximation is the *change of variable formula* or reparameterization of the integral. For example, some approximation methods require that integrals be defined on the unit cube $[0, 1]^k$ and this requires a preliminary transformation. For this, suppose that $T : A \to B$ is a 1–1, onto transformation such that T^{-1} is continuously differentiable and the Jacobian matrix

$$\frac{\partial T^{-1}(y)}{\partial y}$$

satisfies

$$\det\left(\frac{\partial T^{-1}(y)}{\partial y}\right) \neq 0$$

for $y \in B$. We then have the well-known result that

$$\int_A f(x)\,dx = \int_B \left|\det\left(\frac{\partial T^{-1}(y)}{\partial y}\right)\right| f(T^{-1}(y))\,dy. \tag{2.3}$$

Perhaps the simplest example of this is the affine transformation $T(x) = c + Dx$ where $c \in R^k$ and $D \in R^{k\times k}$ is an invertible matrix. We then have that

$$\int_A f(x)\,dx = \int_B |D|^{-1} f\left(D^{-1}(y-c)\right) dy$$

where $B = c + DA = \{y | y = c + Dx, x \in A\}$. Other important transformations are discussed in Exercises 7 and 8.

An important use of transformations is in the *removal of singularities* for the integrand f. The existence of such points commonly leads to difficulties in the approximation of the integral. This is because the basis for virtually all integration techniques is some kind of approximation to the integrand and these approximations are generally smooth or at least don't contain singularities. While integration techniques can be, and have been, devised for functions which contain a particular kind of singularity, typically the simplest approach is to remove the singularity by transforming appropriately. Consider the following simple example.

Example 2.1
Suppose that we want to calculate

$$\int_0^1 \frac{1}{\sqrt{x}}\,dx. \tag{2.4}$$

Of course, $\int_0^1 1/\sqrt{x}\,dx = 2$, but consider the effect of applying a simple approximation directly to (2.4) and to its transformed version

$$2\int_0^1 dy \tag{2.5}$$

which is obtained via the transformation $y = T(x) = \sqrt{x}$. The midpoint rule (see Chapter 5) applied to (2.4) gives the approximations 1.8089 and 1.9395 based on 10 and 100 function evaluations respectively. By contrast, the midpoint rule gives the exact answer when applied to (2.5) with only one function evaluation (and is always exact no matter how many function evaluations are chosen). Note also that when approximating (2.4), we cannot use any method that requires the evaluation of the integrand at $x = 0$.

While unrealistic, Example 2.1 illustrates the importance of considering reparameterizations for integrals as difficult problems can be made much easier. Of course, an arbitrary transformation can make an integral even more difficult to evaluate. For example, sometimes transformations are introduced to make the domain of integration bounded as this is a requirement of a number of integration techniques. If such a transformation is used without understanding its effects on the integrand, there is no guarantee that it will make the problem easier.

Quite often we are confronted with a variable of integration that ranges over $(0, \infty)$ and the integration technique we are contemplating applies to variables with no such restrictions. In such circumstances it is often found that the log transformation is useful. We will see some other common transformations used throughout the text.

The importance of the parameterization of an integral is something that recurs throughout the book. No method is exempt from its effects. We discuss some approaches related to this in later sections. In particular, the following section is connected with this issue.

2.5 Multimodality and shape

By a *mode* of an integrand we mean a point of relative maximum of $|f|$. By the *shape* of an integrand we mean the geometry of its contours. Of course, if we know the shape of an integrand this implies that we know the location of all the modes.

As we will see in this text the shape of an integrand is perhaps the most important information in obtaining a successful approximation. As the dimension rises this kind of information becomes not only more important but generally more difficult to obtain. Perhaps this is the true curse of dimensionality. There is no method currently available that avoids the need for this information if it is to be successful, except perhaps through brute force; i.e. running the algorithm for a "long" time when this is feasible. In other words, if the shape of an integrand is not appropriate to the method being used, then an unsuccessful approximation can arise.

In particular, it is almost always important to know the location of the modes of the integrand. Without this information, an approximation method could miss regions where major contributions to the integral are occurring. How does one obtain the modes of an integrand? When these occur in the interior of A, then numerical methods can be used. Currently, there are no methods which guarantee that all these points are found. In general, iterative algorithms such as Newton–Raphson or steepest descent are used with multiple starting points that in some sense cover A.

Once the modes have been obtained we consider the shape of the integrand in neighborhoods of these points. Some information about the shape of a function at a particular point can be obtained from its *Hessian matrix*; i.e. the matrix of second derivatives evaluated at the mode, or more typically more usefully, from the Hessian matrix of the log of the integrand. The Hessian matrix gives us some idea about how the integrand falls off as we move away from the mode. If the shape of the integrand is approximately elliptical about each mode, then as we will see, there are methods that can lead to successful approximations. On the other hand, the shape may be profoundly skewed and this can lead to problems as this is difficult to detect in high dimensions. While more can be learned about the integrand by looking at higher order derivatives, this becomes complicated as the number of derivatives grows quickly with dimension.

Overall, there is no replacement for studying the shape of an integrand if we want to ensure a successful approximation as brute force integration techniques, that avoid this, typically require too much computation. Currently no methods avoid this fundamental need for information about the behavior of the integrand. Still, even when this information is difficult or even impossible to obtain, we will be required to compute an approximation as part of applying some mathematical theory to a real-world problem. While we must compute approximations, however, we have to be prepared to be humbled on occasion while trying to be as rigorous, in the application of these methods, as possible.

2.6 Error analysis

The analysis of error is an intrinsic part of the application of any approximation method. After all, what value does an approximation have if we have no idea how good it is? In ideal circumstances, we would state the approximation together with a bound on the error in the approximation. However, this is achievable only in rare situations. We identify this as the highest level of error analysis. In our opinion, the goal of any theory of approximation is to develop procedures that produce this information within reasonable computational times.

The lowest level of error analysis arises when little or nothing can be said about error except that in certain problems, where the integral can be precisely evaluated, the approximation technique does well. To a great degree, this is characteristic of the methods discussed in Chapter 4. Still, such methods have value, and historically these methods have led to a number of useful approximations. When given a choice, however, we would prefer a method that provides a higher level of error analysis.

There are a number of methods that give an intermediate level of error analysis. These methods are characterized by error estimates that are based on multiple evaluations of the integrand f. While the error estimates contain error, the idea is that the error can be proven to go to 0 as the number of function evaluations increases. Unfortunately, the number of function evaluations required to ensure a sufficiently small error cannot typically be assessed. A practical strategy in such contexts is to consider approximations and their error estimates for an increasing number of function evaluations and examine how the error estimates change. This is a heuristic strategy which is not guaranteed to work, and therefore we need to be wary about results obtained in this way. The multiple quadrature methods discussed in Chapter 5 and the Monte Carlo methods of Chapters 6 and 7 exhibit this level of error analysis. In fact, this level of error analysis currently characterizes most approaches to high-dimensional integration.

We examine now some basic facts about error that are useful for our future discussions. First of all, we note that there are two sources of error in our approximations; namely, the error intrinsic to the approximation technique, which we call *algorithmic error*, and the error intrinsic to the representation of the problem on a computer, which we refer to as *round-off error*. Primarily, we ignore round-off error in this text, and therefore use "error" as a synonym for algorithmic error. However, this does not mean that round-off error is unimportant. The practitioner should always take steps to try and ensure that round-off error does not destroy the accuracy of an approximation and we discuss this in Section 2.6.2.

2.6.1 ABSOLUTE AND RELATIVE ERROR

Suppose that we wish to approximate a quantity r and to do this we use the approximation r^*. We define the *absolute error* in the approximation by

$$\left|r - r^*\right|$$

and the *relative error* by

$$\frac{|r - r^*|}{|r|}$$

provided that $r \neq 0$. When the value of r is very large or very small, relative error in the approximation is often more relevant than absolute error. We consider now the behavior of absolute and relative errors when we combine approximations to approximate new quantities. Approximations to integrals are often used as components to approximate quantities dependent on their values.

Suppose then that quantities r and s are approximated by r^* and s^* with absolute errors bounded by ϵ_r and ϵ_s respectively. Then the following inequalities give some idea of how absolute errors propagate when we perform elementary operations on the approximations:

$$\begin{aligned}
\left|(r+s) - \left(r^* + s^*\right)\right| &\le \epsilon_r + \epsilon_s \\
\left|rs - r^*s^*\right| &\le |s|\epsilon_r + \left|r^*\right|\epsilon_s \\
\left|\frac{1}{r} - \frac{1}{r^*}\right| &\le \frac{\epsilon_r}{|rr^*|}.
\end{aligned}$$

Note that the absolute error bound in addition propagates simply while the absolute error bounds associated with multiplication and division are more complex. In particular, if r is close to 0, a good approximation to r does not guarantee a good approximation to $1/r$. For a general smooth $f(r, s)$ we have the approximation

$$\begin{aligned} |f(r,s) - f(r^*, s^*)| &\approx \left|(r^* - r)\frac{\partial f(r,s)}{\partial r} + (s^* - s)\frac{\partial f(r,s)}{\partial s}\right| \\ &\le \epsilon_r \left|\frac{\partial f(r,s)}{\partial r}\right| + \epsilon_s \left|\frac{\partial f(r,s)}{\partial s}\right| \end{aligned}$$

where we have used the first two terms in the Taylor expansion of $f(r^*, s^*)$ about (r, s). Note that while we have stated the upper bound in terms of the values of the partial derivatives at (r, s), we could reverse the roles of (r, s) and (r^*, s^*) and obtain an alternative upper bound in terms of (r^*, s^*). Using the alternative upper bound, if we know the values of ϵ_r and ϵ_s, then we have a computable approximate upper bound on the absolute error $|f(r, s) - f(r^*, s^*)|$.

Suppose now that quantities r and s are approximated by r^* and s^* with relative errors bounded by ϵ_r and ϵ_s respectively. We then have as follows:

$$\begin{aligned} \frac{|(r+s) - (r^* + s^*)|}{|r+s|} &\le \frac{|r|}{|r+s|}\epsilon_r + \frac{|s|}{|r+s|}\epsilon_s \\ \frac{|rs - r^*s^*|}{|rs|} &\le \epsilon_r + \frac{|r^*|}{|r|}\epsilon_s \\ \frac{|1/r - 1/r^*|}{|1/r|} &\le \frac{|r|}{|r^*|}\epsilon_r. \end{aligned}$$

With relative error, multiplication and division behave relatively simply while addition is more complicated. In particular, if r and s are of roughly the same magnitude but of differing sign, then the relative error in addition can be disastrously large. This is known as the *cancellation phenomenon* and it is something that should always be guarded against when designing an algorithm as it can destroy the accuracy in a computation. For a general smooth $f(r, s)$ we have the approximation

$$\begin{aligned} \frac{|f(r,s) - f(r^*, s^*)|}{|f(r,s)|} &\approx \frac{\left|(r^* - r)\frac{\partial f(r,s)}{\partial r} + (s^* - s)\frac{\partial f(r,s)}{\partial s}\right|}{|f(r,s)|} \\ &\le \epsilon_r \frac{|r|}{|f(r,s)|}\left|\frac{\partial f(r,s)}{\partial r}\right| + \epsilon_s \frac{|s|}{|f(r,s)|}\left|\frac{\partial f(r,s)}{\partial s}\right| \end{aligned}$$

and we see that the cancellation phenomenon can occur whenever $f(r, s) \approx 0$.

In some circumstances we are able to compute upper and lower bounds for quantities that we are approximating. We can then use these to bound various absolute and relative errors. For example, suppose that we have lower bounds l_r, l_s

and upper bounds u_r, u_s for the unknown quantities r and s respectively. Then we have that $l_r + l_s \leq r + s \leq u_r + u_s$ and therefore $(u_r + u_s) - (l_r + l_s)$ is an absolute error bound for the approximations $u_r + u_s$ and $l_r + l_s$ of $r + s$. Further, if l_s is positive then $l_r/u_s \leq r/s \leq u_r/l_s$ and $u_r/l_s - l_r/u_s$ is an absolute error bound for the approximations u_r/u_s and l_r/l_s of r/s. When l_r is positive then the relative error in u_r, as an approximation of r, is bounded above by $u_r/l_r - 1$. Similarly, we can combine lower and upper bounds on quantities to compute bounds on the absolute and the relative errors of other simple functions of those quantities.

2.6.2 ROUND-OFF ERROR

When approximating integrals the floating point representation of real numbers is used. Nowadays most computers use a binary representation for real numbers and so the real number x is represented as $m2^e$ where, for $a_i \in \{0, 1\}$,

$$e = \pm \sum_{i=0}^{s-1} a_i 2^i,$$

is the *exponent* and, for $b_i \in \{0, 1\}$,

$$m = \pm \sum_{i=1}^{t} b_i 2^{-i}$$

is the *mantissa* such that, whenever $x \neq 0$ then $b_1 = 1$. Here s and t are fixed characteristics of an individual machine. For example, when $s \geq 3$ and $t \geq 3$, then the number 10 is represented using $a_2 = 1$, $a_i = 0$ when $i \neq 2$, $b_1 = b_3 = 1$ and $b_i = 0$ otherwise. We immediately see that the computer represents only a finite set of real numbers because some numbers lie outside the floating point range

$$\left[-\max\left|m2^e\right|, -\min\left|m2^e\right|\right] \cup \left[\min\left|m2^e\right|, \max\left|m2^e\right|\right]$$

and because some numbers in this range cannot be represented exactly in the form $m2^e$.

If in the process of a computation a number x arises such that $|x| > \max|m2^e|$ then we say we have an *overflow* error while if $|x| < \min|m2^e|$ then we have an *underflow* error. Typically, an overflow error causes a computation to stop, as it is clear that the computer is no longer accurately representing the problem. With an underflow error, the computation continues and the resulting value is set equal to 0. Note however that an excessive number of underflow errors can lead to numerical inaccuracies; e.g. if $c < \min|m2^e|$ and $f(x) = c$ on $(0, 1/c)$, then $\int_0^{1/c} f(x)\,dx = 1$ while the integrand, and thus the integral, is identically 0 as far as the computer is concerned. Therefore, when approximating an integral it is important that both overflow and underflow in the evaluation of the integrand be considered and avoided when possible.

The difference between a number x in the floating point range and its approximation by a number of the form $m2^e$ is called the round-off error. There are various methods of determining the approximation $m2^e$ in the floating point range and the particular one used depends upon the machine architecture. For example, suppose we have that

$$x = \pm \left(\sum_{i=1}^{\infty} b_i 2^{-i} \right) 2^e$$

where $b_1 = 1$. Then the *chop* method uses the approximation

$$\pm \left(\sum_{i=1}^{t} b_i 2^{-i} \right) 2^e$$

whereas *symmetric rounding* uses the approximation

$$\begin{cases} \pm \left(\sum_{i=1}^{t} b_i 2^{-i} \right) 2^e & \text{if } b_{t+1} = 0 \\ \pm \left(\sum_{i=1}^{t} b_i 2^{-i} + 2^{-t} \right) 2^e & \text{if } b_{t+1} = 1. \end{cases}$$

Note that the absolute round-off error in the approximation depends upon the size of the number. With symmetric rounding, when $b_{t+1} = 0$ we have that the relative rounding error is

$$\frac{|x| - \left(\sum_{i=1}^{t} b_i 2^{-i} \right) 2^e}{|x|} \leq 2^{-t}$$

and when $b_{t+1} = 1$ the relative rounding error is

$$\frac{\left(\sum_{i=1}^{t} b_i 2^{-i} + 2^{-t} \right) 2^e - |x|}{|x|} \leq 2^{-t}.$$

Therefore, in the case of symmetric rounding, 2^{-t} serves as an upper bound on the relative error in the floating point approximation of any number in the floating point range. This quantity is often referred to as the *machine precision*.

Rounding error is intrinsic to computations with real numbers on the computer. Its effects in integration problems are difficult to assess. Although individually small, an accumulation of rounding errors can destroy the accuracy of an approximate integration and this possibility has to be kept in mind. Particular circumstances where accumulated rounding errors can cause problems include situations where we are summing numbers with similar magnitudes and different signs. This occurs with many integration techniques when the integrand oscillates in sign. Therefore, it is important that this be considered and avoided when possible.

One protection against rounding error is to increase the precision of computation. By this we mean using more than t bits in the mantissa. For example, if we double the number of bits (i.e. increase the number of bits by $t + 1$ as the sign takes up one bit in the single precision) then we have double precision floating point numbers. The machine precision is then $2^{-(2t+1)}$ and the relative error is substantially improved. Many programming languages such as C and FORTRAN contain the ability to work with double precision values and this is almost always recommended. Unfortunately, the ability to increase the precision beyond this is not common.

2.6.3 INTERVAL ANALYSIS

One way of controlling error in computations is through the use of *interval analysis*. By this we mean that an upper and lower bound is provided for each real number in a computation such that the true value of the real number is guaranteed to be in this range. There is a calculus associated with this so that we can compute intervals that contain the exact values of real numbers that arise from computations on such intervals. A good discussion of this approach to controlling and measuring error in computations is found in Moore (1979). We will touch very briefly on some methods that lead to intervals that contain the exact values of integrals. As mentioned above this provides the highest level of error analysis and is the ultimate goal of all approximation. In general, however, this is currently not achievable, at least for integration problems.

2.7 Software

We have already mentioned the packages MAPLE and MATHEMATICA when discussing symbolic integration. Both of these packages also contain methods for obtaining numerical approximations to one-dimensional integrals. When we are confronted with one-dimensional integrals these packages can be very useful as they are easy to use, and in principle, arbitrary accuracy can be achieved. Still, this does not avoid the need for other approximation techniques for this class of integrals, as many circumstances arise when use of these software packages is not suitable.

Both the IMSL (1994) and NAG (1996) libraries of subroutines contain algorithms which are useful in approximating integrals. In general, it is our experience that having access to one of these libraries, or one with a nearly equivalent set of routines, is essential to a practitioner. These packages contain a number of essentials for integrating; e.g. tables of Gauss rules, variate generation routines, etc. These packages also contain many relevant numerical routines. In particular, we mention numerical optimization, convex programming, root-finding and many linear algebra operations like matrix inversion and eigensystem analysis. In addition, these libraries have excellent approximations to various special functions which can be written as integrals; e.g. the distribution functions of various random variables. Therefore, it is important that readers acquaint themselves with the capabilities of these libraries, ensure that they have ready access to them and acquaint themselves with their use. With such libraries, much error-prone programming can be avoided.

There are other packages which can be used for the approximation of integrals. However, we are not aware of other packages that have the coverage of NAG or IMSL.

2.8 Exercises

(1) Use MAPLE or MATHEMATICA to check the result obtained for (2.2).

(2) Verify the integrability of the function $e^{-x^2/2}$ on R^1. Verify the integrability on R^1 of any function of the form $p(x)e^{-x^2/2}$ where p is a polynomial.

(3) Determine the integrability of $\frac{1}{x}e^{-x^2/2}$ on R^1.

(4) Verify the integrability of the function $\left(1+x^2\right)^{-(\lambda+1)/2}$ on R^1 when $\lambda > 0$.

(5) Determine the integrability of the function $p(x)\left(1+x^2\right)^{-(\lambda+1)/2}$ on R^1 when $\lambda > 0$ and p is a polynomial.

(6) Verify the change of variable formula for the affine transformation.

(7) For $x \in R^k$ show that the Jacobian of the transformation $x \to (r, \theta_1, \ldots, \theta_{k-1})$ where

$$
\begin{aligned}
x_1 &= r\cos\theta_{k-1}\cos\theta_{k-2}\cdots\cos\theta_2\cos\theta_1 \\
x_2 &= r\cos\theta_{k-1}\cos\theta_{k-2}\cdots\cos\theta_2\sin\theta_1 \\
x_3 &= r\cos\theta_{k-1}\cos\theta_{k-2}\cdots\cos\theta_3\sin\theta_2 \\
&\vdots \\
x_{k-1} &= r\cos\theta_{k-1}\sin\theta_{k-2} \\
x_k &= r\sin\theta_{k-1}
\end{aligned}
$$

for $r > 0, 0 \le \theta_1 < 2\pi$ and $-\pi/2 < \theta_{k-1} < \pi/2$ for $i = 2, \ldots, k-1$ is given by

$$r^{k-1}\left(\cos\theta_{k-1}\right)^{k-2}\left(\cos\theta_{k-2}\right)^{k-3}\cdots\left(\cos\theta_2\right).$$

The coordinates $(r, \theta_1, \ldots, \theta_{k-1})$ of a point $x \in R^k$ are called *spherical coordinates*. Points with constant r lie on the sphere of radius r in R^k.

(8) For $x \in S_k = \{x | x_i \ge 0, x_1 + \cdots + x_k \le 1\}$; i.e. the *k-simplex*, show that the Jacobian of the transformation $x \to y$, where

$$
\begin{aligned}
x_1 &= y_1 \\
x_2 &= y_2\left(1-y_1\right) \\
x_3 &= y_3\left(1-y_2\right)\left(1-y_1\right) \\
&\vdots \\
x_k &= y_k\left(1-y_{k-1}\right)\cdots\left(1-y_2\right)\left(1-y_1\right),
\end{aligned}
$$

is given by

$$(1 - y_1)^{k-1} (1 - y_2)^{k-2} \cdots (1 - y_{k-1})$$

where $0 \leq y_i \leq 1$ for $i = 1, \ldots, k$. Therefore this transformation transforms integrals over the k-simplex into integrals over the unit cube $[0, 1]^k$.

(9) Use the Taylor series method to obtain approximate upper bounds on the absolute errors in r^*s^* and $1/r^*$ when we approximate r and s with absolute errors bounded by ϵ_r and ϵ_s respectively.

(10) Use the Taylor series method to obtain approximate upper bounds on the relative errors in r^*s^* and $1/r^*$ when we approximate r and s with relative errors bounded by ϵ_r and ϵ_s respectively.

(11) Determine $\min |m2^e|$ and $\max |m2^e|$ for a computer where the exponent is represented with s bits and the normalized mantissa is represented with t bits.

(12) Verify that the machine precision is given by 2^{-t} for a computer where the normalized mantissa is represented with t bits.

3
Algorithms for sampling

3.1 Introduction

Monte Carlo methods are an essential part of any practitioner's toolkit when considering the approximation of integrals. Typically this will require the ability to generate from a wide variety of probability distributions. In fact, a major constraint on the implementation of Monte Carlo methods can be the lack of a suitable algorithm for generation in a given context.

In recent years there has been progress in developing so-called "black-box" methods. The goal behind such algorithms is to provide methods for the construction of efficient generators that require minimal information about the underlying distribution and minimal work on the part of the user.

There are contexts where variates must be generated from literally thousands of unique distributions and therefore it is not possible to spend time developing optimal algorithms for each distribution. On the other hand, there are some common distributions; e.g. the $N(0, 1)$ (standard normal) distribution, where it is worth having the most efficient algorithm possible. Therefore we need both quick-and-easy methods as well as those providing high-quality algorithms.

Our purpose here is to review the primary methods for constructing generators. If a problem is encountered for which a good algorithm is not available, then hopefully these methods will enable practitioners to develop their own algorithms. Still, problems will be encountered for which no method works well and therefore random variate generation continues to be an active and important area of research.

In addition to Devroye (1986), there are other books which deal with random variate generation. In particular, Hammersley and Handscomb (1964), Rubinstein (1981), Bratley, Fox and Schrage (1983), Ripley (1987), Lewis and Orav (1989) and Fishman (1996) are all recommended as they deal with various aspects which we do not cover.

In Section 3.2 we consider the important problem of generating a stream of independently distributed $U(0, 1)$ random variables. This is a very important problem as it underlies all Monte Carlo work. If we do not have a good source of such random variates then the quality of whatever Monte Carlo algorithm we are considering can be severely compromised. In Section 3.3 we discuss the most straightforward mechanism for generating a random variable from an arbitrary distribution given a stream of independent $U(0, 1)$ [uniform on $(0, 1)$] variables;

namely the method of inversion. In Sections 3.4 and 3.6 rejection-based methods for random variate generation are presented. In Section 3.5 the hit-or-miss integration algorithm is presented as it is closely related to rejection-based methods for generation. In Section 3.7 other methods are discussed that have somewhat limited applicability but with some particular important applications. Finally, in Section 3.8 we consider algorithms for generating from multivariate distributions.

3.2 Generating uniform random variables

3.2.1 RANDOMNESS

All Monte Carlo methods are based on having available a method for generating a sequence $u_1, u_2, \ldots$ of values in $[0, 1]$ that can be considered to be a realization of an independent and identically distributed (iid) sequence from a uniform distribution. By this we mean that the sequence $u_1, u_2, \ldots$ has all the properties that an iid $U(0, 1)$ sequence is supposed to have. For example, for any m and for Borel sets $B_1, \ldots, B_m \subseteq [0, 1]$ we want

$$\frac{\#\{(u_{im+1}, \ldots, u_{im+m}) \in B_1 \times \cdots \times B_m, 0 \le i \le n-1\}}{n} \rightarrow \int_{B_1 \times \cdots \times B_m} du_1 \cdots du_m \tag{3.1}$$

as $n \to \infty$. It is worth noting that (3.1) does not mathematically characterize what is meant by an iid uniform sequence. For example, deterministic sequences can be constructed with this property and of course such a sequence lacks the element of unpredictability that we regard as essential for randomness. A more appropriate definition of randomness exists due to Kolmogorov (1963) and Chaitin (1966) but this does not seem to lead to practical algorithms for constructing such sequences; see Li and Vitányi (1993) for an extensive discussion.

With integration problems it is reasonable to ask: do we really need the full properties of a random sequence since we are only interested in approximating an integral and not modeling a random system? This is an issue that we return to in Chapter 5 where the goal of constructing random sequences is given up in favor of constructing *quasi-random sequences* which possess properties that are suitable for the integration problem. However, in our opinion, the existence of such sequences in no way avoids the need to develop sequences that mimic random sequences, so-called *pseudo-random sequences*. In part, history validates this, as the pseudo-random methodology remains the predominant approach. We also note a distinct theoretical advantage of the pseudo-random approach as it makes available all the machinery of probability theory for the error and convergence analysis of our approximations.

3.2.2 PSEUDO-RANDOM SEQUENCES OF UNIFORMS

The quality of any simulation procedure is dependent on the quality of the underlying stream of uniform variates. For if the uniform generator is not very good then the output from the simulation cannot accurately mimic the random system and any analysis based on such an assumption is flawed. Accordingly, considerable work has been invested in developing increasingly better uniform generators. We present some very basic developments on this topic so that the reader is at least familiar with the main issues. More extensive discussion can be found in Knuth (1981), Ripley (1987) and Niederreiter (1992). At the very least, one should be aware of the type of uniform generator used when carrying out a simulation.

Other characteristics, beyond accurately modeling iid $U(0, 1)$ sequences, are also desirable in a generator. For example, because Monte Carlo simulations tend to be computationally time intensive, the generator should be fast. Also, the simulation results should be independent of the particular computing environment, and moreover, the simulation results should be reproducible. By *reproducibility*, we mean that we can restart our algorithm and obtain exactly the same output. Reproducibility is important not only for the verification of our work but also for the debugging of software. Clearly, one way of ensuring reproducibility is to save the uniform stream. However, in contexts where millions, or perhaps billions, of values are used, this is not an attractive option.

Considerations such as these have led researchers away from the use of physical devices to generate random numbers to the development of deterministic mathematical algorithms. Such algorithms cannot produce truly random numbers but the point of view here is to consider a truly random sequence as an idealized mathematical object that, through ingenuity and labor, can be approximated to any degree desired. As simulations become more demanding in their needs, and this is a continuing trend particularly with respect to the dimensionality of the integration problems considered, the uniform generators must be constantly improved. The discussion in Marsaglia (1995) makes many relevant comments concerning physical random number generators and their role in numerical work.

We note first that since computers can only store values of finite accuracy, it is typical to consider uniform generators of random integers in a finite range and then construct approximate $U(0, 1)$ values from these. For suppose that we have a generator capable of producing iid $U\{0, 1, \ldots, M-1\}$ values. Then if x is output from this generator we record $u = x/M$ as the approximate $U(0, 1)$ value. Obviously u is constrained to lie on a grid of mesh length $1/M$ but provided that M is large enough this is typically not a problem.

The most common class of generators is based on a choice of M, a function $f : \{0, 1, \ldots, M-1\}^k \to \{0, 1, \ldots, M-1\}$ and a set of initial or *seed* values $x_1, \ldots, x_k$. The sequence is then generated via

$$x_{i+1} = f(x_{i-k+1}, \ldots, x_i) \tag{3.2}$$

for $i = k, k+1, \ldots$. Although the choice of seed and M are not irrelevant in determining the quality of the generator, attention is primarily focused on the choice of f. Knuth (1981) cites his experience in making arbitrary choices for f and finding that these algorithms have very poor characteristics. Typically, the approach is to base the choice of f on mathematical theory, but in the end, assess

f empirically by testing generated sequences using a battery of statistical tests such as the Chi-square test. See Knuth (1981) and Marsaglia (1985, 1995) for discussion of some of these tests. In particular the DIEHARD battery of tests introduced in Marsaglia (1995) is very thorough and many generators fail these tests. No generator should be accepted for use without undergoing such tests.

There is something that can be proved about generators of the form (3.2) that has implications concerning performance; namely, these generators cycle if we let them run long enough.

Lemma 3.1 There exists $\alpha, \beta \in \aleph$ such that $x_{i+\beta} = x_i$ for all $i > \alpha$.

Proof: There are at most M^k different k-tuples $(x_{i-k+1}, \ldots, x_i)$. Therefore after x_{M^k+k-1} one of the previously occurring k-tuples must occur again. Let α be such that $(x_{\alpha-k+1}, \ldots, x_\alpha)$ is the k-tuple that first reappears in the sequence and let β be the smallest integer such that $\left(x_{\alpha-k+\beta+1}, \ldots, x_{\alpha+\beta}\right) = (x_{\alpha-k+1}, \ldots, x_\alpha)$. Then $x_{\alpha+\beta+j} = x_{\alpha+j}$ for $j = 1, 2, \ldots$ and therefore $x_{i+\beta} = x_i$ for all $i > \beta$.

We would like the smallest such β, hereafter called the *period* of the generator f, to be independent of the initial values $(x_1, \ldots, x_k)$ and also to be as big as possible. Note that the maximum period for such an f is M^k. We have the following result.

Lemma 3.2 There exists an f with maximum period.

Proof: We proceed iteratively. Put $(x_1, \ldots, x_k) = (0, \ldots, 0)$ and for $n > 0$ put x_{n+k} equal to the largest value in $\{0, 1, \ldots, M - 1\}$ such that $(x_{n+1}, \ldots, x_{n+k-1}, x_{n+k})$ has not occurred in the sequence. Stop when there is no such value. An easy argument (Exercise 1) then completes the proof.

When $M = 2$ sequences constructed via the method of Lemma 3.2 are known as *de Bruijn sequences*, and in Popper (1959) this construction is claimed to be the mathematical characterization of randomness. Perhaps because de Bruijn sequences are difficult to analyze mathematically, they have never found common usage as random generators.

3.2.3 LINEAR CONGRUENTIAL GENERATORS

Perhaps the most important class of generators, at least historically, are the *linear congruential generators*. For these generators, $k = 1$ and

$$f : \{0, 1, \ldots, M - 1\} \rightarrow \{0, 1, \ldots, M - 1\}$$

is given by

$$f(x) = (ax + c) \bmod M$$

and so $x_i = (ax_{i-1} + c) \bmod M$. Part of the importance of this class of generators is that they are relatively simple and can be analyzed mathematically so that good choices can be made for the parameters a, b and M.

Clearly we want M to be large. Also suppose that p is a prime such that $p|M$ (i.e. p divides M) and define a new sequence $y_i = x_i \bmod p$. Then for some integers q_i and r_i, we have that $y_i = (ax_{i-1} + c + q_i M) \bmod p = (ay_{i-1} + c + ar_i p + q_i M) \bmod p = (ay_{i-1} + c) \bmod p$ and so the sequence $\{y_i\}$ will not look very random if p is a small prime. Therefore we should choose M so that it only has large prime factors. The following result, due to Hull and Dobell (1962), characterizes those triples (a, c, M) for which the maximum period is obtained. Also see Knuth (1981) for a proof.

Theorem 3.3
The linear congruential generator specified by (a, c, M) has period M if and only if

(i) $(c, M) = 1$; i.e. c and M are relatively prime,
(ii) $p|(a-1)$ for every prime p such that $p|M$,
(iii) $4|(a-1)$ if $4|M$.

Note that Theorem 3.3 is trivially satisfied by $(1, 1, M)$ for any M. Although this choice provides maximum period, it does not guarantee a good generator. Recall that statistical tests should be used to choose amongst generators with good mathematical properties.

Also note that when $c = 0$ then condition (i) of Theorem 3.3 is not satisfied and therefore a *pure multiplicative generator* cannot have full period. Still, because they are slightly simpler, pure multiplicative generators have some appeal and we consider some of their basic properties. For a pure multiplicative generator we have

$$x_i = ax_{i-1} \bmod M = a^2 x_{i-2} \bmod M = \cdots = a^{i-1} x_1 \bmod M$$

and therefore we cannot have the seed value $x_1 = 0$. Further if $d|M$ and $d|x_i$ then $x_{i+1} = ax_i + kM$ for some integer k implies that $d|x_{i+1}, x_{i+2}, \ldots$ and we observe highly nonrandom behavior in the sequence when d is small. Therefore, for a pure multiplicative generator we want $(x_i, M) = 1$ for every i. If $(x_1, M) = 1$ and $(a, M) = 1$, then $x_{i+j} = x_i$ iff $a^{i+j-1} x_1 \bmod M = a^{i-1} x_1 \bmod M$ iff $a^{i+j-1} \bmod M = a^{i-1} \bmod M$ iff $a^j \bmod M = 1$ iff $h|j$ where h is the smallest integer satisfying $a^h \bmod M = 1$ (see Exercise 2).

In the case that $(a, M) = 1$ we say that a belongs to the exponent h modulo M whenever h is the smallest positive integer such that $a^h \bmod M = 1$. Let $\phi(M)$ equal the number of values in $\{1, \ldots, M-1\}$ relatively prime to M; i.e. ϕ is *Euler's totient*. Euler's theorem (see Nivens and Zuckerman 1972) says that if $(a, M) = 1$ then $a^{\phi(M)} \bmod M = 1$. Therefore there is an $h \le \phi(M)$ such that a belongs to the exponent h modulo M. Note that if a belongs to the exponent h modulo M then the period of the generator given by $(a, 0, M)$ is h. Therefore the maximum possible period for a multiplicative generator with $(a, M) = 1$ is $\phi(M)$.

If a belongs to the exponent $\phi(M)$ modulo M then a is called a *primitive root* of M. To attain the maximum possible period we want to choose M and a so that a is a primitive root of M and $\phi(M)$ is large. It can be shown, see Nivens and Zuckerman (1972), that the only integers having primitive roots are $1, 2, 4, p^e, 2p^e$ where p is an odd prime and e is an integer and that there are $\phi(p-1)$ primitive roots of p. Since $\phi(p^e) = p^{e-1}(p-1)$, see Nivens and Zuckerman (1972), these results enumerate the number of primitive roots. There are techniques for finding primitive roots and thus obtaining pure multiplicative generators of maximum period.

Example 3.1
The triple $(7^5, 0, 2^{31}-1)$ corresponds to a very popular pure multiplicative generator. Here $M = 2^{31}-1$ is a prime and $a = 7^5 = 16807$ is a primitive root of M. Therefore this is a full period pure multiplicative generator with period $2^{31}-2$.

We can generalize these developments to *higher order linear congruential generators*. For these we must specify $(a_1, \ldots, a_k, M)$ and initial values $(x_1, \ldots, x_k)$ and then for $i > k$ the generator is given by

$$x_i = (a_1 x_{i-1} + \cdots + a_k x_{i-k}) \bmod M.$$

It can be shown that $(a_1, \ldots, a_k, M)$ can always be chosen so that the maximum period of $M^k - 1$ is attained independent of the initial values; see Niedereitter (1992).

A number of generators can be considered as particular examples of higher order linear congruential generators. For example, *lagged Fibonacci generators* take the form $x_i = (x_{i-r} + x_{i-s}) \bmod M$ for some choice of r, s and M. Lewis and Payne (1973) present *generalized feedback shift register generators* that give uniforms directly via

$$u_i = \sum_{j=1}^{l} \frac{x_{i+k_j}}{M^j}$$

where the x_i's arise from a higher order linear congruential generator and $k_1, \ldots, k_l$ are specified. The properties of these and more general nonlinear congruential generators are studied in Niederreiter (1992).

3.2.4 OTHER GENERATORS

Perhaps the simplest way of creating more general generators with hopefully better characteristics is to combine generators. For example, in MacLaren and Marsaglia (1965), a table of variates is created using one generator and a second generator is then used to select an entry in the table. The entry is returned as the reported value and is replaced in the table with a newly generated variate from the first generator. This procedure is referred to as *shuffling*. In Marsaglia (1985, 1995) many other methods of combining generators are discussed. Also see L'Ecuyer (1988).

In Marsaglia (1995), *multiply-with-carry* generators are introduced. For example, starting with a, c_1 and the seed x_1, we generate

$$x_2 = (ax_1 + c_1) \bmod M$$

but now also compute a new $c_2 = \lfloor (ax_1 + c_1)/M \rfloor$ where $\lfloor \cdot \rfloor$ denotes the *floor* or *greatest integer function*. We then generate $x_3 = (ax_2 + c_2) \bmod M$ and repeat the process. These generators have some particular advantages for computer implementation and can, depending on the choices of (a, M) and (c_1, x_1), give very long periods. There exist notable examples of these generators which pass all of the tests in DIEHARD and are strongly recommended by Marsaglia. More general versions of this type of generator have also been developed.

Marsaglia (1995) includes several files which contain billions of randomly generated bits from various generators. Included are some sequences generated from generators that failed DIEHARD. Interestingly, failures occurred for a number of physically generated sequences. In spite of their failure, Marsaglia suggests a use for such sequences. With any deterministically generated sequence there is always the risk that someone sufficiently clever can "crack" the generator after relatively few observations. Combining such a generator, perhaps via convolution (see Exercise 3), with a physically generated sequence removes such a risk. Any sequence combined via convolution with a truly random sequence produces a random sequence so the risk that the quality of the initial sequence will be diminished seems low. Marsaglia provides a number of bit sequences that are recommended for rigorous Monte Carlo work.

3.3 Inversion methods for random variables

The most direct method to generate from a probability distribution, when it is possible, is to use *inversion*. Suppose then that we have a random variable X with distribution function given by F. Then for $u \in [0, 1]$ define the *inverse distribution function* $F^{-1} : [0, 1] \to R$ by

$$F^{-1}(u) = \inf\{x | u \leq F(x)\}.$$

We note (Exercise 4) that (i) F^{-1} is increasing, (ii) $F(F^{-1}(u)) \geq u$, $F^{-1}(F(x)) \leq x$, (iii) $u \leq F(x)$ iff $F^{-1}(u) \leq x$ and (iv) F^{-1} is left-continuous. We then have the following result.

Theorem 3.4 If $U \sim U(0, 1)$ then $X = F^{-1}(U) \sim F$.

Proof: $P(X \leq x) = P(F^{-1}(U) \leq x)$ by definition and then $P(F^{-1}(U) \leq x) = P(U \leq F(x))$ by property (iii) and this is just $F(x)$.

We provide some examples of this algorithm.

Example 3.2 *Exponential*(θ) *distribution*
We have $F(x) = 1 - \exp(-x/\theta)$ for $x > 0$ and thus $F^{-1}(u) = -\theta \ln(1 - u)$.

Example 3.3 *Bernoulli* (p) *distribution*
Here

$$F(x) = \begin{cases} 0 & \text{if } x < 0 \\ 1 - p & \text{if } 0 \leq x < 1 \\ 1 & \text{if } x \geq 1 \end{cases}$$

and

$$F^{-1}(u) = \begin{cases} 0 & \text{if } u \le 1 - p \\ 1 & \text{if } u > 1 - p. \end{cases}$$

Example 3.4 *Uniform*$\{1, \ldots, K\}$ *distribution*
In this case we have that

$$F(x) = \begin{cases} 0 & \text{if } x < 1 \\ \lfloor x \rfloor / K & \text{if } 1 \le x < K \\ 1 & \text{if } x \ge K. \end{cases}$$

Then $F^{-1}(u) = \lceil Ku \rceil$ where $\lceil \cdot \rceil$ denotes the *ceiling* or *least integer function.*

The difficulty in using the inversion algorithm to generate a random variable X is the computation of $F^{-1}(U)$. Circumstances are rare when we have a nice closed-form expression for this function. One approach in this context is to generate $U \sim U(0, 1)$ and then numerically solve

$$X = \inf\{x | U \le F(x)\}$$

for X. For example, when F is strictly increasing on the support of the distribution we use a root-finding algorithm to solve

$$F(X) = U \tag{3.3}$$

for X. Typically this requires a tractable expression for F which is not always the case. Even when F is tractable, solving (3.3) numerically rarely leads to an efficient algorithm.

Solving (3.3) numerically could be considered a brute force algorithm. In some cases this may be acceptable if efficiency is not a great concern. Suppose then that the distribution of X is concentrated on $[a, b]$ with F strictly increasing there and we determine to use the bisection method to solve (3.3). Letting ϵ denote the specified accuracy of the variable, we proceed iteratively as follows:

1. put $x_{\text{low}} = a$, $x_{\text{high}} = b$, $X_{\text{old}} = b$;
2. generate $U \sim U(0, 1)$;
3. put $X_{\text{new}} = \left(x_{\text{low}} + x_{\text{high}}\right)/2$ and calculate $F_{\text{new}} = F(X_{\text{new}})$;
4. if $|X_{\text{new}} - X_{\text{old}}| \le \epsilon$ then return $X = X_{\text{new}}$ and stop
   else put $X_{\text{old}} = X_{\text{new}}$;
5. if $F_{\text{new}} < U$ then put $x_{\text{low}} = X_{\text{new}}$ and go to 3
   else put $x_{\text{high}} = X_{\text{new}}$ and go to 3.

When the support of the distribution of X is not contained in a compact interval we must modify the algorithm. Typically this means choosing $[a, b]$ to contain the bulk of the probability and modifying the search appropriately if we have $U < F(a)$ or $U > F(b)$. Note that when inversion is feasible for a distribution then inversion can also be used for truncated versions of the distribution.

3.4 Rejection methods for random variables

One of the most useful methods for constructing algorithms for generating from distributions is the *rejection method* or *acceptance/rejection* sampling as it is sometimes called. We present it for densities on R^k but it is typically only feasible when $k = 1$. Suppose then that f is a possibly unnormalized density on R^k from which we wish to sample, g is a density on R^k and that there exists $c > 0$ such that $f \le cg$. Suppose further that we have an algorithm to generate from g. We note that we are interpreting density here in the general sense that it is a density with respect to some support measure for a probability distribution on R^k. The distribution associated with f could be discrete, continuous or a mixture of the two. The fact that the algorithm works even when f is not normalized is of considerable significance since the normalizing constant is sometimes difficult to calculate. The rejection algorithm then proceeds as follows:

1. generate $Y \sim g$ and independently generate $U \sim U(0, 1)$;
2. if $Ucg(Y) > f(Y)$ then go to 1
   else return $X = Y$ and stop.

Step 2 is called the rejection step. We have the following result.

Theorem 3.5 The rejection algorithm stops with probability 1 and the returned X has the distribution corresponding to f.

Proof: Let

$$\begin{aligned} p &= P(Ucg(Y) \le f(Y)) \\ &= E\left[P(Ucg(Y) \le f(Y)|Y)\right] \\ &= E\left[P\left(U \le \frac{f(Y)}{cg(Y)}|Y\right)\right] \\ &= E\left[\frac{f(Y)}{cg(Y)}\right] \\ &= \frac{1}{c}E\left[\frac{f(Y)}{g(Y)}\right] \end{aligned}$$

and note that $E[f(Y)/g(Y)]$ is the inverse of the normalizing constant for f. Then we see that the probability that we accept on a given iteration is $p > 0$ and the probability that the rejection algorithm stops is $\sum_{i=1}^{\infty}(1-p)^{i-1}p = 1$. Further, letting (Y_i, U_i) denote the pair of random variables generated at the i-th iteration

$$\begin{aligned} P(X \in B) &= \sum_{i=1}^{\infty} P\,(Y_i \in B \text{ and we stop on the } i\text{-th iteration}) \\ &= \sum_{i=1}^{\infty} P\,(Y_i \in B \text{ and } U_i cg(Y_i) \le f(Y_i))\,(1-p)^{i-1} \end{aligned}$$

$$= \sum_{i=1}^{\infty} E\left[P\left(Y_i \in B \text{ and } U_i cg(Y_i) \le f(Y_i) | Y_i\right)\right](1-p)^{i-1}$$

$$= \sum_{i=1}^{\infty} E\left[I_B(Y)\frac{f(Y)}{cg(Y)}\right](1-p)^{i-1}$$

$$= E\left[I_B(Y)\frac{f(Y)}{g(Y)}\right] \Big/ E\left[\frac{f(Y)}{g(Y)}\right].$$

Since $E\left[I_B(Y)f(Y)/g(Y)\right]$ is the integral of f over B, this implies that X is distributed with density proportional to f.

We measure the efficiency of a particular rejection algorithm by looking at the distribution of the number of iterations until we accept. Clearly this has expected value

$$\frac{1}{p} = \frac{c}{E[f(Y)/g(Y)]}$$

and therefore rejection sampling is more efficient the closer c is to the inverse of the normalizing constant of f. Therefore, in designing a rejection algorithm we should try to choose a density g and a constant $c > 0$ so that this ratio is as small as possible while ensuring that $f \le cg$.

Sometimes the evaluation of f can be expensive in computational time. The technique of *squeezing*, when it is applicable, can be helpful in such contexts in improving the overall efficiency of the algorithm. Suppose we have functions h_1, h_2 with $h_1(x) \le f(x) \le h_2(x) \le cg(x)$. Further, suppose that h_1 and h_2 are much easier to evaluate than f. Then we can avoid the comparison with f by accepting whenever $Ucg(Y) \le h_1(Y)$ and rejecting whenever $Ucg(Y) > h_2(Y)$. The efficiency of squeezing is measured by the probability of having to make a comparison with f; namely

$$\begin{aligned} P\left(h_1(Y) < Ucg(Y) \le h_2(Y)\right) &= E\left[P\left(h_1(Y) < Ucg(Y) \le h_2(Y) | Y\right)\right] \\ &= E\left[\frac{h_2(Y) - h_1(Y)}{cg(Y)}\right] \\ &= \frac{1}{c} E\left[\frac{h_2(Y) - h_1(Y)}{g(Y)}\right] \end{aligned}$$

which depends on c and the "volume" between h_1 and h_2 over the support of the distribution of Y. Note that the probability is small and squeezing is efficient when the rejection algorithm is poor; i.e. when c is large. Except in very intractable problems; e.g. an f resulting from a very complicated likelihood function in a Bayesian computation, or in situations where we wish to construct very efficient generating algorithms, squeezing does not appear to be very useful.

Example 3.5
Suppose that $f(x) = (x+3)^2(x+1)$ on $[0, 1]$. Then f takes its maximum at 1 and the maximum value is 32. Therefore if g is the density of the $U(0, 1)$ we have that $f \leq cg$ on $[0, 1]$ with $c = 32$. This gives a rejection algorithm. Since $\int_0^1 f(x)\,dx = 229/12 = 19.083$ we have that $1/p = 32/19.083 = 1.677$ and therefore it takes an average of 1.7 iterations until an acceptance occurs.

The rejection algorithm in conjunction with other methods has been ingeniously employed to develop highly efficient generators for a number of important distributions. In particular, we mention the use of rejection sampling by Marsaglia, MacLaren and Bray (1964) to develop an excellent algorithm for the $N(0, 1)$ distribution. This algorithm has been subsequently improved by Kinderman and Ramage (1976). In general, the rejection algorithm is difficult to implement efficiently without considerable study of the density f. The adaptive rejection algorithms which we subsequently discuss can yield very efficient algorithms with little work and are therefore recommended for applications where it does not make sense to spend a lot of time optimizing the generator.

3.5 Hit-or-miss integration

Discussion of the rejection algorithm brings up the first integration technique in the text, sometimes referred to as *hit-or-miss integration*. However, this method is mostly of historical interest. In Chapter 6, we show that better methods are available whenever hit-or-miss integration is applicable.

Suppose now that f is a general integrable function and that we have $|f| \leq cg$ as described in Section 3.4. We write $f = f^+ - f^-$ where $f^+ = \max(0, f)$ and $f^- = \max(0, -f)$. Then we can approximate the integral of f over the support of g by generating n independent pairs (U, Y) and recording

$$p_n = \frac{1}{n}\sum_{i=1}^{n} W_i$$

where

$$\begin{aligned} W_i = {} & I_{\{(u,y):f^+(y)>0,Ucg(y)\leq f^+(y)\}}(U_i, Y_i) \\ & - I_{\{(u,y):f^-(y)>0,Ucg(y)\leq f^-(y)\}}(U_i, Y_i). \end{aligned}$$

Note that if $f^- = 0$ then p_n is the proportion of times that $Ucg(Y) \leq f(Y)$. The estimate of the integral $\int_{R^k} f(x)\,dx$ is then given by cp_n. Let

$$r = P(W_1 = -1) = \frac{1}{c}\int_{R^k} f^-(x)\,dx$$

$$s = P(W_1 = 1) = \frac{1}{c}\int_{R^k} f^+(x)\,dx.$$

We have the following result.

Theorem 3.6 The hit-or-miss algorithm satisfies

$$E\left[cp_n\right] = \int_{R^k} f(x)\,dx$$

and

$$\operatorname{Var}\left[cp_n\right] = \frac{1}{n}\left[c\int_{R^k} |f(x)|\,dx - \left\{\int_{R^k} f(x)\,dx\right\}^2\right].$$

Proof: We have that

$$\begin{aligned} E\left[p_n\right] = E\left[W_1\right] = s - r &= \frac{1}{c}\int_{R^k} f^+(x)\,dx - \frac{1}{c}\int_{R^k} f^-(x)\,dx \\ &= \frac{1}{c}\int_{R^k} f(x)\,dx. \end{aligned}$$

Also $\operatorname{Var}\left[p_n\right] = n^{-1}\operatorname{Var}\left[W_1\right]$ so that

$$\begin{aligned} \operatorname{Var}\left[W_1\right] = E\left[W_1^2\right] - (s-r)^2 &= s + r - (s-r)^2 \\ &= \frac{1}{c}\int_{R^k} |f(x)|\,dx - \frac{1}{c^2}\left\{\int_{R^k} f(x)\,dx\right\}^2 \\ &= \frac{1}{c^2}\left[c\int_{R^k} |f(x)|\,dx - \left\{\int_{R^k} f(x)\,dx\right\}^2\right] \end{aligned}$$

which establishes the result.

Note that the closer c is to

$$\left\{\int_{R^k} f(x)\,dx\right\}^2 \Big/ \int_{R^k} |f(x)|\,dx$$

the more efficient is hit-or-miss integration.

3.6 Adaptive rejection methods

Adaptive rejection methods are amongst the most useful general approaches to constructing generators. However, the methods are not without their limitations. In particular, we must have an expression for the density f and the expression must be tractable in a certain sense which we describe later. Still, many problems can be dealt with using adaptive rejection methods and the construction of good algorithms requires very little input from the user.

The first adaptive generating algorithm was developed by Gilks and Wild (1992) and is applicable to densities that are *log-concave*; i.e. $\ln f$ is concave over its entire convex domain D. Basically, the idea is to use the log-concavity of f to construct upper and lower envelopes to f by constructing piecewise linear upper and lower envelopes to $\ln f$ where the envelopes depend on points $\{x_1, \ldots, x_m\}$ and the values of f and $f^{(1)}$ at these points. The normalized upper envelope serves as g in the rejection algorithm while the lower envelope can be used as a squeezer if desired. The adaptive part of the algorithm occurs when we reject a candidate value generated from g. As a rejection indicates an inadequacy in the upper envelope as an approximation to f, the adaptive step adds the point where the rejection occurs to $\{x_1, \ldots, x_m\}$ and constructs new, more accurate, envelopes.

Actually, there is no need to restrict the approach to the logarithm transformation and more general transformations can be employed. This point is made by Hörmann (1995) but without the adaptive step. In Evans and Swartz (1998a), the class of transformations is broadened, the requirement of concavity over the whole range is dropped and the adaptive step is added. Further, it is pointed out that the transformations are really only needed to handle the tails of distributions and different transformations can be used to handle different regions of the density.

We describe an implementation of the algorithm discussed in Evans and Swartz (1998a, 1998b). The construction of the envelopes in these papers is a little different from that employed in Gilks and Wild (1992) as it does not require continuous envelopes. Dropping continuity has some advantages in terms of the construction process and the analysis of the approximation properties of the envelopes as discussed in Chapter 5. Let T be a 1–1 transformation defined on a subinterval of $[0, \infty)$. Then we say $f : D \to R$ is T-concave on D, a convex subset of R, if $T \circ f$ is concave there. If $T \circ f$ is smooth then f is T-concave if and only if $(T \circ f)^{(2)} = (T^{(2)} \circ f)(f^{(1)})^2 + (T^{(1)} \circ f)f^{(2)} \leq 0$. We say that f is T-convex on D if $-(T \circ f)$ is concave there. It is clear that T-concavity or T-convexity is preserved under location-scale transformations and truncations of the distribution. The inflection points of $T \circ f$, perhaps for several T transformations, is typically the information needed to construct a generator and we note that the transformation can often be taken to be the identity transformation in the interior of the distribution. For example, a general approach is to first find T_l, x_l, T_r, x_r so that f is T_l-concave with T_l increasing on $(-\infty, x_l)$ or T_l-convex with T_l decreasing on $(-\infty, x_l)$, and that f is T_r-concave with T_r increasing on $[x_r, \infty)$ or T_r-convex with T_r decreasing on $[x_r, \infty)$. We then find the inflection points of $T \circ f$ in (x_l, x_r) for some T such as the identity, so that $\{x_1, \ldots, x_m\}$ consists of these inflection points together with x_l and x_r.

Now suppose that f is T-concave or T-convex on the compact interval $[a, b)$. We construct a lower envelope function L and an upper envelope function U to f on $[a, b)$. Let t denote the linear function given by the tangent to $T \circ f$ at a and c denote the linear function given by the chord joining the points $(a, T \circ f(a))$ and $(b, T \circ f(b))$; i.e.

$$t(x) = T \circ f(a) + (T \circ f)^{(1)}(a)(x - a)$$

$$c(x) = T \circ f(a) + \frac{T \circ f(b) - T \circ f(a)}{b - a}(x - a).$$

Then on $[a, b)$ we have

$$L(x) = \begin{cases} T^{-1}(c(x)) & \text{if } \begin{array}{l} T \circ f \text{ concave,} \\ T \text{ increasing} \end{array} \text{ or } \begin{array}{l} T \circ f \text{ convex,} \\ T \text{ decreasing} \end{array} \\ T^{-1}(t(x)) & \text{if } \begin{array}{l} T \circ f \text{ concave,} \\ T \text{ decreasing} \end{array} \text{ or } \begin{array}{l} T \circ f \text{ convex,} \\ T \text{ increasing;} \end{array} \end{cases}$$

$$U(x) = \begin{cases} T^{-1}(t(x)) & \text{if } \begin{array}{l} T \circ f \text{ concave,} \\ T \text{ increasing} \end{array} \text{ or } \begin{array}{l} T \circ f \text{ convex,} \\ T \text{ decreasing} \end{array} \\ T^{-1}(c(x)) & \text{if } \begin{array}{l} T \circ f \text{ concave,} \\ T \text{ decreasing} \end{array} \text{ or } \begin{array}{l} T \circ f \text{ convex,} \\ T \text{ increasing.} \end{array} \end{cases}$$

Notice that evaluation of the envelopes L and U requires that T^{-1} be easy to evaluate. Suppose further that the antiderivatives of $T^{-1} \circ c$ and $T^{-1} \circ t$ are easy to evaluate and that we can easily generate from the density

$$g(x) = \frac{U(x)}{\int_a^b U(x)\, dx}.$$

For example, when $T = \ln$ then $T^{-1} = \exp$ and we can generate from g using inversion because $g(x) \propto \exp(\alpha + \beta x)$ on $[a, b)$ for some α and β. We then have a rejection algorithm for generating from the density proportional to f on $[a, b)$. Note that f need not be normalized for this to work and that the lower envelope $L(x)$ is available for squeezing.

If the support of f can be decomposed into a finite set $[a_i, b_i)$, $i = 1, \ldots, k$ of compact subintervals where f is either T_i-concave or T_i-convex on each $[a_i, b_i)$ then we can construct densities $g_i \propto U_i$ on $[a_i, b_i)$ as above. We then generate using the rejection density $g(x) = \sum_{i=1}^k p_i g_i(x)$ where $p_i = d_i/(d_1 + \cdots + d_k)$ and $d_i = \int_{a_i}^{b_i} U_i(x)\, dx$. We can generate from g using the method of composition (Exercise 10). In other words we first generate i with probability p_i from the discrete distribution on $\{1, 2, \ldots, k\}$ and then generate a variate from the distribution corresponding to g_i.

The only point we haven't covered in developing a rejection algorithm for f is the treatment of the tails. For this we consider only the right-tail as the left-tail can be handled in the same fashion. Suppose then that we have a transformation T and a point x_r such that f is T-concave or T-convex on $[x_r, \infty)$. In the case of T-concavity this leads to the inequality

$$\begin{aligned} T \circ f(x) &\le T \circ f(x_0) + (T \circ f(x_0))^{(1)} (x - x_0) \\ &= T \circ f(x_0) + T^{(1)} \circ f(x_0)\, f^{(1)}(x_0)\,(x - x_0) \end{aligned} \tag{3.4}$$

for all $x, x_0 \in [x_r, \infty)$. In the case of T-convexity we get the inequality

$$\begin{aligned} T \circ f(x) &\ge T \circ f(x_0) + (T \circ f(x_0))^{(1)\prime} (x - x_0) \\ &= T \circ f(x_0) + T^{(1)} \circ f(x_0)\, f^{(1)}(x_0)\,(x - x_0) \end{aligned} \tag{3.5}$$

for all $x, x_0 \in [x_r, \infty)$. As we want the right-hand sides of (3.4) and (3.5) to give upper envelopes to f when we apply T^{-1}, we see that T has to be strictly increasing in the case of (3.4) and strictly decreasing in the case of (3.5). We then obtain the inequality

$$f(x) \leq U(x) = T^{-1}\{T \circ f(x_0) + T^{(1)} \circ f(x_0) f^{(1)}(x_0)(x - x_0)\} \quad (3.6)$$

for all $x, x_0 \in [x_r, \infty)$. This immediately leads to the following result.

Theorem 3.7 (*Envelope inequality*) If the function $f \geq 0$ on $[x_r, \infty)$ and the transformation $T : (0, \infty) \to R$ are such that

(i) f is T-concave with T increasing or

(ii) f is T-convex with T decreasing

then for any $x_0 \in [x_r, \infty)$,

$$\int_{x_r}^{\infty} f(x)\,dx \leq \int_{x_r}^{\infty} T^{-1}\{T \circ f(x_0) + T^{(1)} \circ f(x_0) f^{(1)}(x_0)(x - x_0)\}dx. \quad (3.7)$$

The transformation T must be chosen so that the right-hand side of (3.7) is integrable and can be easily evaluated. As discussed in Evans and Swartz (1998b), this leads to a number of interesting inequalities for distributions. For example, in the context of the $N(0, 1)$ distribution we get the well-known *Mills ratio inequality*

$$\int_{x_r}^{\infty} \frac{1}{\sqrt{2\pi}} \exp\left(-\tfrac{1}{2}x^2\right) dx \leq \frac{1}{\sqrt{2\pi}\,x_r} \exp\left(-\tfrac{1}{2}x_r^2\right)$$

obtained by taking $T = \ln$ and $x_0 = x_r$. In fact, an optimal choice can be made for x_0 which tightens the bound.

The envelope inequality allows us to choose the value x_r so that there is very little probability in the right tail and therefore we do not have to generate from this region very often. When we do have to generate from this tail we use the density

$$\frac{U_r(x)}{\int_{x_r}^{\infty} U_r(x)\,dx}$$

where U_r is the integrand of the right-hand side of (3.7). This density is included to form a mixture component in the composite density g for the rejection algorithm. Obviously we have a similar inequality and generating algorithm for the left tail $(-\infty, x_l)$.

We note that the density g for the rejection algorithm may or may not be good in the sense of Section 3.4. But suppose we are generating from g_i over the interval $[a_i, b_i)$ and we reject the generated value y. Then we improve the upper and lower envelopes by subdividing $[a_i, b_i)$ into subintervals $[a_i, y)$ and $[y, b_i)$ and constructing new envelopes to f based on the subdivision. This adaptive step is a key component in obtaining excellent generators for particular distributions. For example, if we have a distribution for which a highly efficient generator is required, we simply apply the above algorithm and allow the adaptation to continue until the probability of obtaining an acceptance from g is large. As pointed out in Evans and Swartz (1998a) it is possible to get a lower bound on this probability. Denoting the overall lower and upper envelopes to f by L and U respectively, with L defined to be 0 outside of (x_l, x_r), we have

$$\int_{-\infty}^{\infty} L(x)\,dx \leq \int_{-\infty}^{\infty} f(x)\,dx \leq \int_{-\infty}^{\infty} U(x)\,dx$$

and this implies that, for

$$g(x) = \frac{U(x)}{\int_{-\infty}^{\infty} U(x)\,dx},$$

so that $c = \int_{-\infty}^{\infty} U(x)\,dx$, the probability of rejection is

$$1 - p = 1 - \frac{\int_{-\infty}^{\infty} f(x)\,dx}{\int_{-\infty}^{\infty} U(x)\,dx} \leq 1 - \frac{\int_{-\infty}^{\infty} L(x)\,dx}{\int_{-\infty}^{\infty} U(x)\,dx}.$$

Intuitively as we improve the envelopes we have that the integrals of L and U converge to the integral of f. Therefore the probability of accepting a variate goes to 1 as we adapt.

It is reasonable to ask at this point what transformations are available for implementing adaptive rejection sampling. We have already mentioned $T = \ln$ but others are indeed possible. For example, the power transformations defined by $T_p(f) = f^p$ for $p \neq 0$ are a useful class. Here T_p is increasing when $p > 0$, T_p is decreasing when $p < 0$ and $T_p^{-1}(x) = x^{1/p}$ for $x > 0$. Then provided that α and β are such that $\alpha + \beta x \geq 0$ on the relevant interval,

$$\int T_p^{-1}(\alpha + \beta x)\,dx = \begin{cases} \dfrac{1}{\beta}\dfrac{p}{p+1}(\alpha + \beta x)^{(p+1)/p} & \text{if } p \neq 0, -1 \\ \dfrac{1}{\beta}\ln(\alpha + \beta x) & \text{if } p = -1. \end{cases}$$

From this we see that $T_p^{-1}(\alpha + \beta x)$ is a density on $[a, b)$ for every $p \neq 0$ and is a density on $(-\infty, x_l)$ or $[x_r, \infty)$ whenever $p \in (-1, 0)$. In all of these cases the inverse cdf is easily obtained, and therefore it is easy to generate from these distributions via inversion. The requirement that $\alpha + \beta x \geq 0$ on an interval is typically easy to satisfy by requiring that all the critical points of $T \circ f$ are part of $\{x_1, \ldots, x_m\}$.

Example 3.6 $N(0, 1)$ *distribution*
While excellent algorithms exist for the $N(0, 1)$ distribution, Evans and Swartz (1998a) considered an application of the adaptive rejection algorithm to this distribution. Note that the $N(0, 1)$ density is log-concave. Therefore set $T = \ln$ and

choose $\{x_1, \ldots, x_{61}\}$ by dividing each of the intervals $[-4, -1]$, $[-1, 0]$, $[0, 1]$ and $[1, 4]$ into 15 equal-length subintervals. This gives 0.9974 as an initial lower bound on the probability of acceptance. Other than including 0 as an initial point, as it is a critical point of $\ln \circ f$, the x_i's are chosen in what seems a reasonable way to span the support of the distribution. We compare the performance of this algorithm with the IMSL algorithm drnnoa, which is based on developments in Kinderman and Ramage (1976). The accompanying documentation cites this as the fastest normal generator available in that package of routines. Based on 10^6 calls to our routine and to drnnoa, the adaptive rejection algorithm is 51% slower. Also the final lower bound on the acceptance probability is 0.9998. While this does not lead us to advocate the use of the adaptive rejection algorithm for the standard normal distribution, we note that the implementation requires minimal information concerning the normal density.

Example 3.7 *Truncated Student*(λ) *distribution*
This example was also discussed in Evans and Swartz (1998a). There are many algorithms for generating from Student distributions. We note, however, that this statement does not extend to truncated versions of these distributions. We consider generating from the entire Student distribution and then use the truncated envelopes to generate from a truncated Student. The Student (λ) density is proportional to $f(x) = \left(\lambda + x^2\right)^{-\lambda/2-1/2}$ for $\lambda > 0$ and its first and second derivatives are given by

$$f'(x) = -(\lambda+1)\left(\lambda+x^2\right)^{-\lambda/2-3/2}x$$

$$f''(x) = (\lambda+1)\left(\lambda+x^2\right)^{-\lambda/2-5/2}(\lambda x^2 + 2x^2 - \lambda).$$

Then letting $h(x) = \lambda x^2 + 2x^2 - \lambda$ we see that the inflection points of the density occur at $x = \pm\sqrt{\lambda/(\lambda+2)}$ and that the tails are convex. A similar computation shows that the tails cannot be handled by the log transformation as they are also log-convex.

To handle the tails we consider an appropriate choice of p in $h(x) = (T_p \circ f)(x) = (f(x))^p = \left(\lambda + x^2\right)^{(-\lambda/2-1/2)p}$. The first and second derivatives are

$$h^{(1)}(x) = -px(\lambda+1)\left(\lambda+x^2\right)^{(-\lambda/2-1/2)p-1}$$

$$h^{(2)}(x) = p(\lambda+1)[(p\lambda+p+1)x^2 - \lambda]\left(\lambda+x^2\right)^{(-\lambda/2-1/2)p-2}.$$

From $h''(x)$ we see that the tails of f are inevitably T_p-convex whenever $p \in [-1/(\lambda+1), 0)$ and inevitably T_p-convex for $p \in (-\infty, -1/(\lambda+1)) \cup (0, \infty)$. Noting that $-1 < -1/(\lambda+1) < 0$ for all $\lambda > 0$, we have that $T_{-1/(\lambda+1)}$ is decreasing and therefore taking $p = -1/(\lambda+1)$ gives an algorithm for the tails. Moreover, with this choice of p we have that $h''(x) < 0$ for every x and thus the

Student(λ) distribution is $T_{-1/(\lambda+1)}$-convex and this makes for a simple generating algorithm.

As a particular example we consider generating from a Student(0.5) distribution truncated to $(-1, 2)$. The initial envelopes for this distribution are obtained by truncating the initial envelopes constructed for the full Student(0.5) distribution. These in turn were based on dividing each of the intervals $[-4, -1]$, $[-1, 0]$, $[0, 1]$, $[1, 4]$ into 15 equal-length subintervals. This gives an initial lower bound on the probability of acceptance of 0.9991 for the truncated distribution and 0.6776 for the full distribution. After generating 10^4 values from the truncated distribution we obtain the lower bound 0.9992. Similarly, after generating 10^4 values from the full distribution we obtain the lower bound 0.9691 which indicates the usefulness of adaptation. Generation times are immaterial.

A number of other generation problems are discussed in Evans and Swartz (1998a). Further, the methodology can be generalized to using piecewise polynomial envelopes as described in Evans and Swartz (1998b), rather than piecewise linear envelopes. While polynomial envelopes do not lead to meaningful improvements in efficiency over the linear case there are certain problems where it is much easier to implement the polynomial envelopes as the computation of inflection points can sometimes be avoided.

We note that the formation of the envelopes requires knowledge of the derivatives of the density. In fact, this can be avoided and we refer the reader to Exercise 11.

3.7 Specialized methods

There are a number of other approaches to constructing generators that can sometimes lead to excellent algorithms. For example, the *ratio of uniforms* method due to Kinderman and Monahan (1977) can sometimes be useful. For this method, suppose that f is proportional to a density on R^k and let

$$A_f = \left\{(u, v) | v \in R^k, 0 \leq u \leq [f(v/u)]^{1/(k+1)}\right\}.$$

If we can generate (U, V) uniformly on A_f then (Exercise 15) $X = V/U$ is distributed with density proportional to f. If A_f is contained in a compact rectangle then we have an easy algorithm for generating (U, V). Although the method may be inefficient, we simply generate variates (U, V) uniformly on the rectangle until we get a pair in A_f.

The *acceptance-complement* method is based on decomposing a density f as $f = f_1 + f_2$ where $f_1, f_2 \geq 0$. Suppose then that we have density g with $f_1 \leq g$ and that a good generating algorithm for g exists. Suppose further that we can generate from the distribution with density proportional to f_2. Then generate $Y \sim g$, independently generate $U \sim U(0, 1)$ and if

$$U \geq \frac{f_1(Y)}{g(Y)}$$

generate $X \sim f_2/p$ otherwise return $X = Y$. The variate X has unconditional distribution with density proportional to f (Exercise 17). This method is due to Kronmal and Peterson (1981) and also see Devroye (1986).

The *alias method* due to Walker (1974, 1977) is useful whenever we have to generate from a distribution on a finite set of points. It is based on the following result.

Lemma 3.8 Any distribution P_k on $\{1, 2, \ldots, k\}$ can be written as

$$P_k(\{x\}) = \frac{1}{k}\sum_{i=1}^{k} Q_i(\{x\})$$

where Q_i is a probability measure concentrated on two points of $\{1, 2, \ldots, k\}$, one of which is i.

Proof (by induction on k): Let i be such that $P_k(\{i\}) \leq 1/k$ and j be such that $P_k(\{j\}) \geq 1/k$. Define

$$Q_i(\{x\}) = \begin{cases} kP_k(\{i\}) & \text{if } x = i \\ 1 - kP_k(\{i\}) & \text{if } x = j \\ 0 & \text{if } x \neq i, j \end{cases}$$

and

$$P_{k-1}(\{x\}) = \begin{cases} 0 & \text{if } x = i \\ \dfrac{k\,[P_k(\{i\}) + P_k(\{j\})] - 1}{k-1} & \text{if } x = j \\ \dfrac{k}{k-1} P_k(\{x\}) & \text{if } x \neq i, j. \end{cases}$$

Then $P_k(\{x\}) = (1/k)Q_i(\{x\}) + ((k-1)/k)P_{k-1}(\{x\})$ and since P_{k-1} is concentrated on $\{1, 2, \ldots, k\}\backslash\{i\}$ by induction we can write

$$P_{k-1}(\{x\}) = \frac{1}{k-1}\sum_{j=1, j\neq i}^{k} Q_j(\{x\})$$

where each Q_j is concentrated on two points in $\{1, 2, \ldots, k\}\backslash\{i\}$ one of which is j. This completes the result.

Once we have the Q_i the alias algorithm proceeds by generating I uniform on $\{1, 2, \ldots, k\}$ and then generating X from the two-point distribution Q_I. The alias algorithm is the fastest algorithm for generating from a discrete distribution on a finite set as it only requires the generation of a single $U(0, 1)$ variable and one comparison. However, the method requires set-up time to determine the Q_i.

To implement the alias algorithm we need two arrays $R[1], R[2], \ldots, R[k]$ and $A[1], \ldots, A[k]$ where $R[i] = Q_i(\{i\})$ and $A[i]$ contains the point $j \neq i$ where

$Q_i(\{j\}) > 0$. The alias algorithm then proceeds as follows:

1. generate $U \sim U(0, 1)$;
2. $V \leftarrow kU$;
3. $I \leftarrow \lceil V \rceil$;
4. $W \leftarrow I - V$;
5. if $W \leq R[I]$ then return $X \leftarrow I$
   else return $X \leftarrow A[I]$.

In Exercise 19 you are asked to prove that this algorithm does generate a value from the mixture $(1/k) \sum_{i=1}^{k} Q_i(\{x\})$.

There are many possible choices for the arrays $R[i]$ and $A[i]$ for $i \in \{1, \ldots, K\}$. One simple way to construct these is to initially let $R[i] = P(\{i\})$, $A[i] = i$ and then iteratively take away probability from the largest R so as to bring the smallest R up to the value $1/k$; i.e. if $R[i]$ is the smallest and $R[j]$ is the largest, then let $R[j] \leftarrow R[j] - (1/k - R[i])$ and $A[i] \leftarrow j$. In Exercise 20 you are asked to show that this construction method establishes arrays with the appropriate properties.

Another problem of considerable importance is to generate a simple random sample from $\{1, \ldots, N\}$; i.e. generate a subset of size n in such a way that each subset has probability $1/\binom{N}{n}$ of being chosen. A naive way to do this is to repeatedly generate discrete uniform variates from $\{1, \ldots, N\}$ and stop when n unique values have been obtained. Alternatively, we can examine the elements $1, \ldots, N$ sequentially until n have been selected. For example, if k have already been examined and r elements have been chosen, then select the $(k + 1)$-st element with probability $(n - r)/(N - k)$. In Exercise 24 you are asked to show that this algorithm accomplishes simple random sampling. A modification of this due to Bissell (1986) currently gives the best algorithm for simple random sampling. In Pinkham (1987) this algorithm is modified to cover the case when N is unknown.

3.8 Generating from multivariate distributions

For multivariate distributions the typical approach is to reduce the problem to a sequence of one-dimensional generation problems. For example, if $X = (X_1, \ldots, X_k)$ with density f then we can generate from f by generating sequentially

$$X_1 \sim f_1(\cdot)$$
$$X_2 \sim f_2(\cdot|X_1)$$
$$\vdots$$
$$X_k \sim f_k(\cdot|X_1, \ldots, X_{k-1})$$

where $f_i(\cdot|X_1, \ldots, X_{i-1})$ denotes the conditional density of X_i given $X_1, \ldots, X_{i-1}$. Of course, it is only in specialized circumstances that this algorithm works as it can be very difficult to determine the conditionals let alone develop efficient generating algorithms. There are a number of situations, however, where this approach works well.

Example 3.8 *Multinomial*$(n, p_1, \ldots, p_k)$
We have that X_i given $X_1, \ldots, X_{i-1}$ is distributed

$$\text{binomial}\,(n - X_1 - \cdots - X_{i-1}, p_i / (1 - p_1 - \cdots - p_{i-1}))\,;$$

and $X_1 \sim \text{binomial}\,(n, p_1)$ (Exercise 25). Therefore we have an algorithm for the multinomial distribution provided that we have an algorithm for binomial distributions.

Example 3.9 *Dirichlet*$(r_1, \ldots, r_{k+1})$
We have that X_i given $X_1, \ldots, X_{i-1}$ is distributed

$$(1 - X_1 - \cdots - X_{i-1})\text{beta}\,(r_i, r_{i+1} + \cdots + r_k)\,;$$

and $X_1 \sim \text{beta}\,(r_1, r_2 + \cdots + r_{k+1})$ (Exercise 26). Therefore we have an algorithm for the Dirichlet distribution provided that we have an algorithm for beta distributions.

Another method which is useful with certain multivariate distributions is via transformations.

Example 3.10 *Multivariate normal with mean $\mu \in R^k$ and variance $\Sigma \in R^{k \times k}$*
We restrict attention to the situation where Σ is positive definite and then $X \sim N_k\,(\mu, \Sigma)$ has density on R^k given by

$$f(x) = (2\pi)^{-k/2}\,|\Sigma|^{-1/2} \exp\left\{-\tfrac{1}{2}(x - \mu)^t \Sigma^{-1} (x - \mu)\right\}.$$

As is well known, if $Z \sim N_k(0, I)$ and $\Sigma^{1/2}$ is a positive definite square root of Σ then $X \sim \mu + \Sigma^{1/2} Z$. Also $Z \sim N_k(0, I)$ iff $Z_i \sim N(0, 1)$ for each i and $Z_1, \ldots, Z_k$ are statistically independent. Therefore a generator for the $N(0, 1)$ distribution yields a generator for the general $N_k\,(\mu, \Sigma)$ distribution via this transformation.

Example 3.11 *Multivariate Student with λ degrees of freedom, location vector μ and scale matrix Σ*
We restrict attention to the situation where Σ is positive definite and then $X \sim \text{Student}_k\,(\lambda, \mu, \Sigma)$ has density on R^k given by

$$f(x) = \frac{\Gamma(\lambda + k)/2}{\pi^{k/2}\Gamma(\lambda/2)} |\Sigma|^{-1/2} \left\{1 + \frac{(x - \mu)^t \Sigma^{-1} (x - \mu)}{\lambda}\right\}^{-(\lambda+k)/2} \lambda^{-k/2}.$$

As is well known, if $Z \sim N_k(0, I)$ statistically independent of $W \sim$ chi-square(λ) then

$$X \sim \mu + \Sigma^{1/2} \frac{Z}{\sqrt{W/\lambda}}.$$

Therefore a generator for the $N(0, 1)$ distribution and a generator for the chi-square distribution yields a generator for the multivariate Student distribution.

Example 3.12 *Generating random orthogonal matrices*
The space $O(k)$ of $k \times k$ orthogonal matrices is compact and it therefore makes sense to generate $Q \in O(k)$ from the uniform distribution on this set. One method is to generate a sample $z_1, \ldots, z_k$ from the $N_k(0, I)$ distribution, put $Z = (z_1 \cdots z_k) \in R^{k \times k}$ and then compute the QR decomposition $Z = QR$; i.e. apply the Gram–Schmidt process to the columns of Z. It can then be shown that Q is uniformly distributed on $O(k)$; see Stewart (1980) for a proof.

In Stewart (1980), a more efficient algorithm is described. Let $z_1, \ldots, z_{k-1}$ be independently generated such that $z_i \sim N_{k-i+1}(0, I)$. Now put

$$u_i = \frac{z_i}{\|z_i\|} - e_{i1}$$

where $e_{i1} = (1, 0, \ldots, 0)^t \in R^{k-i+1}$ and

$$\begin{aligned} H_{z_i} &= I - 2\frac{u_i u_i^t}{u_i^t u_i} \\ &= I - \frac{1}{\|z_i\|} \frac{z_i z_i^t - \|z_i\| z_i e_{i1}^t - \|z_i\| e_{i1} z_i^t + \|z_i\|^2 e_{i1} e_{i1}^t}{\|z_i\| - e_{i1}^t z_i} \end{aligned}$$

Note that $H_{z_i} \in O(k-i+1)$ and is symmetric. A matrix of the form $I - xx^t/x^t x$ is known as a *Householder transformation* and note that to premultiply a vector by such a matrix does not require the computation of the matrix since

$$\left(I - \frac{xx^t}{x^t x}\right) y = y - \frac{x^t y}{x^t x} x.$$

Now observe that

$$\begin{aligned} H_{z_i} z_i &= z_i - \frac{\|z_i\| z_i - \left(e_{i1}^t z_i\right) z_i - \|z_i\|^2 e_{i1} + \|z_i\| \left(e_{i1}^t z_i\right) e_{i1}}{\|z_i\| - e_{i1}^t z_i} \\ &= \|z_i\| e_{i1} \end{aligned}$$

and therefore the orthogonal transformation H_{z_i} transforms z_i to the first standard basis vector times the length of z_i. Let $Q_{z_i} \in O(k)$ be the orthogonal transformation given by

$$Q_{z_i} = \begin{pmatrix} I & 0 \\ 0 & H_{z_i} \end{pmatrix}$$

and let $Q = Q_{z_1} Q_{z_2} \cdots Q_{z_{k-1}}$. The proof in Stewart (1980) then establishes that Q is uniformly distributed on $O(k)$. This paper also points out that generating Q via the Gram–Schmidt algorithm requires $O(n^3)$ operations while this algorithm requires $O(n^2)$ operations. Also see Anderson, Olkin and Underhill (1987) for a discussion of alternative methods for generating a random orthogonal matrix.

In general, the problem of generating from a multivariate distribution is quite difficult. In Chapter 7 we discuss an approximate method that works for many multivariate distributions.

3.9 Exercises

(1) Complete the proof of Lemma 3.2.

(2) Prove that if $(a, M) = 1$ and $a^j \bmod M = 1 \bmod M$ then $h|j$ where a belongs to the exponent h modulo M.

(3) Suppose that P, Q are probability measures on $Z_m = \{0, 1, \ldots, M-1\}$ with addition carried out modulo M on this space. Let a metric be defined on the set of probability measures on Z_m by $\|P - Q\| = \sum_{i=0}^{m-1} |P(\{i\}) - Q(\{i\})|$. Suppose $X \sim P$, $Y \sim Q$ and then determine $P*Q$, the probability measure of $X+Y$. If U denotes uniform measure on Z_m then prove that $\|P*Q - U\| \leq \min\{\|P-U\|, \|Q-U\|\}$. Note that this supports combining two uniform generators by addition.

(4) Establish properties (i)–(iv) of F^{-1}.

(5) Prove that F is continuous iff F^{-1} is strictly increasing and F is strictly increasing iff F^{-1} is continuous.

(6) Construct an inversion algorithm for the geometric(p) distribution.

(7) Suppose that $P(X = i) = p_i$ for $i = 1, \ldots, K$. Then determine F^{-1}.

(8) Implement an inversion algorithm for the density $f(x) \propto (x+2)^2(x+3)$ on $(0, 1)$.

(9) Suppose that $X = (X_1, \ldots, X_k)$ is distributed with distribution function F and let $F_i(\cdot|X_1, \ldots, X_{i-1})$ denote the conditional distribution function of X_i given $X_1, \ldots, X_{i-1}$. Putting $U = (U_1, \ldots, U_k)$ where $U_i = F_i(\cdot|X_1, \ldots, X_{i-1})$ establish that U is uniformly distributed on $[0, 1]^k$. The transformation $X \to U$ is known as the *probability transform*.

(10) Suppose that random variables X and Y are such that we can easily generate from the distribution for Y and from each of the conditional distributions of X given Y. Show that this gives a generating algorithm for X. This is called the *method of composition*.

(11) Suppose that the function f is concave on the interval $[a, b]$ and suppose that we know the point $c \in [a, b]$ where f takes its maximum. Provide an algorithm for constructing upper envelopes to f on $[a, b]$ based only on evaluations of f and that more closely approximate f as we increase the number of these evaluations.

(12) Construct an adaptive rejection algorithm for the beta (2,2) distribution.

(13) Construct an adaptive rejection algorithm for the beta (0.5,0.5) distribution.

(14) Construct an adaptive rejection algorithm for the distribution with density proportional to $(x+1)^2 \exp\left(-x^2/2\right)$.

(15) Prove that the ratio of uniforms method described in the text generates from the intended distribution.

(16) Apply the ratio of uniforms method to get a generator for the Student(λ) distribution when $\lambda \geq 1$.

(17) Prove that the acceptance-complement method works.

(18) Apply the acceptance-complement method to generating from a Cauchy distribution truncated to the interval $[-1, 1]$.

(19) Prove that the alias algorithm described in the text correctly generates from the mixture

$$P_k(\{x\}) = \frac{1}{k} \sum_{i=1}^{k} Q_i(\{x\}).$$

(20) Establish the correctness of the method discussed in the text for constructing the arrays $R[1], R[2], \ldots, R[k]$ and $A[1], \ldots, A[k]$, necessary for implementation of the alias algorithm.

(21) Prove that if $U_1, U_2 \sim \mathrm{U}(0, 1)$ then $Y_1 = \cos(2\pi U_1)\sqrt{-2\ln U_2}$ and $Y_2 = \sin(2\pi U_1)\sqrt{-2\ln U_2}$ are statistically independent and $N(0, 1)$. This is known as the *Box–Müller* algorithm for the normal.

(22) Establish that if $V = (V_1, V_2)$ is uniformly distributed on $[-1, 1]^2$ then generating such pairs until $V_1^2 + V_2^2 \leq 1$ and returning that V gives a pair uniformly distributed on the unit disk $B^2 = \{(x, y) | x^2 + y^2 \leq 1\}$ in R^2. What is the expected number of pairs generated until an acceptance is obtained?

(23) Suppose V is generated uniformly on the unit disk B^2 (see Exercise 22) and $X = \sqrt{-4\ln \|V\|}\, V$. Show that $X \sim N_2(0, I)$. This algorithm is known as the *polar method* for the normal.

(24) Show that the methods discussed in Section 3.7 for generating a simple random sample from $\{1, \ldots, N\}$ do indeed accomplish this.

(25) Establish the result in Example 3.8.

(26) Establish the result in Example 3.9.

(27) Show that if $Z \sim N_k(0, I)$ then $Z/\|Z\| \sim$ uniform on the unit sphere S^{k-1}.

4
Asymptotic approximations

4.1 Introduction

Asymptotic methods for integrals have long been used to approximate the special functions of mathematical physics that have integral representations. Two of the most common techniques are the Laplace method and the method of steepest descent or saddle-point method. In this chapter we will describe the essence of these methods so that they can be used in applications. For a more complete treatment, see for example, de Bruijn (1981), Murray (1984), Bleistein and Handelsman (1986) and Wong (1989).

One cautionary remark needs to be stated about these methods. In general there is no way to assess the accuracy of a particular approximation. In practice one looks at a number of specific applications of the particular asymptotic approximation where the exact answer can be computed. If the errors are suitably small then one proceeds to use it in a particular case. From a certain point of view this seems unsatisfactory but there is no denying that this has produced approximations of great practical value. We develop some of these here.

Perhaps the greatest value in asymptotic approximations is that they can produce great simplifications in problems whose mathematical characteristics are difficult to describe. Further, often we are concerned with developing software for a class of integration problems that depend upon some other parameter; e.g. sample size in Bayesian integration problems. Characteristic of such problems is that different integration strategies may be needed dependent on the value of the parameter; e.g. asymptotic approximations may be needed when the sample size is large. Therefore asymptotic approximations have practical as well as theoretical value and the practitioner really cannot avoid their use. On the other hand, there are asymptotic approximation techniques that are very complicated. Unless there are no competitors for the approximation, it is hard to justify the use of asymptotic methods in such contexts.

In Section 4.1 we provide details about the mathematics that underlies the reasoning behind asymptotic approximations to integrals. In Section 4.2 we give a detailed description of some of the most common asymptotic techniques for approximating one-dimensional integrals and in Section 4.3 consider some aspects of their multidimensional analogs.

4.2 Asymptotic expansions

All the functions in this section will be defined on some open set $A \subset \mathcal{C}$, where $\mathcal{C}$ denotes the complex plane, and map into $\mathcal{C}$. Note that we will consider $\mathcal{C}$ as a subset of the extended plane $\mathcal{C} \cup \{\infty\}$ which in turn can be identified with the Riemann sphere S with the north pole identified with the point ∞ and $\mathcal{C}$ identified with the remaining part of the sphere under stereographic projection; see Ahlfors (1966) for details. So if A is unbounded then under this identification $\bar{A}$, the closure of A, will contain the point ∞.

For two functions ϕ, ϑ defined on A we write $\phi \sim \vartheta$ at $x_0 \in \bar{A}$, whenever

$$\lim_{x \to x_0} \frac{\phi(x)}{\vartheta(x)} = 1$$

and we say that the functions are *asymptotically equivalent* at x_0. Now let $\|\cdot\|$ denote Euclidean distance in $\mathcal{C}$. If $x_0 \in \bar{A} \cap \mathcal{C}$ then we write $\phi = O(\vartheta)$ at x_0 if there exist constants $c, \delta > 0$ such that

$$|\phi(x)| \leq c|\vartheta(x)| \tag{4.1}$$

whenever $x \in A$ and $0 < \|x - x_0\| < \delta$. If $x_0 \in \bar{A}$ and $x_0 = \infty$ then we write $\phi = O(\vartheta)$ if there exist constants $c, \delta > 0$ such that whenever $x \in A$ and $\|x\| > \delta$ then (4.1) holds. If $x_0 \in \bar{A} \cap \mathcal{C}$ then we write $\phi = o(\vartheta)$ if for each $\epsilon > 0$ there exists $\delta > 0$ such that

$$|\phi(x)| \leq \epsilon|\vartheta(x)| \tag{4.2}$$

whenever $x \in A$ and $0 < \|x - x_0\| < \delta$. If $x_0 \in \bar{A}$ and $x_0 = \infty$ then we write $\phi = o(\vartheta)$ if for each $\epsilon > 0$ there exists $\delta > 0$ such that (4.2) holds whenever $x \in A$ and $\|x\| > \delta$. This order notation is very useful in simplifying certain calculations and is extended in an obvious way to complex-valued functions defined on R^k.

The basic definitions are given below.

Definition 4.1 A sequence $\{\phi_0, \phi_1, \dots\}$ of functions defined on A is an *asymptotic sequence* at $x_0 \in \bar{A}$ if for every n

$$\lim_{x \to x_0} \frac{\phi_{n+1}(x)}{\phi_n(x)} = 0.$$

For example, $\phi_n(x) = x^n$ is an asymptotic sequence at $x_0 = 0$ and $\phi_n(x) = x^{-n}$ is an asymptotic sequence at $x_0 = \infty$. If the function ϕ_0 in Definition 4.1 has a finite limit as $x \to x_0$ then $\lim_{x \to x_0} \phi_n(x) = 0$ for $n > 0$.

Definition 4.2 If the sequence $\{\phi_n : n = 1, 2, \dots\}$ is asymptotic at $x_0 \in \bar{A}$ then

$$\sum_{i=0}^{\infty} a_i \phi_i$$

with $a_i \in C$ is an *asymptotic expansion* for the function f at x_0 if for each n

$$f(x) = \sum_{i=0}^{n} a_i \phi_i(x) + o(\phi_n)$$

at x_0 and we write $f \sim \sum_{i=0}^{\infty} a_i \phi_i$ at x_0.

A definition equivalent to this is to say that for a sequence $\{\phi_n\}$ asymptotic at $x_0 \in \bar{A}$ then $f \sim \sum_{i=0}^{\infty} a_i \phi_i$ at x_0 if for each $n \geq 1$

$$f(x) = \sum_{i=0}^{n-1} a_i \phi_i(x) + O(\phi_n)$$

at x_0 (Exercise 6). Note that the symbol $\sum_{i=0}^{\infty} a_i \phi_i$ need not have any meaning in the sense of being a convergent series for certain values of x and in fact in the most useful applications it is not convergent. The point is that the sum $\sum_{i=0}^{n} a_i \phi_i(x)$ gives an accurate approximation to $f(x)$ when x is in a neighborhood of x_0 of sufficiently small size. While for a given n the approximation improves the closer x gets to x_0 it does not necessarily improve, at a particular x, as we increase n and in fact could get significantly worse. It is true, however, that the error is going to 0 faster as $x \to x_0$ for higher values of n. We will illustrate this in Example 4.3.

If $f(x) = \sum_{i=0}^{\infty} a_i (x - x_0)^i$ in a neighborhood of x_0; i.e. f is analytic at x_0 and so the series is convergent, then this series is also an asymptotic expansion of f at x_0 with respect to the asymptotic sequence $\{(x - x_0)^n\}$. Similarly if $f(x) = \sum_{i=0}^{\infty} a_i (x - x_0)^{-i}$ in a deleted neighborhood of x_0 then this series is an asymptotic expansion of f at x_0 with respect to the asymptotic sequence $\{(x - x_0)^{-n}\}$.

We have the following basic results about asymptotic expansions.

Lemma 4.3 If $f \sim \sum_{i=0}^{\infty} a_i \phi_i$ at x_0 for the asymptotic sequence $\{\phi_n\}$ then the a_n are unique.

Proof: We have that

$$a_0 = \lim_{x \to x_0} \frac{f(x)}{\phi_0(x)}, \quad a_1 = \lim_{x \to x_0} \frac{(f(x) - a_0 \phi_0(x))}{\phi_1(x)},$$

etc.

Note that it could happen that $a_n = 0$ for every n but $f \neq 0$ in a neighborhood of x_0. For example, if $f(x) = \exp(-a|x|^{-m})$ for $a, m > 0$ and $\phi_n(x) = x^n$ then $f \sim \sum_{i=0}^{\infty} 0\, \phi_i$ at 0.

Lemma 4.4 If $f \sim \sum_{i=0}^{\infty} a_i \phi_i$ and $g \sim \sum_{i=0}^{\infty} b_i \phi_i$ at x_0 for the asymptotic sequence $\{\phi_n\}$ at x_0 then $cf + dg \sim \sum_{i=0}^{\infty} (ca_i + db_i)\phi_i$ at x_0.

Proof: Exercise 7.

In particular if $f \sim \sum_{i=0}^{\infty} a_i \phi_i$ and $g \sim \sum_{i=0}^{\infty} 0\,\phi_i$ at x_0 then $f + g \sim \sum_{i=0}^{\infty} a_i \phi_i$ at x_0.

Consider the class $\mathcal{A}(\{\phi_n\}, x_0)$ of functions that have asymptotic expansions with respect to the asymptotic sequence $\{\phi_n\}$ at x_0. We define an equivalence relation on this class via $f \equiv g$ whenever $f - g \sim \sum_{i=0}^{\infty} 0\phi_i$ or equivalently whenever f and g have the same asymptotic expansion. Then Lemma 4.4 establishes that the equivalence classes form a linear space over C. Of course if $f \equiv g$ then $f \sim g$ but not conversely.

4.3 Approximating one-dimensional integrals

Much more can be said about asymptotic expansions in general but we now turn our attention to obtaining asymptotic expansions for functions that can be represented as integrals.

4.3.1 INTEGRATION BY PARTS

One technique that sometimes works for obtaining an asymptotic expansion for an integral is to use *integration by parts*. We illustrate this via an example which demonstrates the main characteristics of the approach. A general discussion of the integration by parts technique and conditions for its applicability are provided in Bleistein and Handelsman (1986).

Example 4.1 $N(0, 1)$ *tail probabilities*
Consider approximating

$$\sqrt{2\pi}\{1 - \Phi(x)\} = \int_x^{\infty} e^{-z^2/2}\, dz \tag{4.3}$$

for large x; i.e. we are approximating a tail probability $1 - \Phi(x)$. For this we put $u(z) = 1/z$, $dv(z) = z\exp(-z^2/2)$ and then by integration by parts (4.3) equals

$$\frac{1}{x} e^{-x^2/2} - \int_x^{\infty} \frac{1}{z^2} e^{-z^2/2}\, dz. \tag{4.4}$$

Then applying integration by parts to the integral in (4.4) with $u(z) = 1/z^3$, $dv(z) = z\exp(-z^2/2)$ we get

$$\frac{1}{x} e^{-x^2/2} - \frac{1}{x^3} e^{-x^2/2} + \int_x^{\infty} \frac{3}{z^4} e^{-z^2/2}\, dz$$

and we see that continuing in this way we get the formal series $\sum_{i=0}^{\infty} a_i \phi_i$ where $a_0 = 1$,

$$a_i = 1 \cdot 3 \cdots (2i - 1)$$

for $i > 0$ and

$$\phi_i(x) = (-1)^i x^{-(2i+1)} e^{-x^2/2}$$

which is easily seen (Exercise 8) to be an asymptotic sequence at ∞. Further we have that

$$\begin{aligned}
&\lim_{x\to\infty}\left|\frac{\sqrt{2\pi}\{1-\Phi(x)\}-\sum_{i=0}^{n}a_i\phi_i(x)}{a_n(2n+1)x^{-(2n+1)}e^{-x^2/2}}\right| \\
&= \lim_{x\to\infty} x^{2n+1}e^{x^2/2}\int_x^\infty \frac{1}{z^{2n+2}}e^{-z^2/2}\,dz \\
&\le \lim_{x\to\infty}\frac{1}{x}\int_x^\infty e^{-(z^2-x^2)/2}\,dz \\
&= \lim_{x\to\infty}\frac{1}{x}\int_x^\infty e^{-(z-x)(z+x)/2}\,dz \\
&\le \lim_{x\to\infty}\frac{1}{x}\int_x^\infty e^{-(z-x)}\,dz \\
&= 0
\end{aligned}$$

so that

$$\left[1+\sum_{i=1}^{\infty}(-1)^i\frac{[1\cdot 3\cdots(2i-1)]}{x^{2i}}\right]\frac{1}{x}\frac{e^{-x^2/2}}{\sqrt{2\pi}} \tag{4.5}$$

is an asymptotic expansion at ∞ for $1-\Phi(x)$ for $x>0$. Notice that for any $x>0$ the series does not converge because

$$\frac{1\cdot 3\cdots(2i-1)}{x^{2i-1}} = x\left(\frac{1}{x^2}\frac{3}{x^2}\cdots\frac{2i-1}{x^2}\right)\to\infty$$

as $i\to\infty$; i.e. the i-th term of the series does not converge to 0 as i increases.

Note that because the sign of the remainder term R_n, i.e.

$$R_n(x) = \frac{1}{\sqrt{2\pi}}(-1)^{n+1}[1\cdot 3\cdots(2n+1)]\int_x^\infty \frac{1}{z^{2n+2}}e^{-z^2/2}\,dz,$$

oscillates we have that the approximations $\sum_{i=0}^{n}a_i\phi_i(x)$ will oscillate above and below the exact value as n increases. The first term in this asymptotic expansion gives a well-known upper bound on the tail probability known as the Mills ratio inequality; namely for $x>0$

$$1-\Phi(x)\le\frac{1}{x}\frac{e^{-x^2/2}}{\sqrt{2\pi}}. \tag{4.6}$$

Further we must have that

$$|R_n(x)|\le\frac{[1\cdot 3\cdots(2i+1)]}{x^{2n+3}}\frac{e^{-x^2/2}}{\sqrt{2\pi}};$$

i.e. the remainder is bounded in magnitude by the first term neglected in the expansion because they have the same sign. This suggests that the optimal place to stop this expansion is at the term of smallest magnitude and of course this depends on x.

Table 4.1. Asymptotic expansion approximations to a tail probability of the standard normal with exact value 0.02275.

n	Approximation
0	0.02700
1	0.02025
2	0.02531
3	0.01898
4	0.03005

Consider this approximation when $x = 2$. The exact value is given by 0.02275 to five decimal places. In Table 4.1 we give the approximations obtained from the first five terms. The best approximation is given by the choice $n = 1$ and after that the approximation deteriorates.

4.3.2 WATSON'S LEMMA

In this section we develop a basic result for determining an asymptotic expansion for the class of integrals of the form

$$\int_0^\infty f(t)e^{-xt}\,dt \tag{4.7}$$

for large values of $x \in R^1$. The integral (4.7) as a function of x is the Laplace transform of f and we suppose we want to approximate this for large x. The following result is a version of what is known as *Watson's lemma.* For this we suppose have available the values of the *gamma function*

$$\Gamma(\nu) = \int_0^\infty t^{\upsilon-1}e^{-t}\,dt$$

for $\nu \in (0, \infty)$.

Theorem 4.5 For the integral (4.7) suppose that $f(t) = t^\lambda g(t)$ for $\lambda > -1$ where g is analytic at 0 with Taylor expansion

$$g(t) = \sum_{i=0}^\infty a_i t^i,$$

$g(0) = a_0 \neq 0$ and $|g(t)| \leq ce^{dt}$ for some $c, d > 0$. Then

$$\int_0^\infty f(t)e^{-xt}\,dt \sim \sum_{i=0}^\infty \frac{a_i\Gamma(\lambda+i+1)}{x^{\lambda+i+1}}$$

as $x \to \infty$.

Proof: Let t_0 be such that the expansion for g holds as a convergent series in $[-t_0, t_0]$. Then

$$\int_0^\infty f(t)e^{-xt}\,dt = \int_0^{t_0} f(t)e^{-xt}\,dt + \int_{t_0}^\infty f(t)e^{-xt}\,dt$$

and for $x > d+1$

$$\begin{aligned}\left|\int_{t_0}^\infty f(t)e^{-xt}\,dt\right| &\leq ce^{-(x-d)t_0}\int_{t_0}^\infty t^\lambda e^{-(x-d)(t-t_0)}\,dt\\ &\leq ce^{-(x-d)t_0}\int_{t_0}^\infty t^\lambda e^{-(t-t_0)}\,dt\\ &= o\left(\frac{1}{x^{m+\lambda+1}}\right)\end{aligned}$$

as $x \to \infty$ for every m since $x^{m+\lambda+1}e^{-(x-d)t_0} \to 0$ as $x \to \infty$ for every m. Therefore an asymptotic expansion for $\int_{t_0}^\infty f(t)e^{-xt}\,dt$ as $x \to \infty$ is given by $\sum_{i=0}^\infty 0x^{-(i+\lambda+1)}$ and Lemma 4.4 implies that we can ignore $\int_{t_0}^\infty f(t)e^{-xt}\,dt$ in deriving the expansion for $\int_0^\infty f(t)e^{-xt}\,dt$.

So we restrict our attention to

$$\int_0^{t_0} f(t)e^{-xt}\,dt = \sum_{i=0}^n a_i\int_0^{t_0} t^{i+\lambda}e^{-xt}\,dt + \int_0^{t_0}\left(\sum_{i=n+1}^\infty a_it^i\right)t^\lambda e^{-xt}\,dt. \tag{4.8}$$

Now observe that when $x > 1$

$$\begin{aligned}\frac{\Gamma(\lambda+i+1)}{x^{\lambda+i+1}} &= \int_0^\infty t^{i+\lambda}e^{-xt}\,dt\\ &= \int_0^{t_0} t^{i+\lambda}e^{-xt}\,dt + \int_{t_0}^\infty t^{i+\lambda}e^{-xt}\,dt\\ &= \int_0^{t_0} t^{i+\lambda}e^{-xt}\,dt + e^{-xt_0}\int_0^\infty (t+t_0)^{i+\lambda}e^{-xt}\,dt\\ &\leq \int_0^{t_0} t^{i+\lambda}e^{-xt}\,dt + e^{-xt_0}\int_0^\infty (t+t_0)^{i+\lambda}e^{-t}\,dt\\ &= \int_0^{t_0} t^{i+\lambda}e^{-xt}\,dt + o\left(\frac{1}{x^{\lambda+m+1}}\right)\end{aligned}$$

as $x \to \infty$ for every m. Therefore, applying Lemma 4.4 to 4.8, we have that

$$\int_0^\infty f(t)e^{-xt}\,dt = \sum_{i=0}^n \frac{a_i\Gamma(\lambda+i+1)}{x^{\lambda+i+1}} + o\left(\frac{1}{x^{\lambda+m+1}}\right) + \int_0^{t_0} r_n(t)t^\lambda e^{-xt}\,dt \tag{4.9}$$

for every m and where $r_n(t) = \sum_{i=n+1}^{\infty} a_i t^i$. We have that $|r_n(t)| \leq bt^{n+1}$ for all $t \in [-t_0, t_0]$ for some constant b which implies

$$\begin{aligned}
\left| \int_0^{t_0} r_n(t) t^\lambda e^{-xt} \, dt \right| &\leq b \int_0^{t_0} t^{\lambda+n+1} e^{-xt} \, dt \\
&\leq b \int_0^{\infty} t^{\lambda+n+1} e^{-xt} \, dt \\
&= \frac{b\Gamma(\lambda + n + 2)}{x^{\lambda+n+2}} \\
&= o\left(\frac{1}{x^{\lambda+n+1}} \right)
\end{aligned}$$

as $x \to \infty$. This, together with Lemma 4.4 and 4.9 establishes the result.

Note that it is immediate from the proof of Theorem 4.5 that the asymptotic expansion is also valid when we replace the infinite upper limit in (4.7) by any finite value say t_1. This is because $\int_{t_0}^{t_1} f(t) e^{-xt} \, dt$ has the 0 asymptotic expansion with respect to $\{x^{-(\lambda+n)}\}$. More general versions of Watson's lemma can be found in the references cited above.

Example 4.2 *Gamma expectations*
The expectation of the function f with respect to the gamma(α, β) distribution is given by

$$\int_0^{\infty} f(t) \frac{1}{\Gamma(\alpha)} \left(\frac{t}{\beta} \right)^{\alpha-1} e^{-t/\beta} \frac{1}{\beta} \, dt.$$

Therefore we have an asymptotic expansion for this integral when β is small from Theorem 4.5, provided of course that f satisfies the conditions of that theorem. For example, if $f(t) = t^{1/2}$ then $\lambda = \alpha - 1/2$, $a_0 = 1$ and so

$$\int_0^{\infty} t^{1/2} \frac{1}{\Gamma(\alpha)} \left(\frac{t}{\beta} \right)^{\alpha-1} e^{-t/\beta} \frac{1}{\beta} \, dt \sim \frac{\beta^{-\alpha}}{\Gamma(\alpha)} \frac{\Gamma(\alpha + 1/2)}{\beta^{-(\alpha+1/2)}} = \frac{\Gamma(\alpha + 1/2)}{\Gamma(\alpha)} \beta^{1/2}$$

which is in fact exact for all β. If $f(t) = \cos t$ then $\lambda = \alpha - 1$, $a_0 = 1$, $a_1 = 0$ and $a_2 = -1/2$ so

$$\begin{aligned}
\int_0^{\infty} \cos t \, \frac{1}{\Gamma(\alpha)} \left(\frac{t}{\beta} \right)^{\alpha-1} e^{-t/\beta} \frac{1}{\beta} \, dt &\sim \frac{\beta^{-\alpha}}{\Gamma(\alpha)} \left[\frac{\Gamma(\alpha)}{\beta^{-\alpha}} - \frac{1}{2} \frac{\Gamma(\alpha + 1)}{\beta^{-(\alpha+1)}} \right] \\
&= 1 - \tfrac{1}{2} \alpha \beta
\end{aligned}$$

as $\beta \to 0$.

4.3.3 LAPLACE'S APPROXIMATION

One of the most important asymptotic approximations is Laplace's approximation. For this suppose that we have an integral of the form

$$\int_a^b h(t) \exp\{-\lambda k(t)\}\, dt \tag{4.10}$$

where $a, b \in R^1 \cup \{-\infty, \infty\}$. The Laplace formula says that, under conditions,

$$\int_a^b h(t) \exp\{-\lambda k(t)\}\, dt \sim h(\hat{t})(2\pi)^{1/2}\{\lambda k^{(2)}(\hat{t})\}^{-1/2} \exp\{-\lambda k(\hat{t})\}$$

as $\lambda \to \infty$ where $\hat{t} \in (a, b)$ is the global minimum of k. The basic idea is that the big contributions to (4.10) occur locally about $\hat{t}$ when λ is large. A modification of the Laplace technique arises when k has a finite number of relative minima in (a, b). Then we divide (a, b) into disjoint subintervals each of which contains a relative minimum, apply the Laplace formula to each of these and sum them. The following is a simple application of the Laplace technique to a well-known function.

Example 4.3 *Stirling's approximation to the gamma function*
The gamma function for real argument $\alpha \in (0, \infty)$ can be written as

$$\begin{aligned}\Gamma(\alpha) &= \int_0^\infty t^{\alpha-1} e^{-t}\, dt \\ &= \int_0^\infty \exp\{(\alpha - 1) \ln t - t\}\, dt \\ &= (\alpha - 1)^\alpha \int_0^\infty \exp\{(\alpha - 1)(\ln(u) - u)\}\, du\end{aligned}$$

where the last equality follows after the transformation $t \to u = t/(\alpha - 1)$. We want to approximate this for large α using Laplace's formula. For this we see that $h(u) \equiv 1$ and $k(u) = -(\ln(u) - u)$ has a global minimum at $u = 1$. Further $k(1) = 1$, $k^{(2)}(1) = 1$ and therefore an application of Laplace's formula yields

$$\Gamma(\alpha) \sim (2\pi)^{1/2}(\alpha - 1)^{\alpha-1/2} e^{-(\alpha-1)}.$$

Using $\Gamma(\alpha) = (\alpha - 1)\Gamma(\alpha - 1)$ this implies that

$$\begin{aligned}\Gamma(\alpha - 1) &\sim (2\pi)^{1/2}(\alpha - 1)^{\alpha-1/2} e^{-(\alpha-1)} \frac{1}{\alpha - 1} \\ &= (2\pi)^{1/2}(\alpha - 1)^{(\alpha-1)-1/2} e^{-(\alpha-1)}\end{aligned}$$

which is the well-known approximation to the gamma function known as *Stirling's approximation*.

In fact it is possible in many contexts to develop more refined approximations; namely to expand (4.10) in powers of $\lambda^{-1/2}$. We develop such an asymptotic expansion here as stated in Theorem 4.6. In particular (4.10) is a corollary of this result. The proof of Theorem 4.6 is somewhat complicated and readers who wish to avoid this can proceed to Section 4.4 where a direct, and somewhat simpler, proof is provided for Laplace's formula.

Theorem 4.6 If k and h are such that

(i) $\int_a^b |h(t)| \exp\{-\lambda k(t)\}\, dt < \infty$ for all $\lambda \geq \lambda_0$;

(ii) there is a point $\hat{t} \in (a, b)$ such that for every $\epsilon > 0$ with $(\hat{t} - \epsilon, \hat{t} + \epsilon) \subset (a, b)$ then

$$\inf\{k(t) - k(\hat{t}) \colon t \in (a, b) \backslash (\hat{t} - \epsilon, \hat{t} + \epsilon)\} > 0;$$

i.e. $\hat{t}$ is an isolated global minimum for k in (a, b);

(iii) $k^{(5)}$ continuous in an interval about $\hat{t}$;

(iv) $h^{(3)}$ exists and is continuous in an interval about $\hat{t}$;

then

$$\int_a^b h(t) \exp\{-\lambda k(t)\}\, dt = (2\pi)^{1/2} \exp\{-\lambda k(\hat{t})\} \left\{ h(\hat{t}) \frac{1}{(\lambda k^{(2)}(\hat{t}))^{1/2}} \right.$$
$$+ \left[h(\hat{t}) \left\{ -\frac{k^{(4)}(\hat{t})}{4!} \mu_4 + \frac{1}{2} \left(\frac{k^{(3)}(\hat{t})}{3!} \right)^2 \mu_6 \right\} - h^{(1)}(\hat{t}) \frac{k^{(3)}(\hat{t})}{3!} \mu_4 + h^{(2)}(\hat{t}) \frac{1}{2!} \right]$$
$$\left. \times \frac{1}{(\lambda k^{(2)}(\hat{t}))^{3/2}} \right\} + o(\lambda^{-3/2}) \tag{4.11}$$

as $\lambda \to \infty$.

Before proving this we state some corollaries that follow immediately from the theorem.

Corollary 4.7

$$\int_a^b h(t) \exp\{-\lambda k(t)\}\, dt = h(\hat{t})(2\pi)^{1/2} \{\lambda k^{(2)}(\hat{t})\}^{-1/2} \exp\{-\lambda k(\hat{t})\} + o(\lambda^{-1/2}).$$

Note that Corollary 4.7 implies but is not implied by Laplace's formula. We have the following result concerning the relative error in the Laplace approximation.

Corollary 4.8

$$\frac{\int_a^b h(t)\exp\{-\lambda k(t)\}\,dt - h(\hat{t})(2\pi)^{1/2}\{\lambda k^{(2)}(\hat{t})\}^{-1/2}\exp\{-\lambda k(\hat{t})\}}{h(\hat{t})(2\pi)^{1/2}\{\lambda k^{(2)}(\hat{t})\}^{-1/2}\exp\{-\lambda k(\hat{t})\}} = O(\lambda^{-1}).$$

Consider now the application of Theorem 4.6 to the gamma function.

Example 4.4 *Higher order approximation to the gamma function*
We determine the next term in the asymptotic expansion for the gamma function noting that $h(t) = 1$ and $k(t) = -(\ln(t) - t)$. All the derivatives of h are 0 and $k^{(3)}(t) = -2/t^3$, $k^{(4)}(t) = 6/t^4$, $\mu_4 = 3$ and $\mu_6 = 15$ so that

$$\Gamma(\alpha) \sim (2\pi)^{1/2}(\alpha-1)^{\alpha}e^{-(\alpha-1)}\left\{(\alpha-1)^{-1/2} + \frac{1}{12}(\alpha-1)^{-3/2}\right\}$$

which can be written as

$$\Gamma(\alpha-1) \sim (2\pi)^{1/2}(\alpha-1)^{\alpha-1}e^{-(\alpha-1)}\left\{(\alpha-1)^{-1/2} + \frac{1}{12}(\alpha-1)^{-3/2}\right\}.$$

It is readily seen that the proof that follows can be modified to prove Corollary 4.7 under the conditions that h have a continuous derivative in an interval about $\hat{t}$ and k have a continuous second derivative in an interval about $\hat{t}$. Further with similar conditions on higher order derivatives of h and k we use the same method of proof to rigorously establish asymptotic expansions for higher orders. In fact we will develop the proof in such a way that it is easy to generalize to obtain such expansions.

There are many other extensions of these results. For example if $h(\hat{t}) = 0$ and $h^{(1)}(\hat{t}) \neq 0$ then the relevant first term for the asymptotic approximation is the second term in (4.11) and we proceed similarly if still higher order derivatives vanish at $\hat{t}$. Many of the conditions we have stated can be substantially weakened and the reader is directed to the references cited earlier in this chapter for more details.

Of some importance for statistical applications is that it is possible to replace $\lambda k(t)$ by $k(t, \lambda)$; i.e. allow k to depend on the parameter λ as well, provided this dependence is of a suitable form. We provide some discussion of this in Section 4.4.

Proof of Theorem 4.6: We need some facts concerning integrals with respect to the normal distribution.

Lemma 4.9 Suppose that the function g is such that for all $\lambda \geq \lambda_0 > 0$

$$\int_{-\infty}^{\infty} |g(t)| \exp\left(-\frac{\lambda}{2}t^2\right) dt < \infty.$$

Then

$$\int_{\lambda^{1/6}}^{\infty} g(t)\exp\left(-\frac{\lambda}{2}t^2\right) dt = o(\lambda^p)$$

as $\lambda \to \infty$ for every $p \in R^1$.

Proof: For $\lambda \geq \lambda_0$ we have that

$$\begin{aligned}\left|\int_{\lambda^{1/6}}^{\infty} g(t)\exp\left(-\frac{\lambda}{2}t^2\right)dt\right| &\leq \int_{\lambda^{1/6}}^{\infty} |g(t)|\exp\left(-\frac{\lambda-\lambda_0}{2}t^2\right)\exp\left(-\frac{\lambda_0}{2}t^2\right)dt\\ &\leq \exp\left(-\frac{\lambda-\lambda_0}{2}\lambda^{1/3}\right)\int_0^{\infty}|g(t)|\exp\left(-\frac{\lambda_0}{2}t^2\right)dt\\ &= o(\lambda^p)\end{aligned}$$

for any $p \in R^1$.

Note that $\lambda^{1/6}$ in Lemma 4.9 was chosen because it is precisely what we need at a point later in the proof of the theorem. Obviously other choices could have been made and the result would remain valid. Also a similar result holds for $\int_{-\infty}^{-\lambda^{1/6}} g(t)\exp(-\lambda t^2/2)\,dt$. We will need the moments of the $N(0, 1)$ distribution; namely,

$$\mu_i = \int_{-\infty}^{\infty} u^i \frac{1}{\sqrt{2\pi}}\exp\left(-\tfrac{1}{2}u^2\right)du$$

which can be easily shown to satisfy $\mu_{2i+1} = 0$ and

$$\mu_{2i} = (2i-1)(2i-3)\cdots 1.$$

Without loss of generality we assume hereafter that $\hat{t} = 0$ and $k(0) = 0$ (by premultiplying (4.10) by $\exp\{\lambda k(\hat{t})\}$ and changing the variable of integration to $(t - \hat{t})$ as this simplifies the notation. We have the following result.

Lemma 4.10 If $\epsilon > 0$ is such that $(-\epsilon, \epsilon) \subset (a, b)$ then

$$\int_{(a,b)\backslash(-\epsilon,\epsilon)} h(t)\exp\{-\lambda k(t)\}\,dt = o(\lambda^p)$$

as $\lambda \to \infty$ for every $p \in R^1$.

Proof: We have that

$$\begin{aligned}&\left|\int_{(a,b)\backslash(-\epsilon,\epsilon)} h(t)\exp\{-\lambda k(t)\}\,dt\right|\\ &\quad= \left|\int_{(a,b)\backslash(-\epsilon,\epsilon)} h(t)\exp\{-(\lambda-\lambda_0)k(t)\}\exp\{-\lambda_0 k(t)\}\,dt\right|\\ &\quad\leq \exp[-(\lambda-\lambda_0)\inf\{k(t)\colon t \in (a,b)\backslash(-\epsilon,\epsilon)\}]\\ &\qquad\times\int_{(a,b)\backslash(-\epsilon,\epsilon)} |h(t)|\exp\{-\lambda_0 k(t)\}\,dt\end{aligned}$$

and using (ii) of Theorem 4.6 establishes the result.

This implies that $\int_a^b h(t)\exp\{-\lambda k(t)\}\,dt$ and $\int_{-\epsilon}^{\epsilon} h(t)\exp\{-\lambda k(t)\}\,dt$ have identical asymptotic expansions with respect to the asymptotic sequence $\lambda^{-1/2}, \lambda^{-3/2}, \ldots$ as $\lambda \to \infty$. Therefore we can restrict our attention hereafter to the interval $(-\epsilon, \epsilon)$ for some choice of $\epsilon > 0$.

Now suppose that ϵ can be chosen so that k has $n \geq 2$ continuous derivatives in the interval $(-\epsilon, \epsilon)$. Then $k^{(1)}(0) = 0$ and $k^{(2)}(0) > 0$. Without loss of generality we assume $k^{(2)}(0) = 1$ since if $f(x) \sim \sum a_i \phi_i(x)$ as $x \to \infty$ then when $\sigma > 0$ we also have $f(\sigma x) \sim \sum a_i \phi_i(\sigma x)$ as $x \to \infty$. By Taylor's theorem,

$$k(t) = \tfrac{1}{2}t^2 + \sum_{i=3}^{n} \frac{k^{(i)}(0)}{i!} t^i + k_n^*(t)t^n$$

for $t \in (-\epsilon, \epsilon)$ where $k_n^*(t) \to 0$ as $t \to 0$. Define

$$\begin{aligned}
k_{n\epsilon}^{*+} &= \inf\{k_n^*(t) \colon t \in (0, \epsilon)\} \\
k_{n\epsilon}^{*-} &= \begin{cases} \inf\{k_n^*(t) \colon t \in (-\epsilon, 0)\} & \text{if } n \text{ even} \\ \sup\{k_n^*(t) \colon t \in (-\epsilon, 0)\} & \text{if } n \text{ odd} \end{cases} \\
k_n^{*+\epsilon} &= \sup\{k_n^*(t) \colon t \in (0, \epsilon)\} \\
k_n^{*-\epsilon} &= \begin{cases} \sup\{k_n^*(t) \colon t \in (-\epsilon, 0)\} & \text{if } n \text{ even} \\ \inf\{k_n^*(t) \colon t \in (-\epsilon, 0)\} & \text{if } n \text{ odd} \end{cases}
\end{aligned}$$

and note that all of these quantities converge to 0 as $\epsilon \to 0$. Then putting

$$\begin{aligned}
E_\epsilon(k, n)(t) &= [k_{n\epsilon}^{*+} I_{(0,\infty)}(t) + k_{n\epsilon}^{*-} I_{(-\infty,0)}(t)] \\
E^\epsilon(k, n)(t) &= [k_n^{*+\epsilon} I_{(0,\infty)}(t) + k_n^{*-\epsilon} I_{(-\infty,0)}(t)]
\end{aligned}$$

we have that

$$\begin{aligned}
L(k, t) &= \sum_{i=3}^{n} \frac{k^{(i)}(0)}{i!} t^i + E_\epsilon(k, n)(t) t^n \\
&\leq k(t) - \tfrac{1}{2}t^2 \leq \\
U(k, t) &= \sum_{i=3}^{n} \frac{k^{(i)}(0)}{i!} t^i + E^\epsilon(k, n)(t) t^n
\end{aligned}$$

for all $t \in (-\epsilon, \epsilon)$. Observe that $E_\epsilon(k, n)(t)$ and $E^\epsilon(k, n)(t)$ are bounded functions of t; e.g. $|E^\epsilon(k, n)(t)| \leq \max\{|k_n^{*+\epsilon}|, |k_n^{*-\epsilon}|\}$.

Also suppose that ϵ has been chosen so that h has $m \geq 0$ continuous derivatives in $(-\epsilon, \epsilon)$ and $h(0) \neq 0$. Then we can write

$$h(t) = \sum_{i=0}^{m} \frac{h^{(i)}(0)}{i!} t^i + h_m^*(t) t^m$$

for $t \in (a, b)$ where $h_m^*(t) \to 0$ as $t \to 0$. Define $E_\epsilon(h, m)$ and $E^\epsilon(h, m)$ as we did for k and then we have

$$\begin{aligned} L(h,t) &= \sum_{i=0}^{m} \frac{h^{(i)}(0)}{i!} t^i + E_\epsilon(h,m)(t)t^m \\ &\leq h(t) \leq \\ U(h,t) &= \sum_{i=0}^{m} \frac{h^{(i)}(0)}{i!} t^i + E^\epsilon(h,m)(t)t^m \end{aligned}$$

for all $t \in (-\epsilon, \epsilon)$.

We have shown in Lemma 4.10 that in the asymptotic expansion for (4.10) we can restrict our attention to arbitrarily small intervals about 0. In fact the next result shows that the size of these intervals can shrink as $\lambda \to \infty$. This is an essential step in rigorously proving the validity of the expansion.

Lemma 4.11 Let $\epsilon(\lambda) = \lambda^{-1/3}$. Then

$$\int_{\epsilon(\lambda)}^{\epsilon} h(t) \exp\{-\lambda k(t)\}\, dt + \int_{-\epsilon}^{-\epsilon(\lambda)} h(t) \exp\{-\lambda k(t)\}\, dt = o(\lambda^p)$$

as $\lambda \to \infty$ for every $p \in R^1$.

Proof: Since $2k(t)/t^2 \to 1$ as $t \to 0$ there exists $\epsilon > 0$, $A > 0$ such that $k(t) \geq At^2/2$ for all $t \in (-\epsilon, \epsilon)$. Also $|h(t)| \leq B$ on $(-\epsilon, \epsilon)$ for some $B > 0$ so, for λ sufficiently large, we obtain

$$\begin{aligned} \left| \int_{\epsilon(\lambda)}^{\epsilon} h(t) \exp\{-\lambda k(t)\}\, dt \right| &\leq B \int_{\epsilon(\lambda)}^{\epsilon} \exp\{-\lambda k(t)\}\, dt \\ &\leq B \int_{\epsilon(\lambda)}^{\infty} \exp\left(-\frac{A\lambda}{2} t^2\right) dt \\ &= \frac{\sqrt{2\pi}\, B}{\sqrt{A}} \frac{1}{\sqrt{\lambda}} \int_{\sqrt{A\lambda}\epsilon(\lambda)}^{\infty} \frac{1}{\sqrt{2\pi}} \exp\left(-\tfrac{1}{2}t^2\right) dt \\ &\leq \frac{\sqrt{2\pi}\, B}{\sqrt{A}} \frac{1}{\sqrt{\lambda}} \left\{ \frac{1}{\sqrt{A\lambda}\epsilon(\lambda)} \frac{e^{-A\lambda\epsilon^2(\lambda)/2}}{\sqrt{2\pi}} \right\} \\ &= o(\lambda^p) \end{aligned}$$

as $\lambda \to \infty$, for every $p \in R^1$ since $\lambda\epsilon^2(\lambda) = \lambda^{1/3}$ and where the last inequality is an application of the Mills ratio inequality derived in (4.6). Obviously the same result holds for the integral over $(-\epsilon, -\epsilon(\lambda))$.

We henceforth restrict our attention to the integral $\int_{-\epsilon(\lambda)}^{\epsilon(\lambda)} h(t) \exp\{-\lambda k(t)\}\, dt$ with $\epsilon(\lambda) = \lambda^{-1/3}$.

We have the following inequalities

$$\begin{aligned}
&\int_{-\epsilon(\lambda)}^{\epsilon(\lambda)} L(h,t)\exp\{-\lambda U(k,t)\}\exp\left(-\frac{\lambda}{2}t^2\right)dt \\
&\quad\le \int_{-\epsilon(\lambda)}^{\epsilon(\lambda)} h(t)\exp\{-\lambda k(t)\}\,dt \\
&\quad\le \int_{-\epsilon(\lambda)}^{\epsilon(\lambda)} U(h,t)\exp\{-\lambda L(k,t)\}\exp\left(-\frac{\lambda}{2}t^2\right)dt. \qquad (4.12)
\end{aligned}$$

Consider the left side of (4.12) and note that $U(k,t)\to 0$ as $t\to 0$. Using the Taylor's expansion for e^x about 0, we obtain

$$e^{-\lambda U(k,t)} = \sum_{j=0}^{N}\frac{(-1)^j}{j!}\{\lambda U(k,t)\}^j + r_N\{\lambda U(k,t)\}\{\lambda U(k,t)\}^N$$

where $r_N(x)\to 0$ as $x\to 0$. Note that $r_N(x)$ is bounded for all x in any bounded interval. We then have that the left-hand side of (4.12) equals

$$\int_{-\epsilon(\lambda)}^{\epsilon(\lambda)} L(h,t)\left[\sum_{j=0}^{N}\frac{(-1)^j}{j!}\{\lambda U(k,t)\}^j + r_N\{\lambda U(k,t)\}\{\lambda U(k,t)\}^N\right] \times \exp\left(-\frac{\lambda}{2}t^2\right)dt. \qquad (4.13)$$

Notice that the right-most term of the integrand in (4.13) is the $N(0,\lambda^{-1})$ density times $(2\pi/\lambda)^{1/2}$. In (4.13) make the transformation $t\to u=\sqrt{\lambda}t$ to obtain

$$\begin{aligned}
&\lambda^{-1/2}\int_{-\sqrt{\lambda}\epsilon(\lambda)}^{\sqrt{\lambda}\epsilon(\lambda)} L(h,u/\sqrt{\lambda}) \\
&\times\left[\sum_{j=0}^{N}\frac{(-1)^j}{j!}\{\lambda U(k,u/\sqrt{\lambda})\}^j + r_N\{\lambda U(k,u/\sqrt{\lambda})\}\{\lambda U(k,u/\sqrt{\lambda})\}^N\right] \\
&\times\exp\left(-\tfrac{1}{2}u^2\right)du. \qquad (4.14)
\end{aligned}$$

We get that the right side of (4.12) is similarly equal to

$$\begin{aligned}
&\lambda^{-1/2}\int_{-\sqrt{\lambda}\epsilon(\lambda)}^{\sqrt{\lambda}\epsilon(\lambda)} U(h,u/\sqrt{\lambda}) \\
&\times\left[\sum_{j=0}^{N}\frac{(-1)^j}{j!}\{\lambda L(k,u/\sqrt{\lambda})\}^j + r_N\{\lambda L(k,u/\sqrt{\lambda})\}\{\lambda L(k,u/\sqrt{\lambda})\}^N\right] \\
&\times\exp\left(-\tfrac{1}{2}u^2\right)du. \qquad (4.15)
\end{aligned}$$

We consider expanding (4.14) and (4.15) and integrating term-by-term.

Observe that $\lambda U(k, u/\lambda)$ is uniformly bounded for all

$$u \in \left(-\sqrt{\lambda}\epsilon(\lambda), \sqrt{\lambda}\epsilon(\lambda)\right) = (-\lambda^{1/6}, \lambda^{1/6})$$

and all λ such that $\epsilon(\lambda) \leq \epsilon$ (this follows by noting that the absolute value of each term in $\lambda U(k, u/\lambda)$ is bounded by a constant times a term of the form $\lambda\lambda^{i/6}\lambda^{-i/2} = \lambda^{1-i/3}$ where $i \geq 3$). Therefore there is a $B > 0$ such that

$$\left|r_N\left\{\lambda U(k, u/\sqrt{\lambda})\right\}\right| \leq B$$

for all $u \in (-\sqrt{\lambda}\epsilon(\lambda), \sqrt{\lambda}\epsilon(\lambda))$ and all λ such that $\epsilon(\lambda) \leq \epsilon$. Now put

$$p(u, \lambda) = \sum_{i=3}^{n} \frac{|k^{(i)}(0)|}{i!}|u|^i \lambda^{3/2-i/2} + \max\left\{\left|k_n^{*+\epsilon}\right|, \left|k_n^{*-\epsilon}\right|\right\} |u|^n \lambda^{3/2-n/2}.$$

Then

$$\begin{aligned}
&\lambda^{-1/2}\left|\int_{-\sqrt{\lambda}\epsilon(\lambda)}^{\sqrt{\lambda}\epsilon(\lambda)} r_N\left\{\lambda U(k, u/\sqrt{\lambda})\right\}\left\{\lambda U(k, u/\sqrt{\lambda})\right\}^N \exp\left(-\tfrac{1}{2}u^2\right)du\right| \\
&\quad\leq B\lambda^{-(N+1)/2}\int_{\sqrt{\lambda}\epsilon(\lambda)}^{\sqrt{\lambda}\epsilon(\lambda)} p^N(u, \lambda)\exp\left(-\tfrac{1}{2}u^2\right)du \\
&\quad\leq B\lambda^{-(N+1)/2}\int_{-\infty}^{\infty} p^N(u, \lambda)\exp\left(-\tfrac{1}{2}u^2\right)du \\
&\quad= o(\lambda^q) \qquad (4.16)
\end{aligned}$$

as $\lambda \to \infty$ whenever $q > -(N+1)/2$. A similar result holds for all the other integrals arising from expanding (4.14), when the integrand involves

$$r_N\left\{\lambda U(k, u/\sqrt{\lambda})\right\}\left\{\lambda U(k, u/\sqrt{\lambda})\right\}^N$$

though of course the lower bounds on q will be lower. Also a similar result holds for all the integrals arising from expanding (4.15) that involve

$$r_N\left\{\lambda L(k, u/\sqrt{\lambda})\right\}\left\{\lambda L(k, u/\sqrt{\lambda})\right\}^N.$$

This will allow us to ignore these integrals when developing asymptotic expansions to order $-N/2$. Note that if we show that the left and right-hand sides of (4.12) have the same asymptotic expansion in terms of the asymptotic sequence $\lambda^{-1/2}, \lambda^{-1}, \lambda^{-3/2}, \ldots$ as $\lambda \to \infty$. Then this must also be the asymptotic expansion of (4.10).

Consider the left-hand side of (4.12). We now obtain the first three terms in the asymptotic expansion. For this we take $m = 3, n = 5$ and $N = 3$ where these are determined to ensure that the integrals of the integrands that involve error terms are $o(\lambda^{-2})$ and therefore have no effect on the first three coefficients. Note that while we are free to choose N, m is prescribed by h and n is prescribed by k. To determine the coefficients in the expansion, expand the integrand in (4.14) to obtain

$$\frac{1}{\lambda^{1/2}}\left[h^{(0)}(0)+h^{(1)}(0)\frac{u}{\lambda^{1/2}}+\frac{h^{(2)}(0)}{2}\frac{u^2}{\lambda}+\frac{h^{(3)}(0)}{3!}\frac{u^3}{\lambda^{3/2}}\right.$$
$$\left.+E_\epsilon(h,3)\left(\frac{u}{\lambda^{1/2}}\right)\frac{u^3}{\lambda^{3/2}}\right]$$
$$\times\left[1-\lambda\left\{\frac{k^{(3)}(0)}{3!}\frac{u^3}{\lambda^{3/2}}+\frac{k^{(4)}(0)}{4!}\frac{u^4}{\lambda^2}+\frac{k^{(5)}(0)}{5!}\frac{u^5}{\lambda^{5/2}}\right.\right.$$
$$\left.\left.+E^\epsilon(k,5)\left(\frac{u}{\lambda^{1/2}}\right)\frac{u^5}{\lambda^{5/2}}\right\}+\frac{\lambda^2}{2}(\cdot)^2-\frac{\lambda^3}{3!}(\cdot)^3+r_3(\cdot)\lambda^3(\cdot)^3\right]e^{-u^2/2}$$

and we have used the notation

$$(\cdot)=\left\{\frac{k^{(3)}(0)}{3!}\frac{u^3}{\lambda^{3/2}}+\frac{k^{(4)}(0)}{4!}\frac{u^4}{\lambda^2}+\frac{k^{(5)}(0)}{5!}\frac{u^5}{\lambda^{5/2}}+E^\epsilon(k,5)\left(\frac{u}{\lambda^{1/2}}\right)\frac{u^5}{\lambda^{5/2}}\right\}.$$

After expanding this we integrate the sum term-by-term using Lemma 4.10 to replace any integrals of the form

$$\int_{-\sqrt{\lambda}\epsilon(\lambda)}^{\sqrt{\lambda}\epsilon(\lambda)} u^l \exp\left(-\tfrac{1}{2}u^2\right)du=\int_{\lambda^{1/6}}^{\lambda^{1/6}} u^l \exp\left(-\tfrac{1}{2}u^2\right)du,$$

for any nonnegative integer l, by

$$\int_{-\infty}^{\infty} u^l \exp\left(-\tfrac{1}{2}u^2\right)du=\sqrt{2\pi}\,\mu_p.$$

Note that many of these integrals can be ignored because they will be premultiplied by powers of $\lambda^{-1/2}$ higher than 3. Because of the choice of m, n and N we see that we can ignore any integrands that involve error terms; i.e. terms that involve the functions E_ϵ, E^ϵ and r_3. Notice that this implies that the asymptotic expansions for the left and right-hand sides of (4.12), to the order $-3/2$, are identical as the expansions of the integrands only differ in the error terms. Taking all of this into account, and replacing 0 by $\hat{t}$, and rescaling by $(k^{(2)}(\hat{t}))^{1/2}$, we obtain the result stated in Theorem 4.6.

4.3.4 SADDLE-POINT APPROXIMATIONS

Sometimes called the *method of steepest descent*, the *saddle-point approximation* is concerned with approximating integrals of the form

$$\int_C h(z)\exp\{-\lambda k(z)\}\,dz \tag{4.17}$$

when $\lambda > 0$ is large, where C is a path in $\mathcal{C}$; i.e. there is a piecewise continuously differentiable $s : [0, 1] \to \mathcal{C}$ such that $C = \{s(t) : 0 \le t \le 1\}$ and $s^{(1)}(t) \neq 0$ where it is defined, k is an analytic function in a region A containing C and h is continuous there. Note that (4.17) equals

$$\int_0^1 h(s(t))\exp\{-\lambda k(s(t))\}s^{(1)}(t)\,dt.$$

The saddle-point approximation to (4.17) is given by (4.18) in the following result.

Theorem 4.12 Suppose that

(i) $k = u + iv$ is an analytic function and h is a continuous function in a region $A \subset \mathcal{C}$;

(ii) there exists a point $z_0 = x_0 + iy_0 \in A$ such that $k^{(1)}(z_0) = 0, k^{(2)}(z_0) \neq 0$ and $h(z_0) \neq 0$;

(iii) C is a path in $\mathcal{C}$ that can be deformed into a path C_0 satisfying $z_0 \in C_0$, $\inf\{u(z) - u(z_0) : z \in C_0, |z - z_0| > \epsilon\} > 0$,

$$\int_{C_0} |h(z)\exp\{-\lambda k(z)\}|\,dz < \infty$$

for all $\lambda \ge \lambda_0$ and, in a neighborhood of z_0, the path C_0 coincides with the path where u increases the fastest.

Then

$$\int_C h(z)\exp\{-\lambda k(z)\}\,dz \sim h(z_0)(2\pi)^{1/2}\{\lambda|k^{(2)}(z_0)|\}^{-1/2}\exp\{-\lambda k(z_0) + i\alpha\} \tag{4.18}$$

as $\lambda \to \infty$ where if $s(t) = x(t) + iy(t)$ parameterizes C_0 with $s(t_0) = z_0$ then α satisfies

$$\exp\{i\alpha\} = \left[\{x^{(1)}(t_0)\}^2 + \{y^{(1)}(t_0)\}^2\right]^{-1/2}\begin{pmatrix} x^{(1)}(t_0) \\ y^{(1)}(t_0) \end{pmatrix}.$$

Before we prove Theorem 4.12 we develop a bit of intuition. For any path C let $-C$ denote the path with the same image as C but traversed in the opposite sense. We assume that we can find a path $C_0 \subset A$, such that $z_0 \in C_0$ and $C - C_0$ is a simple closed path. By Cauchy's theorem (see for example, Ahlfors (1966) or Silverman (1974) for this and other results in complex analysis referred to here) we have that

$$\int_C h(z)\exp\{-\lambda k(z)\}\,dz = \int_{C_0} h(z)\exp\{-\lambda k(z)\}\,dz.$$

So we can replace C in (4.17) by C_0 and this is what we mean when we say we *deform* C to C_0. The *Cauchy–Riemann equations*

$$\begin{aligned}\frac{\partial u(x,y)}{\partial x} &= \frac{\partial v(x,y)}{\partial y}\\ \frac{\partial u(x,y)}{\partial y} &= -\frac{\partial v(x,y)}{\partial x}\end{aligned} \tag{4.19}$$

imply that

$$\frac{\partial u(x_0,y_0)}{\partial x} = \frac{\partial u(x_0,y_0)}{\partial y} = 0$$

so that (x_0, y_0) is a critical point of u. Since u is a harmonic function, by the maximum modulus principle, u cannot take a maximum or minimum at (x_0, y_0) and therefore this point must be a saddle point of u and similarly, also a saddle point of v. For this reason z_0 is called a saddle point of k. Since z_0 is a saddle point of u it is possible to choose the path C_0 so that (x_0, y_0) is a local minimum of u along C_0. If (x_0, y_0) is also an isolated, global minimum of u along C_0; i.e. if $\epsilon > 0$ then $\inf\{u(z) - u(z_0) \colon z \in C_0, |z - z_0| > \epsilon\} > 0$, then we might expect the greatest contribution to $\int_{C_0} h(z)\exp\{-\lambda k(z)\}\,dz$ to occur in a small subpath about (x_0, y_0). One of several difficulties with this argument is that we have to worry about the behavior of

$$\exp\{-\lambda i v(x,y)\} = \cos(\lambda v(x,y)) + i\sin(\lambda v(x,y))$$

which can be highly oscillatory near (x_0, y_0) as λ grows large. The idea is to choose C_0 so that (x_0, y_0) is a global minimum of u along C_0 and so that v is locally constant at (x_0, y_0) along C_0. In this case the effect of $\exp\{-\lambda i v(x,y)\}$ is just $\exp\{-\lambda i v(x_0, y_0)\}$ which can be factored through the integral. Effectively, the saddle-point method prescribes a path C_0 with these characteristics. Of course there will not always be such a C_0.

There are many generalizations to the saddle-point method and we refer the reader to the above references for a discussion of these. In particular we can develop a saddle-point approximation at z_0 when $k^{(1)}(z_0) = \cdots = k^{(m)}(z_0) = 0$ and $k^{(m+1)}(z_0) \neq 0$. Also it is clear that, just as with the Laplace approximation, we can extend the technique to obtain an asymptotic expansion for the integral in terms of $\lambda^{-1/2}, \lambda^{-1}, \lambda^{-3/2}, \ldots$ when appropriate conditions are met. The proof of Theorem 4.12 requires some knowledge of complex analysis. Readers without this background can proceed directly to the examples.

Proof of Theorem 4.12: We presume first that we have replaced the path C by the path C_0. We have that

$$k^{(2)}(z_0) = \frac{\partial^2 u(x_0, y_0)}{\partial x^2} - i\frac{\partial^2 u(x_0, y_0)}{\partial x \partial y} \neq 0.$$

Then the Cauchy–Riemann equations (4.19) imply that

$$\frac{\partial^2 u(x_0, y_0)}{\partial y^2} = -\frac{\partial^2 u(x_0, y_0)}{\partial x^2};$$

i.e. u is harmonic, and therefore

$$\det U(x_0, y_0) = \det \begin{pmatrix} \dfrac{\partial^2 u(x_0, y_0)}{\partial x^2} & \dfrac{\partial^2 u(x_0, y_0)}{\partial x \partial y} \\ \dfrac{\partial^2 u(x_0, y_0)}{\partial x \partial y} & \dfrac{\partial^2 u(x_0, y_0)}{\partial y^2} \end{pmatrix} = -|k^{(2)}(z_0)|^2 \neq 0.$$

Further, it is easy to show that the eigenvalues of $U(x_0, y_0)$ equal

$$\pm\left[\left(\frac{\partial^2 u(x_0, y_0)}{\partial x^2}\right)^2 + \left(\frac{\partial^2 u(x_0, y_0)}{\partial x \partial y}\right)^2\right]^{1/2} = \pm|k^{(2)}(z_0)|.$$

Consider the path B through (x_0, y_0) along which u increases the fastest as we move away from this point. Observe that the tangent direction at $(x, y) \neq (x_0, y_0)$ on B is given by $\nabla u(x, y)$ whenever $\nabla u(x, y) \neq (0, 0)$. Also we have that $\nabla u(x_0, y_0) = (0, 0)$. Note that without the minus sign in front of k in (4.17) we would look for the contour where u decreases the fastest and so the saddle-point method is sometimes called the method of steepest descent. Then by the uniqueness property for analytic functions there exists $c > 0$ and a subpath B_ϵ along B, of length ϵ containing (x_0, y_0) such that

$$\nabla u(x, y) \neq (0, 0)$$

whenever $(x, y) \in B_\epsilon \backslash \{(x_0, y_0)\}$ (otherwise u has partial derivatives equal to 0 on a set with a limit point which implies, via the Cauchy–Riemann equations, that $k^{(1)}$ is 0 on a set with a limit point and therefore identically 0 on A). By (iii) we can choose ϵ so that $B_\epsilon = C_{0\epsilon} = \{s(t)\colon t_1 < t < t_2\}$ is also a subpath of C_0.

From the Cauchy–Riemann equations we have

$$\begin{pmatrix} \dfrac{\partial u(x, y)}{\partial x} & \dfrac{\partial u(x, y)}{\partial y} \end{pmatrix}^t \begin{pmatrix} \dfrac{\partial v(x, y)}{\partial x} \\ \dfrac{\partial v(x, y)}{\partial y} \end{pmatrix} = 0. \tag{4.20}$$

Now noting that the tangent vector to the path $C_{0\epsilon}$ at $(x, y) \in C_{0\epsilon} \backslash \{(x_0, y_0)\}$ is of the form $c(x, y)\nabla u(x, y)$, for some function c, and that $\nabla u(x_0, y_0) = 0$, we obtain

$$\begin{aligned}\frac{dv(s(t))}{dt} &= \begin{pmatrix}\dfrac{\partial v(x(t), y(t))}{\partial x}\\ \dfrac{\partial v(x(t), y(t))}{\partial y}\end{pmatrix}^t \begin{pmatrix}x^{(1)}(t)\\ y^{(1)}(t)\end{pmatrix}\\ &= c(x(t), y(t)) \begin{pmatrix}\dfrac{\partial v(x(t), y(t))}{\partial x}\\ \dfrac{\partial v(x(t), y(t))}{\partial y}\end{pmatrix}^t \begin{pmatrix}\dfrac{\partial u(x(t), y(t))}{\partial x}\\ \dfrac{\partial u(x(t), y(t))}{\partial y}\end{pmatrix}\\ &= 0\end{aligned}$$

for $t \in (t_1, t_2)$ which in turn implies that $v(s(t))$ is constant for $t \in (t_1, t_2)$.

Now

$$\begin{aligned}\left|\int_{C_0\setminus C_{0\epsilon}} h(z)\exp\{-\lambda k(z)\}\,dz\right| &= \left|\int_{[0,1]\setminus(t_1,t_2)} h(s(t))s^{(1)}(t)\exp\{-\lambda k(s(t))\}\,dt\right|\\ &\le \int_{[0,1]\setminus(t_1,t_2)} |h(s(t))s^{(1)}(t)|\exp\{-\lambda u(s(t))\}\,dt\\ &= o(\lambda^p)\end{aligned}$$

as $\lambda \to \infty$ for every $p \in R^1$ just as in Lemma 4.10. Therefore we can restrict our attention to

$$\begin{aligned}&\int_{C_{0\epsilon}} h(z)\exp\{-\lambda k(z)\}\,dz\\ &= \exp\{-\lambda i v(z_0)\}\int_{(t_1,t_2)} h(s(t))s^{(1)}(t)\exp\{-\lambda u(s(t))\}\,dt\\ &= \exp\{-\lambda i v(z_0)\}\left[\int_{(t_1,t_2)} \mathrm{Re}\{h(s(t))s^{(1)}(t)\}\exp\{-\lambda u(s(t))\}\,dt\right.\\ &\qquad \left. +i\int_{(t_1,t_2)} \mathrm{Im}\{h(s(t))s^{(1)}(t)\}\exp\{-\lambda u(s(t))\}\,dt\right] \qquad (4.21)\end{aligned}$$

and apply the Laplace approximation to both integrals. It is an easy exercise to show that if $a_\lambda \sim A_\lambda$ and $b_\lambda \sim B_\lambda$ then $a_\lambda + ib_\lambda \sim A_\lambda + iB_\lambda$ so we can combine these approximations to approximate (4.17). Therefore we only need to obtain the value of $s^{(1)}(t_0)$ to apply the Laplace approximation to each integral in (4.21).

The direction of steepest increase of u at (x_0, y_0) is given by the unit eigenvector of $U(x_0, y_0)$ associated with its largest eigenvalue. We havc shown that the largest eigenvalue is $|k^{(2)}(z_0)|$. Further it is straightforward to show that

$$\frac{d^2u(s(t_0))}{dt^2} = \begin{pmatrix} x^{(1)}(t_0) \\ y^{(1)}(t_0) \end{pmatrix}^t \begin{pmatrix} \dfrac{\partial^2 u(x_0, y_0)}{\partial x^2} & \dfrac{\partial^2 u(x_0, y_0)}{\partial x \partial y} \\ \dfrac{\partial^2 u(x_0, y_0)}{\partial x \partial y} & \dfrac{\partial^2 u(x_0, y_0)}{\partial y^2} \end{pmatrix} \begin{pmatrix} x^{(1)}(t_0) \\ y^{(1)}(t_0) \end{pmatrix}.$$

This implies that

$$\frac{d^2u(s(t_0))}{dt^2} = \left[\left\{x^{(1)}(t_0)\right\}^2 + \left\{y^{(1)}(t_0)\right\}^2\right] \left|k^{(2)}(z_0)\right|$$

since

$$s^{(1)}(t_0) = \begin{pmatrix} x^{(1)}(t_0) \\ y^{(1)}(t_0) \end{pmatrix}$$

points in the direction of an eigenvector of $U(x_0, y_0)$ associated with the eigenvalue $|k^{(2)}(z_0)|$ (as a simple example, consider $u(x, y) = x^2 - y^2$, $(x_0, y_0) = (0, 0)$ and note that a path of maximum increase from (x_0, y_0) is given by $s(t) = (t, 0)^t$ which has $s^{(1)}(t) = \nabla u(t, 0)/2$ whenever $t \neq 0$ and $s^{(1)}(0) = e_1$ which is an eigenvector of $U(x_0, y_0)$ associated with its largest eigenvalue 1).

Computing the Laplace approximations to the integrals in (4.21) and adding them together gives

$$\begin{aligned} &\int_C h(z) \exp\{-\lambda k(z)\}\, dz \\ &\quad \sim h(z_0)(2\pi)^{1/2}\left\{\lambda\left|k^{(2)}(z_0)\right|\right\}^{-1/2} \exp\{-\lambda k(z_0)\} \\ &\qquad \times s^{(1)}(t_0)\left[\left\{x^{(1)}(t_0)\right\}^2 + \left\{y^{(1)}(t_0)\right\}^2\right]^{-1/2}. \\ &\quad = h(z_0)(2\pi)^{1/2}\left\{\lambda\left|k^{(2)}(z_0)\right|\right\}^{-1/2} \exp\{-\lambda k(z_0) + i\alpha\} \end{aligned}$$

where

$$\exp\{i\alpha\} = s^{(1)}(t_0)\left[\left\{x^{(1)}(t_0)\right\}^2 + \left\{y^{(1)}(t_0)\right\}^2\right]^{-1/2}$$

and α is a direction of steepest ascent (there are only two and which is used is determined by which sense we pass through z_0 along C_0). This completes the proof.

Examples of the use of the saddle-point approximation We consider some applications of the saddle-point approximation.

Example 4.5 *Inversion formulae*
Suppose that $X_1, X_2, \ldots, X_n$ is a sample from a distribution with density function f on R^1 with characteristic function (Fourier transform)

$$c(t) = \int_{-\infty}^{\infty} e^{itx} f(x)\, dx$$

for $t \in R^1$ and $\int_{-\infty}^{\infty} |c(t)|\, dt < \infty$. Further, suppose that we have a relatively simple expression for c. Then by Fourier inversion, see for example, Durrett (1991), we have that

$$f(x) = \frac{1}{2\pi} \int_{-\infty}^{\infty} e^{-itx} c(t)\, dt.$$

The random variable

$$\bar{X} = \frac{1}{n} \sum_{i=1}^{n} X_i$$

has characteristic function $c^n(t/n)$ and density function at w given by

$$\begin{aligned} f_n(w) &= \frac{1}{2\pi} \int_{-\infty}^{\infty} e^{-ity} c^n(t/n)\, dt \\ &= \frac{1}{2\pi} \int_{-\infty}^{\infty} \exp\{n \ln c(t/n) - ity\}\, dt. \end{aligned}$$

Putting $z = t/n$ this becomes

$$f_n(w) = \frac{n}{2\pi} \int_{-\infty}^{\infty} \exp\{n(\ln c(z) - izw)\}\, dz$$

and the integral is in the form of (4.17) with $\lambda = n, h(z) = 1$ and $k(z) = -(\ln c(z) - izw)$. Then

$$k^{(1)}(z) = -\left[\frac{c^{(1)}(z)}{c(z)} - iw\right]$$

and setting this equal to 0 gives the *saddle-point equation*

$$\frac{c^{(1)}(z)}{c(z)} = iw$$

with a saddle point being a solution z_0. Note that if the first moment of f exists then

$$c^{(1)}(z) = i \int_{-\infty}^{\infty} e^{izx} x f(x)\, dx.$$

Also we have that

$$k^{(2)}(z) = -\left[\frac{c^{(2)}(z)}{c(z)} - \left\{\frac{c^{(1)}(z)}{c(z)}\right\}^2\right] = \kappa^{(2)}(z)$$

where $\kappa(z) = \log c(z)$ is called the *cumulant generating function.* We can then write the saddle-point approximation as

$$f_n(w) \sim \frac{\sqrt{n}}{\sqrt{2\pi}} \left[\kappa^{(2)}(z_0)\right]^{-1/2} \exp\{n(\varkappa(z_0) - iz_0 w) + i\alpha\} \tag{4.22}$$

provided of course that the conditions of Theorem 4.12 apply. Under certain conditions we have that $\alpha = 0$ and the formula simplifies.

Example 4.6 *Normal sample mean*
The saddle-point approximation is not really needed in this example but it does illustrate how the technique is to be employed. It also demonstrates that the saddle-point approximation can be remarkably accurate. We compute and justify, in this example, the specific form of (4.22).

Suppose that

$$f(x) = \frac{1}{\sqrt{2\pi}} \exp\left(-\tfrac{1}{2}x^2\right);$$

i.e. f is the density of the $N(0, 1)$ distribution and we want the density of $\bar{X}$ based on a sample of n from this distribution at the point w. Then

$$c(t) = \exp\left(-\tfrac{1}{2}t^2\right)$$

and

$$\begin{aligned} k(z) &= \ (\ln c(z) - izw) \\ &= \frac{z^2}{2} + izw \\ &= \left(\frac{x^2}{2} - \frac{y^2}{2} - wy\right) + i(xy + wx) \\ &= u(x, y) + iv(x, y). \end{aligned}$$

The saddle-point equation is

$$\frac{c^{(1)}(z)}{c(z)} = -z = iw$$

which has solution $z_0 = -iw$ and

$$k(-iw) = \tfrac{1}{2}w^2.$$

Then $v(0, -w) = 0$ and $v(x, y) = 0$ if and only if $x = 0$ or $y = -w$ which are both lines going through the saddle point. Since $u(0, y) = -(y + w)^2/2 + w^2/2$ and $u(x, -w) = x^2/2 + w^2/2$ we see that the path of maximum increase for u at the saddle point is along the line $y = -w$.

For the integral

$$\begin{aligned} f_n(w) &= \frac{n}{2\pi}\int_{-\infty}^{\infty} \exp\{n(\ln c(z) - izw)\}\, dz \\ &= \frac{n}{2\pi}\int_{-\infty}^{\infty} \exp\left(-\frac{n}{2}z^2 - inzw\right) dz \end{aligned}$$

it is clear that

$$\frac{n}{2\pi}\int_{-\infty}^{\infty} \exp\left(-\frac{n}{2}z^2 - inzw\right) dz \sim \frac{n}{2\pi}\int_{-|w|}^{|w|} \exp\left(-\frac{n}{2}z^2 - inzw\right) dz$$

as $n \to \infty$ and then by Cauchy's theorem

$$\frac{n}{2\pi}\int_{-|w|}^{|w|} \exp\left(-\frac{n}{2}z^2 - inzw\right) dz = \frac{n}{2\pi}\int_{C_0} \exp\left(-\frac{n}{2}z^2 - inzw\right) dz$$

for a contour C_0. If $w = 0$ then C_0 is the interval $[-|w|, |w|]$ traversed from left to right. If $w < 0$ then C_0 is composed of the line segment from the point $(-|w|, -w)$ to the point $(|w|, -w)$ and the two vertical segments joining $(-|w|, -w)$ to $(-|w|, 0)$ and $(|w|, -w)$ to $(|w|, 0)$ and traversed clockwise. If $w > 0$ then C_0 is similar to the case $w < 0$ but lies in the lower half-plane. Notice that C_0 passes through the saddle point in each of these cases and that in each case $\alpha = 0$. Also the integrals on the vertical strips are $O(\exp\{-nw^2/2\})$ as $n \to \infty$ and therefore can be ignored.

Calculating we obtain

$$k^{(2)}(z) = -\left[\frac{c^{(2)}(z)}{c(z)} - \left\{\frac{c^{(1)}(z)}{c(z)}\right\}^2\right]\Bigg|_{z=-iw} = 1.$$

Therefore applying the saddle-point approximation (4.18) we obtain

$$f_n(w) \sim (2\pi)^{-1/2}(n)^{1/2} \exp\left(-\frac{n}{2}w^2\right)$$

which is the exact value in this case.

Many more examples can be found in the references given at the start of this chapter. In recent years the saddle-point approximation has found many uses in statistical applications. In particular see Reid (1988), Barndorff-Nielsen and Cox (1989, 1994) and Jensen (1995) for a discussion of such contexts.

4.3.5 EDGEWORTH EXPANSIONS

A technique related to the saddle-point method is commonly used in statistical problems when we want to approximate the distribution of a sum of iid (independent and identically distributed) random variables. For this suppose that

$X_1, X_2, \ldots, X_n$ is a sample from a distribution with density function f on R^1 with mean 0, variance 1 and characteristic function c. Further suppose that this distribution has its first m moments. Then c has its first m derivatives at 0 and we can write

$$c(t) = 1 - \frac{t^2}{2} + \sum_{j=3}^{m} \mu_j \frac{(it)^j}{j!} + t^m c_R(t)$$

where μ_j is the j-th moment and $c_R(t) \to 0$ as $t \to 0$; see Durrett (1991). This implies that we can expand the cumulant generating function $\kappa(t) = \ln c(t)$ about 0 to obtain

$$\kappa(t) = -\frac{t^2}{2} + \sum_{j=3}^{m} \kappa_j \frac{(it)^j}{j!} + t^m \kappa_R(t)$$

where κ_j is the j-th cumulant of the distribution and $\kappa_R(t) \to 0$ as $t \to 0$. We can obtain the cumulants from the moments, and vice versa, using the expansion $\ln(1+x) = x - x^2/2 + x^3/3 - \cdots$ with

$$x = -\frac{t^2}{2} + \sum_{j=3}^{m} \mu_j \frac{(it)^j}{j!} + t^m c_R(t)$$

and then equating coefficients of t with the expansion given above for κ. Note that $\sqrt{n}\bar{X}$ has characteristic function given by $c^n(t/\sqrt{n})$ and thus the cumulant generating function is given by

$$n \ln c(t/\sqrt{n}) = -\frac{t^2}{2} + \sum_{j=3}^{m} \kappa_j \frac{(it)^j}{j!} \left(\frac{1}{\sqrt{n}}\right)^{j-2} + n \left(\frac{t}{\sqrt{n}}\right)^m \kappa_R\left(\frac{t}{\sqrt{n}}\right).$$

Suppose now that we want to approximate

$$F_n(b) - F_n(a) = P(a < \sqrt{n}\bar{X} \le b).$$

Provided that $\int_{-\infty}^{\infty} |c(t)|\, dt < \infty$ we can write, see Durrett (1991),

$$\begin{aligned} F_n(b) &- F_n(a) \\ &= \frac{1}{2\pi} \int_{-\infty}^{\infty} \frac{e^{-ita} - e^{-itb}}{it} c^n\left(\frac{t}{\sqrt{n}}\right) dt \\ &= \frac{1}{\sqrt{2\pi}} \int_{-\infty}^{\infty} \frac{e^{-ita} - e^{-itb}}{it} \exp\left\{\sum_{j=3}^{m} \kappa_j \frac{(it)^j}{j!} \left(\frac{1}{\sqrt{n}}\right)^{j-2} \right. \\ &\qquad \left. + n \left(\frac{t}{\sqrt{n}}\right)^m \kappa_R\left(\frac{t}{\sqrt{n}}\right)\right\} \varphi(t)\, dt \qquad (4.23) \end{aligned}$$

where φ denotes the standard normal density.

The expression (4.23) leads to an asymptotic expansion for $F_n(b) - F_n(a)$ in terms of the asymptotic sequence $n^{-1/2}, n^{-1}, \ldots$ as $n \to \infty$. We establish this here with $m = 5$ to obtain the first three terms of the expansion. This also gives the general methodology for establishing higher order expansions. Now write

$$e^x = 1 + x + \frac{x^2}{2} + \frac{x^3}{3!} + x^3 u(x)$$

where $u(x) \to 0$ as $x \to 0$. Putting

$$x = \sum_{j=3}^{5} \kappa_j \frac{(it)^j}{j!} \left(\frac{1}{\sqrt{n}}\right)^{j-2} + n \left(\frac{t}{\sqrt{n}}\right)^5 \kappa_R \left(\frac{t}{\sqrt{n}}\right)$$

and using the expansion for e^x with this x, and collecting powers of $n^{-1/2}$, we obtain

$$\begin{aligned}
&\exp\left\{\sum_{j=3}^{5} \kappa_j \frac{(it)^j}{j!} \left(\frac{1}{\sqrt{n}}\right)^{j-2} + n \left(\frac{t}{\sqrt{n}}\right)^5 \kappa_R \left(\frac{t}{\sqrt{n}}\right)\right\} \\
&= 1 + \kappa_3 \frac{(it)^3}{3!} \frac{1}{n^{1/2}} + \left\{\kappa_3^2 \frac{(it)^6}{(3!)^2} + \kappa_4 \frac{(it)^4}{4!}\right\} \frac{1}{n} \\
&\quad + \left\{\kappa_3^3 \frac{(it)^9}{(3!)^3} + 2\kappa_3\kappa_4 \frac{(it)^7}{3!4!} + \kappa_5 \frac{(it)^5}{5!}\right\} \frac{1}{n^{3/2}} + r_n(t) \\
&= 1 + q_n(t) + r_n(t) \qquad (4.24)
\end{aligned}$$

where $r_n(t)$ contains terms involving $n^{-2}, n^{-5/2}$ and so on. We plug (4.24) into (4.23) and integrate term by term.

For this we need

$$\frac{d^j}{dx^j} \varphi(x) = (-1)^j H_j(x) \varphi(x)$$

where H_j is the *Hermite polynomial* of degree j as defined by the recursion formula $H_{i+1}(x) = x H_i(x) - i H_{i-1}(x)$ with $H_0(x) = 1$, $H_1(x) = x$. The Hermite polynomials satisfy

$$\int_{-\infty}^{\infty} H_i(x) H_j(x) \varphi(x)\, dx = 0$$

when $i \neq j$ and note that this implies that the odd degree Hermite polynomials have expectation equal to 0 with respect to the $N(0, 1)$ distribution. See Jackson (1941) for more details about orthogonal polynomials. Since the characteristic function of the $N(0, 1)$ distribution is given by $e^{-t^2/2}$ we also have that

$$\begin{aligned}\frac{d^j}{dx^j}\varphi(x) &= \frac{d^j}{dx^j}\frac{1}{2\pi}\int_{-\infty}^{\infty} e^{-itx}e^{-t^2/2}\,dt \\ &= (-1)^j\frac{1}{\sqrt{2\pi}}\int_{-\infty}^{\infty}(it)^j e^{-itx}\varphi(t)\,dt.\end{aligned}$$

Therefore we conclude that for $j \geq 0$

$$\int_{-\infty}^{\infty}(it)^j e^{-itx}\varphi(t)\,dt = \sqrt{2\pi}\,H_j(x)\varphi(x). \tag{4.25}$$

We also have, by the inversion theorem, that

$$\Phi(b) - \Phi(a) = \frac{1}{\sqrt{2\pi}}\int_{-\infty}^{\infty}\frac{e^{-ita} - e^{-itb}}{it}\varphi(t)\,dt. \tag{4.26}$$

Then (4.25) and (4.26) permit us, after substituting (4.24) into (4.23), to evaluate every integral in closed form except of course for the remainder term. Carrying this out we obtain that (4.23) equals

$$[\Phi(b) - \Phi(a)] + [P_n(b)\varphi(b) - P_n(a)\varphi(a)] + R_n(a, b)$$

where

$$\begin{aligned}P_n(x) &= \frac{1}{\sqrt{2\pi}}\int_{-\infty}^{\infty}\frac{-e^{-itx}}{it}q_n(t)\varphi(t)\,dt \\ &= -\left\{\frac{\kappa_3}{3!}H_2(x)\right\}\varphi(x)\frac{1}{n^{1/2}} - \left\{\frac{\kappa_3^2}{(3!)^2}H_5(x) + \frac{\kappa_4}{4!}H_3(x)\right\}\varphi(x)\frac{1}{n} \\ &\quad - \left\{\frac{\kappa_3^3}{(3!)^3}H_8(x) + \frac{2\kappa_3\kappa_4}{3!4!}H_6(x) + \frac{\kappa_5}{5!}H_4(x)\right\}\varphi(x)\frac{1}{n^{3/2}}\end{aligned}$$

and

$$R_n(a, b) = \frac{1}{\sqrt{2\pi}}\int_{-\infty}^{\infty}\frac{e^{-ita} - e^{-itb}}{it}r_n(t)\varphi(t)\,dt.$$

Therefore we approximate $F_n(b) - F_n(a)$ by

$$[\Phi(b) - \Phi(a)] + [P_n(b)\varphi(b) - P_n(a)\varphi(a)] \tag{4.27}$$

and this is a valid asymptotic expansion to order $n^{-3/2}$ if we can show that $R_n(a, b) = o(n^{-3/2})$ as $n \to \infty$.

We have the following result.

Theorem 4.13 If a random variable X has a density with respect to Lebesgue measure, $E|X|^5 < \infty$ and its characteristic function c satisfies $\int_{-\infty}^{\infty}|c(t)|\,dt < \infty$ then $R_n(a, b) = o(n^{-3/2})$ uniformly in a and b.

Proof: We adapt the proof found in Feller (1971). We have that

$$\begin{aligned}
&|R_n(a, b)| \\
&= |F_n(b) - F_n(a) - [\Phi(b) - \Phi(a)] - [P_n(b)\varphi(b) - P_n(a)\varphi(a)]| \\
&= \left| \frac{1}{2\pi} \int_{-\infty}^{\infty} \frac{e^{-ita} - e^{-itb}}{it} \left\{ c^n\left(\frac{t}{\sqrt{n}}\right) - e^{-t^2/2} - q_n(t)e^{-t^2/2} \right\} dt \right| \\
&\leq \frac{1}{\pi} \int_{-\infty}^{\infty} \frac{1}{|t|} \left| c^n\left(\frac{t}{\sqrt{n}}\right) - e^{-t^2/2} - q_n(t)e^{-t^2/2} \right| dt \qquad (4.28)
\end{aligned}$$

and note that this is independent of a and b. For $\delta > 0$ we can write

$$\begin{aligned}
&\frac{1}{\pi} \int_{-\infty}^{\infty} \frac{1}{|t|} \left| c^n\left(\frac{t}{\sqrt{n}}\right) - e^{-t^2/2} - q_n(t)e^{-t^2/2} \right| dt \\
&\leq \frac{1}{\pi} \int_{t \in (-\delta\sqrt{n}, \delta\sqrt{n})} \frac{1}{|t|} \left| c^n\left(\frac{t}{\sqrt{n}}\right) - e^{-t^2/2} - q_n(t)e^{-t^2/2} \right| dt \\
&\quad + \frac{1}{\pi} \int_{t \notin (-\delta\sqrt{n}, \delta\sqrt{n})} \frac{1}{|t|} \left| c\left(\frac{t}{\sqrt{n}}\right) \right|^n dt \\
&\quad + \frac{1}{\pi} \int_{t \notin (-\delta\sqrt{n}, \delta\sqrt{n})} \frac{1}{|t|} |1 + q_n(t)| e^{-t^2/2}\, dt. \qquad (4.29)
\end{aligned}$$

Now we show that there exists positive $\tau_\delta < 1$ such that for all $t \notin (-\delta, \delta)$, $|c(t)| \leq \tau_\delta$. For every characteristic function c we have that $|c(t)| \leq 1$ for all t. If $c(t) = \pm 1$ for some $t \neq 0$ then the distribution corresponding to c would have a discrete point which is not the case here. Since c is continuous (by the dominated convergence theorem) $c(t)$ cannot get arbitrarily close to ± 1 for $t \notin (-\delta, \delta)$. Further the integrability of c implies that $c(t) \to 0$ as $t \to \pm\infty$. Then we let $\tau_\delta = \sup\{c(t) \colon t \notin (-\delta, \delta)\}$. This implies that (4.29) is bounded above by

$$\begin{aligned}
&\frac{1}{\pi} \int_{t \in (-\delta\sqrt{n}, \delta\sqrt{n})} \frac{1}{|t|} \left| c^n\left(\frac{t}{\sqrt{n}}\right) - e^{-t^2/2} - q_n(t)e^{-t^2/2} \right| dt \\
&\quad + \frac{\tau_\delta^{n-1}}{\pi} \int_{t \notin (-\delta, \delta)} \frac{|c(t)|}{|t|} dt + \frac{1}{\pi} \int_{\delta\sqrt{n}}^{\infty} \frac{1}{t} |1 + q_n(t)| e^{-t^2/2}\, dt \\
&\quad - \frac{1}{\pi} \int_{-\infty}^{-\delta\sqrt{n}} \frac{1}{t} |1 + q_n(t)| e^{-t^2/2}\, dt. \qquad (4.30)
\end{aligned}$$

The second integral in (4.30) is $o(n^{-p/2})$ as $n \to \infty$ for every $p > 0$ and so can be ignored. For n sufficiently large the Mills ratio inequality gives

$$\int_{\delta\sqrt{n}}^{\infty} \frac{1}{t} e^{-t^2/2}\, dt \leq \frac{e^{-n\delta^2/2}}{(\delta\sqrt{n})^2}$$

and for $j > 2$, integration by parts gives

$$\left(\frac{1}{n^{1/2}}\right)^k \int_{\delta\sqrt{n}}^{\infty} t^j e^{-t^2/2}\, dt$$
$$= \left(\frac{1}{n^{1/2}}\right)^k \left\{(\delta\sqrt{n})^{j-1} e^{-n\delta^2/2} + (j-1)\int_{\delta\sqrt{n}}^{\infty} t^{j-2} e^{-t^2/2}\, dt\right\}$$

which shows that the third integral in (4.30) is also $o(n^{-p/2})$ as $n \to \infty$ for all $p > 0$ and therefore can be ignored. By a similar argument the fourth integral is also $o(n^{-p/2})$ as $n \to \infty$ for all $p > 0$.

Finally we consider the first integral in (4.30). Put

$$x(t) = \sum_{j=3}^{5} \kappa_j \frac{(it)^j}{j!}$$

and

$$w_n(t) = \sum_{j=0}^{3} \frac{\{nx(t/\sqrt{n})\}^j}{j!}$$

and note that $w_n(t)$ is the first four terms in the expansion of $\exp(nx(t/\sqrt{n}))$. We have that

$$\frac{1}{\pi}\int_{t\in(-\delta\sqrt{n},\delta\sqrt{n})} \frac{1}{|t|}\left|c^n\left(\frac{t}{\sqrt{n}}\right) - e^{-t^2/2} - q_n(t)e^{-t^2/2}\right| dt$$
$$= \frac{1}{\pi}\int_{t\in(-\delta\sqrt{n},\delta\sqrt{n})} \frac{1}{|t|}\left|\exp\left\{n \ln c\left(\frac{t}{\sqrt{n}}\right)\right\} - e^{-t^2/2} - q_n(t)e^{-t^2/2}\right| dt$$
$$\leq \frac{1}{\pi}\int_{t\in(-\delta\sqrt{n},\delta\sqrt{n})} \frac{1}{|t|}\left|\exp\left\{n\kappa\left(\frac{t}{\sqrt{n}}\right) + \tfrac{1}{2}t^2\right\} - w_n(t)\right| e^{-t^2/2} dt$$
$$+ \frac{1}{\pi}\int_{t\in(-\delta\sqrt{n},\delta\sqrt{n})} \frac{1}{|t|}|w_n(t) - 1 - q_n(t)|\, e^{-t^2/2} dt. \tag{4.31}$$

The second integral in (4.31) is easily seen to be $o(n^{-3/2})$ as $q_n(t)$ was formed by collecting terms in $w_n(t)$ and the remaining terms in $w_n(t) - q_n(t)$ involve powers of $n^{-1/2}$ that are 4 or higher and each term integrates finitely over $(-\infty, \infty)$. For the first integral in (4.31) we will use

$$|e^\alpha - 1 - \beta - \beta^2/2! - \beta^3/3!|$$
$$\leq |e^\alpha - e^\beta| + |e^\beta - 1 - \beta - \beta^2/2! - \beta^3/3!|$$
$$\leq \{|\alpha - \beta| + (1/4!)\beta^4\}e^\gamma \tag{4.32}$$

for $\gamma \geq \max\{|\alpha|, |\beta|\}$ where for the second inequality we apply the mean value theorem to $e^\alpha - e^\beta$ and Taylor's theorem with the Lagrange form for the remainder to e^β. Now choose δ so that for $t \in (-\delta, \delta)$, $|\kappa(t)+t^2/2| \le t^2/4$ and $|x(t)| \le t^2/4$. Put

$$\alpha = n\kappa(t/\sqrt{n}) + \tfrac{1}{2}t^2$$

and

$$\beta = nx(t/\sqrt{n})$$

and note that for $t \in (-\delta\sqrt{n}, \delta\sqrt{n})$ we have $|\alpha| \le t^2/4$ and $|\beta| \le t^2/4$. Then, using (4.32) with these choices for α and β and putting $\gamma = t^2/4$, we have that the first integral in the right-hand side of (4.31) is bounded above by

$$\frac{1}{\pi}\int_{t\in(-\delta\sqrt{n},\delta\sqrt{n})} \frac{1}{|t|}\left|n\kappa(t/\sqrt{n}) + \tfrac{1}{2}t^2 - nx(t/\sqrt{n})\right| e^{-t^2/4}\,dt$$
$$+ \frac{1}{4!\pi}\int_{t\in(-\delta\sqrt{n},\delta\sqrt{n})} \frac{1}{|t|}\left|nx(t/\sqrt{n})\right|^4 e^{-t^2/4}\,dt. \tag{4.33}$$

The second integral in (4.33) is $o(n^{-3/2})$. Further for the first integral in (4.33), if $\epsilon > 0$ then we can choose δ^* such that $0 < \delta^* < \delta$ such that for all $t \in (-\delta^*, \delta^*)$

$$\left|\kappa(t) + \tfrac{1}{2}t^2 - x(t)\right| \le \epsilon t^5$$

and therefore for all $t \in (-\delta^*\sqrt{n}, \delta^*\sqrt{n})$

$$\left|n\kappa(t/\sqrt{n}) + \tfrac{1}{2}t^2 - nx(t/\sqrt{n})\right| \le \frac{\epsilon t^5}{n^{3/2}}.$$

This establishes the result since

$$\limsup_{n\to\infty} n^{3/2}|R_n(a,b)| \le \epsilon\left(\int_{-\infty}^{\infty} t^5 e^{-t^2/4}dt\right)$$

and the right-hand side can be made arbitrarily small by an appropriate choice of ϵ.

While we have obtained an asymptotic expansion for the difference $F_n(b) - F_n(a)$ we might prefer to obtain an approximation for $F_n(b)$ and the natural candidate for this is

$$\Phi(b) + P_n(b)\varphi(b). \tag{4.34}$$

The validity of this approximation, under conditions as stated in Theorem 4.13, can be established using methods discussed in Feller (1971).

It is possible to obtain valid expansions under more general conditions than those stated in Theorem 4.13 so this is not the strongest result. Also we can obtain Edgeworth expansions for the density of $\sqrt{n}\bar{X}$ exactly as above and we leave this as an exercise. We conclude with an example.

Example 4.7 *Distribution of the sample mean when sampling from the $U(0, 1)$ distribution*
Suppose that $U_1, U_2, \ldots, U_n$ is a sample from the $U(0, 1)$ distribution. Then $E[U_i] = 1/2$, $\text{Var}[U_i] = 1/12$ and $E[U_i^m] = 1/(m+1)$. Now put

$$X_i = \frac{U_i - 1/2}{\sqrt{1/12}}.$$

Then $X_1, X_2, \ldots, X_n$ is a sample from a distribution with mean 0 and variance 1 and

$$E[X_i^3] = E\left[\left(\frac{U_i - 1/2}{\sqrt{1/12}}\right)^3\right] = 0.$$

Since $\kappa_3 = \mu_3$ (when a random variable has mean 0) we have that up to order $n^{-1/2}$

$$F_n(b) - F_n(a) \sim \Phi(b) - \Phi(a).$$

Also we have that

$$\begin{aligned} E[X_i^4] &= E\left[\left(\frac{U_i - 1/2}{\sqrt{1/12}}\right)^4\right] \\ &= \left(\frac{1}{12}\right)^2 E\left[U_i^4 - 2U_i^3 + \frac{3}{2}U_i^2 - \frac{1}{2}U_i + \frac{1}{16}\right] \\ &= \left(\frac{1}{12}\right)^2 \left[\frac{1}{5} - 2\frac{1}{4} + \frac{3}{2}\frac{1}{3} - \frac{1}{2}\frac{1}{2} + \frac{1}{16}\right] \\ &= \left(\frac{1}{12}\right)^2 \frac{1}{80} \end{aligned}$$

and since $\kappa_4 = \mu_4 - 3\mu_2^2$ (when a random variable has mean 0) then

$$\kappa_4 = \left(\frac{1}{12}\right)^2 \frac{1}{80} - 3.$$

Therefore up to order n^{-1} we have that

$$F_n(b) - F_n(a) \sim \Phi(b) - \Phi(a) + \frac{\kappa_4}{4!}\{-H_3(b)\varphi(b) + H_3(a)\varphi(a)\}\frac{1}{n}$$

where $H_3(x) = -x^3 + 3x$. We leave it as an exercise to determine the $n^{-3/2}$ term.

We have presented here Edgeworth expansions for the distribution function of the sample mean. The basis of the method is inversion of the characteristic function or, equivalently, the cumulant function of the sample mean. In fact, we only used a finite expansion for the cumulant function. Therefore if we have a finite expansion for the cumulant function for a real-valued function $W(X_1, X_2, \ldots, X_n)$ of the sample we can employ the Edgeworth expansion technique to approximate probabilities and expect that rigorous results like Theorem 4.13 will hold under similar conditions. See McCullagh (1987), Barndorff-Nielsen and Cox (1989) and Ghosh (1991) for more discussion of the Edgeworth expansion technique.

4.4 Approximating multidimensional integrals

We now consider the approximation of multidimensional integrals. In effect the asymptotic methods used here are just generalizations of those developed for one-dimensional integrals. The methods are basically the same for developing asymptotic expansions but the complexity is greatly increased. In fact some of the expressions become so complicated that calculating these is a computational problem in itself; see for example, McCullagh (1987) for a discussion of the complexities involved with computing multivariate Edgeworth expansions. In such contexts symbolic calculation programs become essential; see for example, Heller (1991). The additional complexity arises partly from the multi-indexing associated with multivariable Taylor expansions. We refer the reader to the references cited at the beginning of this chapter for more on asymptotic approximations to multidimensional integrals. We concentrate here on the Laplace method and its applications in statistical contexts.

We develop the Laplace formula in the multivariate context under very weak conditions. Suppose then that we want to approximate an integral of the form

$$\int_A h(t)e^{-\lambda k(t)}dt \tag{4.35}$$

where A is an open subset of R^k and $\lambda > 0$ is a real-valued parameter. The following gives the multidimensional version of the Laplace formula.

Theorem 4.14 The asymptotic equivalence

$$\int_A h(t)e^{-\lambda k(t)}dt \sim h(\hat{t})(2\pi)^{k/2}|\lambda K(\hat{t})|^{-1/2}\exp\{-\lambda k(\hat{t})\} \tag{4.36}$$

as $\lambda \to \infty$ holds under the following conditions:

(i) A is open;

(ii) $\int_A |h(t)|e^{-\lambda k(t)}dt < \infty$ for all $\lambda \geq \lambda_0$;

(iii) there exists $\hat{t} \in A$ such that for every $\epsilon > 0$,

$$\inf\{k(t) - k(\hat{t}) \mid t \in A, \|t - \hat{t}\| \geq \epsilon\} > 0;$$

i.e. $\hat{t}$ is an isolated global minimum of k;

(iv) h is continuous in a neighborhood of $\hat{t}$ with $h(\hat{t}) \neq 0$;

(v) $k \in C^2(A)$ and $K(\hat{t}) = \left(\partial^2 k(\hat{t})/\partial t_i \partial t_j\right)$ is positive definite.

Proof: The proof given here follows Wong (1989), Exercise 5, page 507. Without loss of generality assume $k(\hat{t}) = 0$. Let $B_\epsilon(\hat{t})$ be the open ball of radius ϵ centered at $\hat{t}$ and note that

$$\int_A h(t)e^{-\lambda k(t)}\,dt = \int_{B_\epsilon(\hat{t})} h(t)e^{-\lambda k(t)}\,dt + \int_{A\backslash B_\epsilon(\hat{t})} h(t)e^{-\lambda k(t)}\,dt.$$

Then

$$\begin{aligned}\left|\int_{A\backslash B_\epsilon(\hat{t})} h(t)e^{-\lambda k(t)}\,dt\right| &\leq \int_{A\backslash B_\epsilon(\hat{t})} |h(t)|e^{-(\lambda-\lambda_0)k(t)-\lambda_0 k(t)}\,dt \\ &\leq \exp\left\{-(\lambda-\lambda_0)\inf\left\{k(t) \mid t \in A, \|t-\hat{t}\| \geq \epsilon\right\}\right\} \\ &\quad \times \int_{A\backslash B_\epsilon(\hat{t})} |h(t)|e^{-\lambda_0 k(t)}\,dt \\ &= o(\lambda^p)\end{aligned}$$

as $\lambda \to \infty$ for every $p \in R^1$ and we can ignore this term hereafter.

From Taylor's Theorem, with Young's form for the remainder, and using $k(\hat{t}) = 0$, $\nabla k(\hat{t}) = 0$, we can write

$$\begin{aligned}k(t) &= \tfrac{1}{2}(t-\hat{t})'K(\hat{t})(t-\hat{t}) + \tfrac{1}{2}r(t)\|t-\hat{t}\|^2 \\ &= \tfrac{1}{2}(t-\hat{t})'\{K(\hat{t}) + r(t)I\}(t-\hat{t})\end{aligned}$$

where r is continuous and $r(t) \to 0$ as $t \to \hat{t}$. Let $h^\epsilon = \sup\{h(t)|t \in B_\epsilon(\hat{t})\}$, $h_\epsilon = \inf\{h(t)|t \in B_\epsilon(\hat{t})\}$ and note that $h_\epsilon, h^\epsilon \to h(\hat{t})$ as $\epsilon \to 0$. Therefore we have that

$$h_\epsilon \int_{B_\epsilon(\hat{t})} e^{-\lambda k(t)}\,dt \leq \int_{B_\epsilon(\hat{t})} h(t)e^{-\lambda k(t)}\,dt \leq h^\epsilon \int_{B_\epsilon(\hat{t})} e^{-\lambda k(t)}\,dt.$$

Putting $r^\epsilon = \sup\{|r(t)||t \in B_\epsilon(\hat{t})\}$, we have that

$$\tfrac{1}{2}(t-\hat{t})'\{K(\hat{t}) - r^\epsilon I\}(t-\hat{t}) \leq k(t) \leq \tfrac{1}{2}(t-\hat{t})'\{K(\hat{t}) + r^\epsilon I\}(t-\hat{t})$$

for all $t \in B_\epsilon(\hat{t})$. Now note that we can choose ϵ small enough so that the matrix $K(\hat{t}) - r^\epsilon I$ is positive definite; e.g. make sure that r^ϵ is smaller than the smallest eigenvalue of $K(\hat{t})$ which we know is positive. Therefore we have that

$$\begin{aligned}&h_\epsilon \int_{B_\epsilon(\hat{t})} \exp\left\{-\frac{\lambda}{2}(t-\hat{t})'(K(\hat{t}) + r^\epsilon I)(t-\hat{t})\right\} dt \\ &\leq \int_{B_\epsilon(\hat{t})} h(t)e^{-\lambda k(t)}\,dt \\ &\leq h^\epsilon \int_{B_\epsilon(\hat{t})} \exp\left\{-\frac{\lambda}{2}(t-\hat{t})'(K(\hat{t}) - r^\epsilon I)(t-\hat{t})\right\} dt. \qquad (4.37)\end{aligned}$$

Now consider

$$\frac{\int_{B_\epsilon(\hat{t})} \exp\left\{-(\lambda/2)(t-\hat{t})'(K(\hat{t})+r^\epsilon I)(t-\hat{t})\right\} dt}{(2\pi)^{k/2}|\lambda K(\hat{t})|^{-1/2}}$$
$$= \frac{|K(\hat{t})+r^\epsilon I|^{-1/2}}{|K(\hat{t})|^{-1/2}} P(\|X-\hat{t}\| \le \epsilon)$$

where $X \sim N_k(\hat{t}, \lambda^{-1}(K(\hat{t})+r^\epsilon I)^{-1})$. Note that $\lim_{\lambda\to\infty} P(\|X-\hat{t}\| \le \epsilon) = 1$. Using a similar argument on the right-hand side of (4.37) we have that

$$\begin{aligned} h_\epsilon \frac{|K(\hat{t})+r^\epsilon I|^{-1/2}}{|K(\hat{t})|^{-1/2}} &\le \liminf_{\lambda\to\infty} \frac{\int_{B_\epsilon(\hat{t})} h(t)e^{-\lambda k(t)}dt}{(2\pi)^{k/2}|\lambda K(\hat{t})|^{-1/2}} \\ &\le \limsup_{\lambda\to\infty} \frac{\int_{B_\epsilon(\hat{t})} h(t)e^{-\lambda k(t)}dt}{(2\pi)^{k/2}|\lambda K(\hat{t})|^{-1/2}} \\ &\le h^\epsilon \frac{|K(\hat{t})-r^\epsilon I|^{-1/2}}{|K(\hat{t})|^{-1/2}}. \end{aligned} \tag{4.38}$$

Now note that we can make the left- and right-hand sides of (4.38) as close as we like by an appropriate choice of ϵ. This establishes the theorem.

Accordingly we are effectively approximating (4.35) by

$$(2\pi)^{k/2}|\lambda K(\hat{t})|^{-1/2} \exp\{-\lambda k(\hat{t})\}$$

times the expectation of h with respect to the $N_k(\hat{t}, (\lambda K(\hat{t}))^{-1})$ distribution as $\lambda \to \infty$. Intuitively Laplace's formula is saying that as $\lambda \to \infty$, i.e. as the variance of the multivariate normal goes to 0, most of the contribution to (4.35) is occurring near the maximum of $\exp\{-\lambda k(t)\}$. If k and h are assumed to have more derivatives, then more terms in an asymptotic expansion for (4.35) in powers of $\lambda^{-1/2}$ can be obtained as in Section 4.3.3 for the one-dimensional context.

One obvious necessity for the application of Laplace's formula is the existence of the parameter λ such that the integral in question can be written in the form (4.35). This is not always possible but the use of this approximation can still make sense as we see in the following example.

Example 4.8 *Bayesian asymptotics*
Recall from Example 1.3 that in Bayesian inference problems we are required to evaluate ratios of integrals of the form

$$R(m) = \frac{\int_\Omega m(\theta) f_\theta(x)\pi(\theta)\,d\theta}{\int_\Omega f_\theta(x)\pi(\theta)\,d\theta} = \frac{I(m)}{I(1)} \tag{4.39}$$

where $I(m) = \int_\Omega m(\theta) f_\theta(x)\pi(\theta)\,d\theta$. For the numerator we might write $h(\theta) = m(\theta)\pi(\theta)$ and

$$k(\theta, x) = -\frac{1}{\lambda} \log f_\theta(x) \tag{4.40}$$

where $\lambda = \lambda(x)$ is a measure of the size of the data set x such that $\lambda \to \infty$ as the size of the data set grows. Typically we are interested in approximations to (4.39) as the data set grows. We cannot apply Theorem 4.14, however, because of the dependence of k and λ on x. Note that we have suppressed the dependence in the notation but it is important to remember this. It is typical, however, for $k(\theta, x)$ to converge almost surely to some value $k(\theta)$ as $\lambda \to \infty$. Also the maximum $\hat{\theta}(x)$ of $f_\theta(x)$, which is also the minimum of $k(\cdot, x)$, typically converges almost surely. Under these circumstances it is at least intuitively plausible that, with these choices, a formal application of Laplace's method to each integral in (4.39) will lead to a reasonable approximation. This point was made in Lindley (1961, 1980) and it has been subsequently made rigorous in, for example, Kass, Tierney and Kadane (1990).

There is a degree of arbitrariness in this way of formulating the problem, however, as we can also write $h(\theta) = m(\theta)$ and

$$k(\theta, x) = -\frac{1}{\lambda}[\log f_\theta(x) + \log \pi(\theta)]; \tag{4.41}$$

i.e. including the prior with the likelihood $f_\theta(x)$, in which case $\hat{\theta}(x)$ is the posterior mode. If (4.41) is used then we note that, as λ becomes large, the effect of the prior disappears and the behavior of k is dominated by the likelihood so the intuition described above is again applicable. Throughout the remainder of this example we use (4.41).

We consider now the formal application of Laplace's method in the Bayesian context; i.e. the application of the methodology without providing the necessary conditions to ensure that the asymptotic expansions are valid to the order claimed. We refer the reader to the references cited earlier for the details. An extensive review of the use of the Laplace method in Bayesian inference is also given in Kass, Tierney and Kadane (1988).

For the integral $I(m)$ we let $\hat{I}(m)$ denote the estimate obtained by applying Laplace's formula to it using the posterior mode as $\hat{\theta}(x)$; i.e.

$$\hat{I}(m) = m(\hat{\theta}(x))(2\pi)^{k/2}|\lambda K(\hat{\theta}(x), x)|^{-1/2} \exp\{-\lambda k(\hat{\theta}(x), x)\}$$

where

$$K(\theta, x) = \left(\frac{\partial^2 k(\theta, x)}{\partial \theta_i \partial \theta_j} \right).$$

Notice that $\hat{I}(1)$ gives an estimate of the inverse normalizing constant. We then approximate $R(m)$ by

$$\hat{R}_1(m) = \frac{\hat{I}(m)}{\hat{I}(1)} = m(\hat{\theta}(x)).$$

We obtain a formal asymptotic error order for $\hat{R}_1(m)$, and also generalizations of this estimate, by performing a number of operations on the integrals $I(m)$ and $I(1)$. These operations are fairly standard and so we describe them without worrying about providing justification for their validity. For a function g let $S_i(\theta; g, \hat{\theta})$ denote the first i terms in the Taylor expansion of g about $\hat{\theta}$ and $S_{i,\infty}(\theta; g, \hat{\theta})$ denote the series from the $(i+1)$-st term on, so that $S_{0,\infty}(\theta; g, \hat{\theta})$ is the full series. Then the operations are as follows.

(i) Replace $m(\theta)$ by $S_{0,\infty}(\theta; m, \hat{\theta})$ and replace $k(\theta, x)$ by $S_{0,\infty}(\theta; k(\cdot, x), \hat{\theta})$.

(ii) Write

$$\begin{aligned}
&\exp\{-\lambda k(\theta, x)\}\\
&\quad = \exp\left\{-\lambda S_3(\theta; k(\cdot, x), \hat{\theta})\right\} \exp\left\{-\lambda S_{i,\infty}(\theta; k(\cdot, x), \hat{\theta})\right\}\\
&\quad = \exp\left\{-\lambda k(\hat{\theta}, x) - \tfrac{1}{2}(\theta - \hat{\theta})^t K(\hat{\theta}, x)(\theta - \hat{\theta})\right\}\\
&\qquad \times \sum_{j=0}^{\infty} \frac{(-1)^j \lambda^j S_{i,\infty}^j(\theta; k(\cdot, x), \hat{\theta})}{j!}\\
&\quad = \exp\left\{-\lambda k(\hat{\theta}, x) - \tfrac{1}{2}(\theta - \hat{\theta})^t K(\hat{\theta}, x)(\theta - \hat{\theta})\right\}\\
&\qquad \times S_{0,\infty}(-\lambda S_{i,\infty}(\theta; k(\cdot, x), \hat{\theta}), \exp, 0)
\end{aligned}$$

where we have used $\left(\partial k(\hat{\theta}(x), x)/\partial \theta_i\right) = 0$.

(iii) Reorganize the terms in $S_{0,\infty}(-\lambda S_{i,\infty}(\theta; k(\cdot, x), \hat{\theta}), \exp, 0)$ and in the product $S_{0,\infty}(\theta; m, \hat{\theta}) S_{0,\infty}(-\lambda S_{i,\infty}(\theta; k(\cdot, x), \hat{\theta}), \exp, 0)$ so that they are in increasing powers of $\lambda^{-1/2}$ and denote these sums by $1 + L_\infty(\lambda, \theta, \hat{\theta})$ and $m(\hat{\theta}) + U_\infty(\lambda, \theta, \hat{\theta})$ respectively (actually, as we will see, we only need finitely many terms in these sums). Make the change of variable $\theta \to v = (\lambda K(\hat{\theta}, x))^{1/2}(\theta - \hat{\theta})$ in the integrals. This gives that

$$R(m) = \frac{\int_\Omega (m(\hat{\theta}) + U_\infty(\lambda, \hat{\theta} + (\lambda K(\hat{\theta}, x))^{-1/2} v, \hat{\theta})) \psi(v)\, dv}{\int_\Omega (1 + L_\infty(\lambda, \hat{\theta} + (\lambda K(\hat{\theta}, x))^{-1/2} v, \hat{\theta})) \psi(v)\, dv} \tag{4.42}$$

where ψ is the $N_k(0, I)$ density function. Note that

$$U_\infty(\lambda, \hat{\theta} + (\lambda K(\hat{\theta}, x))^{-1/2} v, \hat{\theta}) = O(\lambda^{-1/2})$$

and

$$L_\infty(\lambda, \hat{\theta} + (\lambda K(\hat{\theta}, x))^{-1/2} v, \hat{\theta}) = O(\lambda^{-1/2})$$

as $\lambda \to \infty$.

(iv) Evaluate the integrals in (4.42) by taking expectations term-by-term (these integrals can be evaluated in closed form as moments of the standard multivariate normal). Write the result as

$$\frac{m(\hat{\theta}) + U_\infty(\lambda, \hat{\theta})}{1 + L_\infty(\lambda, \hat{\theta})} = (m(\hat{\theta}) + U_\infty(\lambda, \hat{\theta}))(1 - L_\infty(\lambda, \hat{\theta}) + L_\infty^2(\lambda, \hat{\theta}) - \cdots) \quad (4.43)$$

where we have used the exapnsion for $(1+x)^{-1}$ about 0 which is formally valid since $L_\infty(\lambda, \hat{\theta}) = O(\lambda^{-1})$ (Exercise 22).

(v) Write the right-hand side of (4.43) in terms of increasing powers of $\lambda^{-1/2}$ and stop when the desired error order is obtained.

Note that because the odd moments of the $N_k(0, I)$ distribution are 0, this implies that $L_\infty(\lambda, \hat{\theta}) = O(\lambda^{-1})$ and $U_\infty(\lambda, \hat{\theta}) = O(\lambda^{-1})$ (Exercise 22). This implies that the formal absolute error in $\hat{R}_1(m)$ is $O(\lambda^{-1})$ and when m is bounded away from 0 the relative error in $\hat{R}_1(m)$ is $O(\lambda^{-1})$ (Exercise 22). An estimate with formal error equal to $O(\lambda^{-2})$ is obtained by taking the first two terms in the final expansion. This is equal to

$$\hat{R}_2(m) = \hat{R}_1(m) - \frac{1}{2\lambda}\sum_{i,j}\Bigg[k^{(i,j)}(\hat{\theta}(x), x) \times \left\{\frac{\partial m(\hat{\theta}(x))}{\partial \theta_i}\sum_{r,s} k^{(r,s)}(\hat{\theta}(x), x)\frac{\partial^3 k(\hat{\theta}(x), x)}{\partial\theta_r\partial\theta_s\partial\theta_j} - \frac{\partial^2 m(\hat{\theta}(x))}{\partial\theta_i\partial\theta_j}\right\}\Bigg] \quad (4.44)$$

where $(K(\hat{\theta}(x), x))^{-1} = (k^{(i,j)}(\hat{\theta}(x), x))$ (Exercise 23). Under certain conditions, it can be proved that these asymptotic expansions hold rigorously; for example, see Kass, Tierney and Kadane (1990).

In Tierney and Kadane (1986) this technique is modified for the case when $m > 0$ by taking

$$k^*(\theta, x) = -(1/\lambda)[\log f_\theta(x) + \log \pi(\theta) + \log m(\theta)]$$

and $h = 1$ in the numerator and finding $\theta^*(x)$ which maximizes $-k^*(\theta, x)$. The first order estimate is

$$R^*(m) = \frac{|K^*(\theta^*(x), x)|^{-1/2}\exp\{-\lambda k^*(\theta^*(x), x)\}}{|K(\hat{\theta}(x), x)|^{-1/2}\exp\{-\lambda k(\hat{\theta}(x), x)\}}. \quad (4.45)$$

Tierney and Kadane (1986) argue formally, as above, that when m is bounded away from 0, then the absolute and relative errors in $R^*(m)$ are in fact $O(\lambda^{-2})$ because of a cancellation between the numerator and denominator (Exercise 24).

The R^* approximation requires only second order derivatives and two maximizations. This is a distinct advantage over $\hat{R}_2$ which requires third order derivatives. The R^* approximation is called the *fully exponential* Laplace method for obvious reasons.

The use of R^* is restricted to positive functions. Several alternative approaches are possible for an arbitrary m. Tierney, Kass and Kadane (1989) recommend using the fully exponential approach to approximate $R(\exp\{sm(\theta)\})$, the moment generating function of m, for several values of s about 0 and then numerically differentiating this to get an estimate of $R(m)$. They suggest using a symmetric difference quotient with s equal to ± 0.01 times the first order approximate standard deviations $[\hat{R}_2(m^2) - \hat{R}_1^2(m)]^{1/2}$. Higher moments and cumulants of m can also be estimated in this way.

There is another approximation that is commonly used in Bayesian statistics. This is also based upon the approximation of the posterior density $\pi(\theta|x)$ by a $N_k(\hat{\theta}(x), (\lambda K(\hat{\theta}(x), x))^{-1})$ distribution where $\hat{\theta}(x)$ is the posterior mode. The expected value of m under this multivariate normal distribution is then used to approximate $R(m)$. We will refer to this approximation as the *asymptotic normality approximation* and denote it by $\hat{R}(m)$. In Chen (1985) conditions are given for the asymptotic normality of a sequence of posterior distributions and this gives a partial justification for this approximation via weak convergence. It is easy to see that these conditions also establish the validity of Laplace's formula for the normalizing constant. With mild regularity conditions, the continuity of m and the convergence of $\hat{\theta}(x)$ as $\lambda \to \infty$, the proof in Chen (1985) can be adjusted to establish the validity of Laplace's formula for $I(m)$ as well. The conditions in this paper are stated for any sequence of densities. In general it is possible to state that these conditions will hold with probability 1 for a particular sequence of posterior densities; see Kass, Tierney and Kadane (1990) for a more extensive discussion of this issue.

A formal analysis establishes that the asymptotic error in the asymptotic normality approximation is $O(\lambda^{-1/2})$ (Exercise 25). Accordingly we would expect the Laplace approximations to perform much better than the asymptotic normal approximation. On the other hand, the Laplace approximations require some regularity conditions on the function m whereas nothing is required in the normal approximation beyond integrability; e.g. the Laplace approximation does not apply if m is an indicator function.

Example 4.9 *Bayesian linear model*
For this example we consider a Bayesian analysis of a linear statistical model which requires that integrals of dimension 10 be approximated. The statistical model that we use is specified as follows. Suppose that we observe the data (X, y) where $X \in R^{45\times 9}$ has $x_{ij} = 1$ for $5(j-1)+1 \le i \le 5j$ and 0 otherwise, and $y \in R^{45}$. The statistical model is then specified by the equation

$$\mathbf{y} = X\beta + \sigma z$$

where $\beta \in R^9, \sigma \in (0, \infty)$ and the error $z \in R^{45}$ is a sample of 45 from a distribution in the family $\{\text{Student}^*(\nu): \nu \in (2, \infty)\}$ where $\text{Student}^*(\nu)$ denotes the $\text{Student}(\nu)$ distribution standardized to have variance 1; i.e. the density is given by

$$g_\nu(z) = \frac{\Gamma\left(\frac{\nu+1}{2}\right)}{\Gamma\left(\frac{\nu}{2}\right)\Gamma\left(\frac{1}{2}\right)}\left(1 + \frac{z^2}{\nu - 2}\right)^{-(\nu+1)/2} \frac{1}{\sqrt{\nu - 2}}$$

for $-\infty < z < \infty$. Therefore we have nine location parameters β_i, $1 \le i \le 9$, a scale parameter σ and a shape parameter ν. In an application all these parameter values are unknown but we will treat ν as being known here and make the assignment $\nu = 3$. The restriction in the degrees of freedom for the Student distribution ensures that all response variables y_i have a mean and variance. We note that this model corresponds to five independent observations from each of the nine distributions specified by (β_i, σ) for $i = 1, \ldots, 9$. We denote the j-th observation from the i-th distribution by y_{ij}.

We place the improper prior $\pi(\beta, \sigma) = 1/\sigma$ on (β, σ). Then the posterior density for the 10-dimensional parameter (β,σ) is proportional to

$$\prod_{i=1}^{9}\prod_{j=1}^{5} \frac{1}{\sigma} g_\nu\left(\frac{y_{ij} - \beta_1 x_{i1} - \cdots - \beta_9 x_{i9}}{\sigma}\right)$$

and we note that this is a reasonably complicated function. Actually in this particular case some integrations can be carried out in closed form but we choose to ignore this here as this is not characteristic of such problems. The integrals we are approximating are therefore 10-dimensional. It makes sense to make the transformation from σ to $\log \sigma$ so that the variable in the integration is unconstrained in R^1. This more closely conforms to the behavior of an approximating multivariate normal density which, as we have seen, is closely associated with the approximations of this section. With this transformation, and letting $\theta = (\theta_1, \ldots, \theta_9, \theta_{10}) = (\beta_1, \ldots, \beta_9, \log \sigma)$, we see that we want to approximate integrals of the form

$$I(m) = \int_{-\infty}^{\infty} \cdots \int_{-\infty}^{\infty} m(\theta_1, \ldots, \theta_{10}) \exp\{-9\, n\, \theta_{10}\}$$
$$\times \prod_{i=1}^{9}\prod_{j=1}^{n} g_\nu\left(\frac{y_{ij} - \theta_i}{\exp(\theta_{10})}\right) d\theta_1 \cdots d\theta_{10}$$

where n serves as the asymptotic parameter λ. Recall that $n = 5$ in this example.

To obtain a specific numerical example we constructed a data set by fixing values for the parameters and then generating the z_{ij}. For this we specified $\beta_i = 0$

for every $i \neq 1$, $\beta_1 = 1.8$, $\sigma = 1$ and generated the z_{ij} from a Student*(3) distribution. The values obtained are listed in Table 4.2.

We now consider numerical approximations to the ratio of integrals $R(m)$ for various choices of m using some of the Laplace methods and the asymptotic normality approximation discussed in Example 4.8. The maximizations were of necessity done numerically using the IMSL routine DUMIAH. In Table 4.3 we give the estimates $\hat{R}_1(\theta_i)$, $\hat{R}(\theta_i^2)$, $\hat{R}_1(\theta_i^2)$ and $R^*(\theta_i^2)$ as well as the exact values. In this case we have that $\hat{R}(\theta_i) = \hat{R}_1(\theta_i)$. The normal approximation provides reasonable accuracy except that it fails significantly in the approximation of $R(\theta_{10})$ and $R(\theta_{10}^2)$. The $\hat{R}_1(\theta_i)$ approximations are reasonable except for $i = 4$ and $i = 10$ where the relative errors are 61% and 218%. The $\hat{R}_1(\theta_i^2)$ approximations are generally not as good as the $\hat{R}_1(\theta_i)$ approximations as the relative errors averaged 44% over the 10 estimates. The $R^*(\theta_i^2)$ approximations are better than the $\hat{R}_1(\theta_i^2)$ approximations in every coordinate and are quite good. In this case the asymptotic normal approximations can be obtained immediately once we have carried out the maximization and calculated the inverse Hessian. For more complicated functions m we will have to use the methods of Chapters 6 and 7 to calculate the normal approximations; i.e. generate samples from the particular normal distribution. For more discussion of this example see Evans and Swartz (1995a) and Genz *et al.* (1996).

Overall, the methods we have discussed under Bayesian asymptotics are easy to use and in many cases produce the kind of accuracy necessary for statistical applications. We note that in this example we have been applying these in a context where rigorous asymptotic results have not been justified and, to a certain extent, this seems characteristic of many applications of this methodology. When they are applicable they are easy to implement and the computations are so fast that computing time is an irrelevant factor. Their weaknesses lie in a dependence on a fairly regular structure; e.g. a single dominant mode and symmetry. In particular we need the normal approximation to be at least reasonable. As is easily seen the approximate normality is highly dependent on the parameterization of a problem.

Table 4.2. The 45 generated z values in Example 4.9.

i	$j=1$	$j=2$	$j=3$	$j=4$	$j=5$
1	0.1518	1.0856	0.5915	−0.0684	−0.3781
2	−0.6383	0.1866	−0.0095	0.4953	0.2616
3	2.8333	1.0984	−0.4761	−0.3833	1.1289
4	0.0420	0.1740	−0.5817	0.4472	−0.0711
5	−0.6161	−0.5138	−1.3213	0.0432	0.0725
6	−0.8166	0.9349	−0.1565	−0.1723	−0.1744
7	−0.8785	0.02807	−0.1104	−1.1933	0.9740
8	−0.6106	−0.5486	−0.5446	−0.4029	−1.300
9	0.2896	0.7551	−0.1500	0.7462	−3.0065

Table 4.3. Estimates computed using the Laplace methods and the asymptotic normality approximation in Example 4.9 when $\nu = 3$.

i	$R(\theta_i)$	$\hat{R}_1(\theta_i)$	$R(\theta_i^2)$	$\hat{R}(\theta_i^2)$	$\hat{R}_1(\theta_i^2)$	$\hat{R}^*(\theta_i^2)$
1	2.043	2.018	4.263	4.141	4.073	4.313
2	0.095	0.116	0.081	0.061	0.014	0.062
3	0.696	0.883	0.721	0.914	0.780	0.747
4	0.018	0.029	0.069	0.046	0.001	0.039
5	−0.426	−0.408	0.270	0.231	0.166	0.273
6	−0.147	−0.172	0.107	0.082	0.030	0.094
7	−0.287	−0.271	0.212	0.177	0.073	0.214
8	−0.650	−0.629	0.488	0.437	0.396	0.505
9	0.319	0.370	0.206	0.208	0.137	0.200
10	−0.073	−0.232	0.033	0.079	0.054	0.033

Another disadvantage is the lack of error estimates for a particular approximation as simply increasing the number of terms in an asymptotic expansion does not necessarily increase the accuracy of the approximation.

4.5 Exercises

(1) Prove that if $\phi = o(\vartheta)$ then $\phi = O(\vartheta)$.

(2) Prove that if $\phi_i = O(\vartheta_i)$ at x_0 for $i = 1, \ldots, n$ then $\sum_{i=1}^n a_i\phi_i = O\left(\sum_{i=1}^n |a_i||\vartheta_i|\right)$ at x_0. Show that this is also true for o.

(3) Prove that if $\phi_i = O(\vartheta_i)$ at x_0 for $i = 1, \ldots, n$ then $\prod_{i=1}^n \phi_i = O\left(\prod_{i=1i}^n \vartheta_i\right)$. Show that this is also true when we replace O by o.

(4) Prove that if $\{\phi_n\}$ is an asymptotic sequence at x_0 and $\lim_{x\to x_0} \phi_0$ is finite then $\lim_{x\to x_0} \phi_n(x) = 0$ for $n > 0$.

(5) Show that $\phi_n(x) = x^{-n} \exp\left(-x^2/2\right)$ is an asymptotic sequence at ∞.

(6) Prove that if the sequence $\{\phi_n\}$ is asymptotic at $x_0 \in C \cup \{\infty\}$ then $f \sim \sum_{i=0}^\infty a_i\phi_i$ at x_0 if and only if

$$f(x) = \sum_{i=0}^{n-1} a_i\phi_i(x) + O(\phi_n)$$

at x_0.

(7) Prove Lemma 4.4.

(8) Prove that

$$\phi_i(x) = (-1)^i x^{-(2i+1)} e^{-x^2/2}$$

is an asymptotic sequence at ∞.

(9) Establish that for $f, g \in \mathcal{A}(\{\phi_n\}, x_0)$ then $f \equiv g$ implies $f \sim g$ at x_0 but not conversely.

(10) Prove that if $f(x) \sim \sum_{i=0}^{\infty} a_i x^{-i}$ at ∞ and $g(x) \sim \sum_{i=0}^{\infty} a_i x^{-i}$ at ∞, with these expansions holding in the same domain A, then $f(x)g(x) \sim \sum_{i=0}^{\infty} c_i x^{-i}$ at ∞ where

$$c_i = a_0 b_i + a_1 b_{i-1} + \cdots + a_i b_0.$$

(11) Use repeated integration by parts to determine an asymptotic expansion for the function

$$\mathrm{Ei}(x) = \int_x^{\infty} \frac{e^{-t}}{t}\, dt$$

as $x \to \infty$.

(12) Establish the following properties of the gamma function: (i) for $\nu > 1$, $\Gamma(\nu) = (\nu - 1)\Gamma(\nu - 1)$, (ii), $\Gamma(1) = 1$, (iii) for integer $n \geq 0$, $\Gamma(n + 1) = n!$ and (iv) $\Gamma(1/2) = \sqrt{\pi}$.

(13) Prove that if $f(x) \sim \sum a_n \phi_n(x)$ as $x \to \infty$ for the asymptotic sequence $\{\phi_n\}$ then for $\sigma > 0$ the asymptotic expansion $f(\sigma x) \sim \sum a_n \phi_n(\sigma x)$ holds.

(14) Prove that $\mu_{2i} = (2i - 1)(2i - 3) \cdots 3 \cdot 1$, $\mu_{2i+1} = 0$ for the $N(0, 1)$ distribution.

(15) Determine the terms of order x^{-2} and $x^{-5/2}$ for (4.10).

(16) Apply the results of Exercise 15 to the gamma function.

(17) Show that if $a_\lambda \sim A_\lambda$ and $b_\lambda \sim B_\lambda$ then $a_\lambda + i b_\lambda \sim A_\lambda + i B_\lambda$.

(18) Construct the saddle-point approximation to the density of the sample mean for a sample of n from the standard exponential distribution; i.e. $f(x) = e^{-x}$ for $x > 0$.

(19) Determine the $n^{-3/2}$ term in Example 4.7.

(20) Develop the Edgeworth expansion for the density function for the situation discussed in Section 4.3.5.

(21) Calculate the first term in the Edgeworth expansion for the distribution function when the sample comes from an exponential (θ) distribution.

(22) Establish formally that $L_\infty(\lambda, \hat{\theta}) = O(\lambda^{-1})$ and $U_\infty(\lambda, \hat{\theta}) = O(\lambda^{-1})$ in (4.43) and therefore that the absolute error in $\hat{R}_1(m)$ is $O(\lambda^{-1})$ and that the relative error in $\hat{R}_1(m)$ is $O(\lambda^{-1})$ when m is bounded away from 0.

(23) Using the formal expansion technique described in Example 4.8 verify (4.44) and that the formal error is $O(\lambda^{-2})$.

(24) Establish formally that the absolute and relative errors in $R^*(m)$ in (4.45) are $O(\lambda^{-2})$.

(25) Establish formally that the absolute error in $\hat{R}(m)$ is $O(\lambda^{-1/2})$.

(26) Verify the numerical results obtained in Example 4.9.

5
Multiple quadrature

5.1 Introduction

Methods of approximate integration based on *quadrature* are historically the oldest of the integration techniques. In fact, the term quadrature stems from the ancient Greek problem of "squaring the circle". The squaring of a circle is the process of inscribing or circumscribing a circle with rectangles (or polygons) of known area for the purpose of estimating π (i.e. the area of a circle with unit radius). Engels (1980) provides a brief history of the development of quadrature.

Not only has quadrature been around for a long time, its theory encompasses some very elegant mathematics. With advances in computing power, research in quadrature continues today for higher dimensional problems; this area of study is typically referred to as multiple quadrature or sometimes as *cubature*. Unfortunately, even though quadrature is an effective technique in low dimensional problems, it is not as effective in higher dimensions. We restrict our attention here to only a small fraction of the myriad of quadrature techniques, those which we believe are especially useful. For detailed accounts of quadrature, see the books by Engels (1980), Davis and Rabinowitz (1984) and Krommer and Ueberhuber (1998).

In Section 5.2 we give the basic definition and make some general observations about multiple quadrature rules. In Section 5.3 we discuss the basic rules used for one dimensional integrals. In Section 5.4 we present a simple method for obtaining approximations to some one dimensional integrals that provides exact error bounds. In Section 5.5 we consider applying one dimensional rules to higher dimensional integration problems via product rules. In Section 5.6 monomial rules are developed for integration problems of dimension greater than 1. In Section 5.7 a relatively new method for deriving multiple quadrature rules, known as Bayesian quadrature, is presented. In Section 5.8 some of the theory of quasi-random integration is developed. In Section 5.9 lattice rules are introduced and in particular good lattice point rules are discussed. Section 5.10 deals with the various methods that have been developed for adaptive quadrature.

5.2 Multiple quadrature rules

Consider the approximation of the integral

$$I = \int_A f(x)\,dx. \tag{5.1}$$

A *multiple quadrature rule of order n* takes the form

$$\hat{I}_n = \sum_{i=1}^{n} w_i f(x_i) \tag{5.2}$$

where the w_i are known as the *weights* and the x_i are known as the *points* of the rule. The weights and points are fixed values and depend on n, although we have suppressed the dependence. There are various schemes for devising the sets

$$R_n = \{(x_i, w_i) : i = 1, \ldots, n\}$$

depending upon what properties we want the rule to possess. Basically when we talk about a rule we mean a sequence of sets $\{R_n\}$ constructed according to some criterion. For example, a common criterion is to make the R_n *exact*; i.e. (5.2) equals (5.1), for some class of functions that increases with n.

A first observation is that if a rule has been developed for a region A and an integral is given by

$$I = \int_B f(x)\, dx$$

where $B = T(A)$ for some smooth, one-to-one $T : A \to R^k$, then we can transform the integral to

$$I = \int_A f(T(x)) \left| \det \frac{\partial T(x)}{\partial x} \right| dx$$

and apply R_n to the new integrand. Equivalently, we can transform the rule $\{(x_i, w_i)\}$ to be applicable over B using the rule

$$R_n^* = \left\{ \left(T(x_i), w_i \left| \det \frac{\partial T(x_i)}{\partial x} \right| \right) : i = 1, \ldots, n \right\}$$

on this region.

Whatever criterion is employed to construct $\{R_n\}$, accompanying any such rule is a theorem which establishes that (5.2) converges to (5.1) as $n \to \infty$ typically for any Riemann integrable function. This is what justifies the rule as providing meaningful approximations. In practice, increasing n generally improves an approximation although this is not guaranteed. Note that in increasing n we cannot make use of previously computed approximations; i.e. $\hat{I}_{n_0}$ for $n_0 < n$ as, in general the R_n don't nest. On the other hand use is made of these values by looking at absolute error estimates

$$|\hat{I}_{n_0} - \hat{I}_n| \tag{5.3}$$

or the relative error estimates

$$\frac{|\hat{I}_{n_0} - \hat{I}_n|}{|\hat{I}_n|} \tag{5.4}$$

which provide an indication as to whether or not the approximation has converged. Although more refined exact error bounds can often be established for multiple quadrature rules, these are not often practically useful and (5.3) and (5.4) are what is commonly used. Often (5.3) or (5.4) is computed for several choices of n and n_0 and we stop increasing n when this estimate is suitably small. Of course, this heuristic error estimate is not guaranteed to give accurate approximations.

Another way of improving an approximation is through *compounding* the rule. This is applicable when A is a bounded region that can be iteratively partitioned into subregions where the rule can be applied. For example, suppose $A = \prod_{i=1}^{k}[a_i, b_i]$ is a bounded rectangle and the rule R_n is devised for $[0, 1]^k$. Then we can subdivide each $[a_i, b_i]$ into m equal-length subintervals to give m^k subrectangles, apply the rule R_n (after a suitable location-scale transformation) in each of these and then sum the approximations to get the compound approximation. Let us denote this compound approximation by $\hat{I}_n^m$. We then have the following result for rectangles.

Theorem 5.1 If f is Riemann integrable on $A = \prod_{i=1}^{k}[a_i, b_i]$, R_n integrates constants exactly on A and we partition A into m^k subrectangles by subdividing each $[a_i, b_i]$ into m equal-length subintervals, then

$$\hat{I}_n^m \to \int_A f(x)\,dx$$

as $m \to \infty$.

Proof: We have that $\hat{I}_n^m = \sum_{i=1}^{n} w_i S_i^m$ where S_i^m is a sum over all subrectangles. The j-th term in S_i^m is equal to the ratio of the volume of the j-th subrectangle to the volume of A all times the function f evaluated at the transformation of the point x_i and of course this point lies inside the j-th subrectangle. Accordingly each S_i^m is a Riemann sum for $\int_A f(x)\,dx$ and therefore $S_i^m \to \int_A f(x)\,dx$ as $m \to \infty$ since the mesh size of the partition goes to 0 as $m \to \infty$. Therefore

$$\lim_{m\to\infty} \hat{I}_n^m = \left(\sum_{i=1}^{n} w_i\right) \int_A f(x)\,dx \Big/ \int_A dx$$

which establishes the result as $\sum_{i=1}^{n} w_i = \int_A dx$.

While we have proved the result for rectangles it is clear that it is also true for more general regions like simplices.

Compounding probably represents the most typical application of multiple quadrature; i.e. we choose n and then increase m while observing quantities such as (5.3) and (5.4) as m increases and stopping when the error estimate becomes suitably small. Compounding has some limitations as it only applies to rules applicable to bounded regions, may require computation of entirely new function evaluations for each m and, perhaps most significantly, it suffers from the curse of dimensionality, as we have to apply the rule in m^k subrectangles. For example, when the dimension is $k = 15$ and m is a modest 10, we require $n10^{15}$ function evaluations and this is prohibitively large.

For low dimensions, however, such as $k = 1, 2$ and 3, compounding is an effective strategy in many situations. This together with an appropriate choice of

R_n can lead to highly accurate approximations and a great deal of confidence in our results.

We mention another strategy when applying multiple quadrature rules. When presented with an integral (5.1) we can sometimes modify this in some fashion so as to render the integral more amenable to a particular multiple quadrature rule. We have already discussed reparameterizations and recall Example 2.1 where we used a transformation to remove a singularity. Basically, multiple quadrature rules are designed to give very precise results; e.g. exact results, for some class of functions. Therefore when considering using a particular rule we attempt to modify the integral so that the integrand is in some sense more like a function in the class. For example, if the integrand has a singularity and we are using a rule appropriate for polynomials, then removing the singularity via a transformation seems like a good idea because polynomials do not have singularities.

Many multiple quadrature rules are designed to be suitable when the integrand is of the form, or well approximated by, pw on A where p is a low degree polynomial and w is a nonnegative *weight function*. We require that w have a finite integral on A and such that $\text{supp}(f) \subset \text{supp}(w)$ where $\text{supp}(f) = cl\{x : f(x) \neq 0\}$ is the support of f. For convenience we assume that the integral of w over A is 1 so that w is a probability density and in fact we will require, in addition, that w have a number of moments. We then write (5.1) as

$$I = \int_A f(x)\,dx = \int_A g(x)w(x)\,dx \tag{5.5}$$

where $g = f/w$. The idea is that we choose a suitable w so that g is well approximated by a low degree polynomial. The multiple quadrature rule then takes the form

$$\hat{I}_n = \sum_{i=1}^{n} w_i g(x_i). \tag{5.6}$$

In essence, the weight function has been "integrated out of the integrand" leaving a simpler function. Notice that (5.5) is the expectation of g with respect to the probability density function w.

There is another way of dealing with integrands that we feel are approximately of the form pw for some polynomial p and weight function w. For this note that the integral can always be transformed to the unit cube, at least in principle, via the probability transformation based on w which is discussed in Exercise 3.9.9. This approach also effectively integrates out w and if we use a rule for the unit cube that is exact for constants then Theorem 5.1 applies; i.e. this transformation permits compounding which is not available in general with weighted integration rules. For an arbitrary weight function this can be difficult to implement, however, typically because the inverse transformation is difficult to compute. This *reweighting* by a density w is a method also applicable to the Monte Carlo methods for approximating integrals that we discuss in Chapters 6 and 7.

5.3 Interpolatory rules on R^1

Interpolatory rules represent a broad class of quadrature rules. These rules are characterized by the way the weights are determined; namely by the values of the integrals of certain functions prescribed by their values at some points $\{x_1, \ldots, x_n\}$ where $x_1 < x_2 < \cdots < x_n$. Suppose then that we want to approximate (5.1) where $A \subset R^1$ and the points of the rule are prescribed as $\{x_1, \ldots, x_n\} \subset A$.

We consider polynomial interpolation. Define the i-th *Lagrange interpolating polynomial* of degree $n-1$ with respect to $\{x_1, \ldots, x_n\}$ by

$$l_i(x) = \frac{(x - x_1) \cdots (x - x_{i-1})(x - x_{i+1}) \cdots (x - x_n)}{(x_i - x_1) \cdots (x_i - x_{i-1})(x_i - x_{i+1}) \cdots (x_i - x_n)}$$

for $i = 1, \ldots, n$. Note that $l_i(x_i) = 1$ and $l_i(x_j) = 0$ if $j \neq i$. Now suppose that we want a rule that calculates

$$\int_A p(x)w(x)\, dx$$

exactly for all polynomials p where $\deg(p) \leq n-1$ and w is a probability density on A. Of course we must restrict our attention to weight functions w that satisfy

$$\int_A x^i w(x)\, dx < \infty$$

for $i = 1, \ldots, n-1$; i.e. weight functions that have their first $n-1$ moments. Now put

$$w_i = \int_A l_i(x)w(x)\, dx$$

for $i = 1, \ldots, n$. We call the rule R_n constructed according to this scheme the *Lagrange interpolation rule* or *polynomial interpolation rule* based on $\{x_1, \ldots, x_n\}$. We have the following result.

Theorem 5.2 For the Lagrange interpolation rule based on $\{x_1, \ldots, x_n\}$

$$\hat{I}_n = \sum_{i=1}^{n} w_i p(x_i) = \int_A p(x)w(x)\, dx$$

for every polynomial p satisfying $\deg(p) \leq n-1$; i.e. it integrates exactly every polynomial of degree less than n.

Proof: We have that

$$p(x) = \sum_{i=1}^{n} p(x_i) l_i(x) \tag{5.7}$$

because the left and right-hand sides are equal at the points $\{x_1, \ldots, x_n\}$ and a polynomial of $\deg(p) \leq n-1$ is completely determined by its values at n points. Therefore

$$\int_A p(x)w(x)\,dx = \sum_{i=1}^{n} p(x_i) \int_A l_i(x)w(x)\,dx = \sum_{i=1}^{n} w_i p(x_i)$$

and the result is established.

So in essence a Lagrange interpolation rule approximates a function f by an interpolating polynomial $\sum_{i=1}^{n} f(x_i)l_i(x)$, interpolating at the points $\{x_1, \ldots, x_n\}$, and then integrates this to obtain $\sum_{i=1}^{n} w_i f(x_i)$ as the approximation to

$$\int_A f(x)w(x)\,dx.$$

Note that the functions $\{l_1, \ldots, l_n\}$ serve as a basis for the linear space of all polynomials of degree $n-1$ or less, with (5.7) being the unique representation of the polynomial p with respect to this basis. Also, we need only evaluate the weights once, as they are applicable to any function f. We discuss several specific classes of interpolation rules as determined by the choice of the points $\{x_1, \ldots, x_n\}$ and the weight function w.

Other possibilities exist for the interpolation procedure. For example, we can interpolate the function f at the points $\{x_1, \ldots, x_n\}$ by the piecewise linear, continuous function on $[x_1, x_n]$ given by

$$\sum_{i=1}^{n-1} \left\{ f(x_i) + \frac{f(x_{i+1}) - f(x_i)}{x_{i+1} - x_i}(x - x_i) \right\} I_{(x_i, x_{i+1})}(x)$$

which is a degree-one *spline* approximation to f. Suppose that $A = [a, b]$ and $x_1 = a$, $x_n = b$. Integrating this with respect to w, we obtain the rule

$$\sum_{i=1}^{n} w_i f(x_i)$$

where

$$w_1 = \int_a^{x_2} \left(1 - \frac{x - a}{x_2 - a}\right) w(x)\,dx,$$

$$w_i = \int_{x_i}^{x_{i+1}} \left(1 - \frac{x - x_i}{x_{i+1} - x_i}\right) w(x)\,dx + \int_{x_{i-1}}^{x_i} \frac{x - x_{i-1}}{x_i - x_{i-1}} w(x)\,dx$$

when $1 < i \leq n-1$ and

$$w_n = \int_{x_{n-1}}^{b} \frac{x - x_{n-1}}{b - x_{n-1}} w(x)\,dx.$$

We call this rule the *spline interpolation rule* of degree one based on $\{x_1, \ldots, x_n\}$. Clearly this rule integrates polynomials of degree one or lower exactly. Higher order spline rules can be obtained by using higher degree splines and integrating as above; see Thisted (1988) and Gautschi (1997) for more discussion of splines.

As another example of an interpolating procedure, we consider *Hermite interpolation*. Here we write $\pi(x) = (x - x_1)\cdots(x - x_n)$ and put

$$p(x) = \sum_{i=1}^{n} f(x_i)\left\{1 - (x - x_i)\frac{\pi^{(2)}(x_i)}{\pi^{(1)}(x_i)}\right\} l_i^2(x) + \sum_{i=1}^{n} f^{(1)}(x_i)(x - x_i)l_i^2(x) \tag{5.8}$$

which is a polynomial of degree $2n - 1$ and satisfies $p(x_i) = f(x_i)$, $p^{(1)}(x_i) = f^{(1)}(x_i)$ for $i = 1, \ldots, n$. Note that if f is a polynomial of degree $2n - 1$ or less then $p = f$. Integrating to obtain the weights gives a *Hermite rule*, of the form

$$\sum_{i=1}^{n} w_i f(x_i) + \sum_{i=1}^{n} w_i^{(1)} f^{(1)}(x_i)$$

which is exact for all polynomials of degree $2n - 1$ or less. Note that Hermite interpolation at n points $x_1, \ldots, x_n$ arises from Lagrange interpolation at $2n$ points $x_1, \ldots, x_n, x_{n+1}, \ldots, x_{2n}$ by taking limits as $x_{n+i} \to x_i$ for $i = 1, \ldots, n$ (Exercise 5). See Davis (1975) for a general discussion of interpolation and a description of other bases for constructing integration rules.

We consider some general results for polynomial interpolation rules when A is a bounded interval. Since we can always transform a rule on $[0, 1]$ to an arbitrary interval by a location-scale transformation we consider only rules on the unit interval. We begin by providing the following upper bound for the absolute error in a polynomial interpolation rule.

Theorem 5.3 If $f \in C^n[0, 1]$ and R_n is a polynomial interpolation rule then

$$\left|\int_0^1 f(x)w(x)\,dx - \sum_{i=1}^{n} w_i f(x_i)\right| \le \frac{\sup\left\{|f^{(n)}(t)| : t \in [0, 1]\right\}}{n!} \int_0^1 \prod_{i=1}^{n} |x - x_i| w(x)\,dx.$$

Proof: Consider the function of t

$$g(t) = f(t) - \sum_{i=1}^{n} f(x_i)l_i(t) - \frac{f(x) - \sum_{i=1}^{n} f(x_i)l_i(x)}{\prod_{i=1}^{n}(x - x_i)} \prod_{i=1}^{n}(t - x_i)$$

where $x \in [0, 1]$, $x \neq x_i$ for $i = 1, \ldots, n$. Then $g(x_i) = 0$ for $i = 1, \ldots, n$ and $g(x) = 0$. Therefore, by Rolle's theorem, there is a point $\zeta(x) \in (\min(x, x_1), \max(x_n, x))$ such that

$$\begin{aligned} 0 &= g^{(n)}(\zeta(x)) \\ &= f^{(n)}(\zeta(x)) - n!\frac{f(x) - \sum_{i=1}^{n} f(x_i)l_i(x)}{\prod_{i=1}^{n}(x - x_i)}. \end{aligned} \tag{5.9}$$

This implies that

$$f(x) - \sum_{i=1}^{n} f(x_i)l_i(x) = \frac{f^{(n)}(\zeta(x))}{n!} \prod_{i=1}^{n}(x - x_i)$$

for all $x \in [0, 1]$ where we define $\zeta(x_j)$ as a point in $[0, 1]$, which exists by the continuity of $f^{(n)}$, satisfying

$$\begin{aligned}
&f^{(n)}(\zeta(x_j)) \\
&\quad = \lim_{x \to x_j} n! \frac{f(x) - \sum_{i=1}^{n} f(x_i)l_i(x)}{\prod_{i=1}^{n}(x - x_i)} \\
&\quad = n! \lim_{x \to x_j} \left\{ \frac{f(x) - f(x_j)}{\prod_{i=1}^{n}(x - x_i)} - \frac{f(x_j)(l_j(x) - 1)}{\prod_{i=1}^{n}(x - x_i)} - \frac{\sum_{i \neq j} f(x_i)l_i(x)}{\prod_{i=1}^{n}(x - x_i)} \right\} \\
&\quad = \frac{n!}{\prod_{i \neq j}(x_j - x_i)} \left\{ f^{(1)}(x_j) - \sum_{i=1}^{n} f(x_i)l_i^{(1)}(x_j) \right\}. \qquad (5.10)
\end{aligned}$$

Then from the continuity of the right-hand side of (5.9), when it is defined at the x_j by (5.10), we have that $f^{(n)}(\zeta(x))$ is a continuous function of $x \in [0, 1]$. Therefore

$$\begin{aligned}
&\left| \int_0^1 f(x)w(x)\,dx - \sum_{i=1}^{n} w_i f(x_i) \right| \\
&\quad = \left| \int_0^1 \left\{ f(x) - \sum_{i=1}^{n} l_i(x)f(x_i) \right\} w(x)\,dx \right| \\
&\quad = \left| \int_0^1 \frac{f^{(n)}(\zeta(x))}{n!} \prod_{i=1}^{n}(x - x_i)w(x)\,dx \right| \\
&\quad \leq \int_0^1 \frac{|f^{(n)}(\zeta(x))|}{n!} \prod_{i=1}^{n} |x - x_i| w(x)\,dx \\
&\quad \leq \frac{\sup\{|f^{(n)}(t)| : t \in [0, 1]\}}{n!} \int_0^1 \prod_{i=1}^{n} |x - x_i| w(x)\,dx.
\end{aligned}$$

For this bound to be useful we need to be able to calculate $\sup\{|f^{(n)}(t)| : t \in [0, 1]\}$ and also require the value $\int_0^1 \prod_{i=1}^{n} |x - x_i| w(x)\,dx$. Notice, however, that

$$\int_0^1 \prod_{i=1}^{n} |x - x_i| w(x)\,dx \leq 1.$$

An alternative, and at times more useful, bound is commonly available. This is obtained by using the error in the Taylor expansion for f rather than the interpolation error used in Theorem 5.3. We have the following results where $x_+^p = (\max\{x, 0\})^p$.

Theorem 5.4 If $f \in C^{p+1}[0, 1]$ and R_n is a rule that integrates all polynomials of degree p or less then

$$\int_0^1 f(x)w(x)\,dx - \sum_{i=1}^n w_i f(x_i) = \int_0^1 f^{(p+1)}(t)K(t)\,dt$$

where

$$K(t) = \frac{1}{p!}\left\{\int_0^1 (x-t)_+^p\, w(x)\,dx - \sum_{i=1}^n w_i (x_i - t)_+^p\right\}.$$

Proof: By Taylor's theorem with the integral form for the remainder we can write

$$\begin{aligned} f(x) &= \sum_{i=0}^p \frac{f^{(i)}(0)}{i!}x^i + \frac{1}{p!}\int_0^x (x-t)^p f^{(p+1)}(t)\,dt \\ &= \sum_{i=0}^p \frac{f^{(i)}(0)}{i!}x^i + \frac{1}{p!}\int_0^1 (x-t)_+^p f^{(p+1)}(t)\,dt. \end{aligned}$$

Therefore, since the rule integrates all polynomials of degree p or lower,

$$\begin{aligned} &\int_0^1 f(x)w(x)\,dx - \sum_{i=1}^n w_i f(x_i) \\ &= \int_0^1 \frac{1}{p!}\left\{\int_0^1 (x-t)_+^p f^{(p+1)}(t)\,dt\right\} w(x)\,dx \\ &\quad - \sum_{i=1}^n w_i \frac{1}{p!}\int_0^1 (x_i - t)_+^p f^{(p+1)}(t)\,dt \\ &= \int_0^1 f^{(p+1)}(t)\frac{1}{p!}\left\{\int_0^1 (x-t)_+^p\, w(x)\,dx - \sum_{i=1}^n w_i (x_i - t)_+^p\right\} dt \\ &= \int_0^1 f^{(p+1)}(t)K(t)\,dt. \end{aligned}$$

The function K is called the *Peano kernel* for R_n. We have the following corollaries.

Corollary 5.5 If $f \in C^{p+1}[0, 1]$ and R_n is a rule that integrates all polynomials of degree p or less then

$$\left|\int_0^1 f(x)w(x)\,dx - \sum_{i=1}^n w_i f(x_i)\right| \le \sup\{|f^{(p+1)}(t)| : t \in [0, 1]\} \int_0^1 |K(t)|\,dt$$

Corollary 5.6 If $f \in C^{p+1}[0, 1]$, R_n is a rule that integrates all polynomials of degree p and whose Peano kernel is of constant sign in $[0, 1]$ then

$$\int_0^1 f(x)w(x)\,dx - \sum_{i=1}^{n} w_i f(x_i) = f^{(p+1)}(t_0) \int_0^1 K(t)\,dt$$

for some $t_0 \in [0, 1]$.

Proof: Suppose without loss of generality that K is nonnegative. Then we have that

$$\begin{aligned} &\inf\left\{ f^{(p+1)}(t) : t \in [0, 1] \right\} \int_0^1 K(t)\,dt \\ &\leq \int_0^1 f^{(p+1)}(t)K(t)\,dt \\ &\leq \sup\left\{ f^{(p+1)}(t) : t \in [0, 1] \right\} \int_0^1 K(t)\,dt. \end{aligned}$$

Since $f^{(p+1)}$ is continuous, the result follows by the intermediate value theorem.

It is rare that the error bounds given in Theorem 5.3 and Corollary 5.5 are useful. This is because we need to evaluate $\sup\left\{|f^{(n)}(t)| : t \in [0, 1]\right\}$ which may be as difficult as evaluating the integral. Of course, we can always substitute an upper bound for this quantity whenever it is available.

We want to establish the convergence of these rules as $n \to \infty$. Convergence depends, however, on how the sequence of sets $\{x_1, \ldots, x_n\}$ is constructed. Obviously, very poor choices can be made that do not lead to the convergence of $\sum_{i=1}^{n} w_i f(x_i)$ to $\int_A f(x)w(x)\,dx$. Surprisingly, as we will see in the next section, even seemingly reasonable choices, may not lead to convergence.

We note, however, that even if a rule does not converge as $n \to \infty$ all interpolatory rules with respect to a constant weight function converge under compounding as $m \to \infty$ when $A = [a, b]$ by Theorem 5.1. The following result gives the rate of convergence of the rule under compounding.

Theorem 5.7 If a rule R_n, with respect to a constant weight function on $[a, b]$, has exact error of the form

$$E_1 = c(b-a)^{r+1} f^{(r)}(t_0)$$

for some constant c, some r and some $t_0 \in [a, b]$, where $f^{(r)}$ is continuous, then the error E_m in the rule after compounding m times satisfies

$$\lim_{m\to\infty} m^r E_m = c(b-a)^r \left\{ f^{(r-1)}(b) - f^{(r-1)}(a) \right\}.$$

Proof: We apply the exact error expression on each subinterval

$$I_i = \left[a + (b-a)\frac{(i-1)}{m}, a + (b-a)\frac{i}{m}\right]$$

to obtain

$$E_m = \sum_{i=1}^{m} c \left\{\frac{(b-a)}{m}\right\}^{r+1} f^{(r)}(t_{0i})$$

where $t_{0i} \in I_i$. Then

$$m^r E_m = c(b-a)^r \sum_{i=1}^{m} \frac{(b-a)}{m} f^{(r)}(t_{0i})$$

and the result follows because

$$\sum_{i=1}^{m} \frac{(b-a)}{m} f^{(r)}(t_{0i})$$

is a Riemann sum for $\int_a^b f^{(r)}(t)\,dt$.

While Theorem 5.7 gives the asymptotic error term explicitly it is somctimes just noted that the error in the compounded rule is $O(1/m^r)$. For example, suppose that Corollary 5.6 applies with $w = 1$. Then if we apply this rule to the interval $[a, b]$ we get that the error in the rule is

$$E_1 = c(b-a)^{p+2} f^{(p+1)}(t_0)$$

where $t_0 \in [a, b]$ and

$$c = \int_0^1 K(t)\,dt$$

and so the error under compounding is $O(1/m^{p+1})$.

5.3.1 RECTANGLE RULES

There are two *rectangle rules* on [0, 1] given by $R_0 = \{(0, 1)\}$ and $R_1 = \{(1, 1)\}$; i.e. we approximate $\int_0^1 f(t)\,dt$ by $f(0)$ and $f(1)$ respectively. These rules integrate constants exactly. The Peano kernel for R_0 is given by $K(t) = 1 - t$ and for R_1 it is given by $K(t) = -t$. Notice that in both cases K is of constant sign so that Corollary 5.6 is applicable whenever $f \in C^1[0, 1]$.

5.3.2 THE MIDPOINT RULE

The midpoint rule approximates $\int_0^1 f(t)\,dt$ by $f(1/2)$; i.e. $R = \{(1/2, 1)\}$. Clearly this rule exactly integrates all polynomials of degree one or less. The Peano kernel is given by

$$K(t) = \tfrac{1}{2}(t-1)^2 - \left(\tfrac{1}{2} - t\right)_+$$

which is of constant sign and so Corollary 5.6 is applicable whenever $f \in C^2[0, 1]$. Note that this implies a certain superiority for the midpoint rule over the rectangle rules. Even though they are all based upon one function evaluation the midpoint rule gives a higher rate of convergence under compounding for functions in C^2.

5.3.3 NEWTON–COTES RULES

The *Newton–Cotes rules* in $[0, 1]$ are based on choosing $x_i = (i-1)/(n-1)$ for $i = 1, \ldots, n$ and then setting

$$w_i = \int_0^1 l_i(x)\,dx;$$

i.e. we interpolate using $n \geq 2$ equidistant points with $x_1 = 0$ and $x_n = 1$. When the endpoints 0 and 1 are omitted, we obtain open as opposed to closed Newton–Cotes rules. The midpoint rule is an example of an open rule. The open rules are useful whenever we are integrating a function which is not defined at the endpoints. Recall, however, our comments about transforming integrands so that the behavior of the transformed integrand is more like the class of integrands for which the rule is designed.

When $n = 2$ we have the *trapezoid rule* which is given by

$$\tfrac{1}{2}\{f(0) + f(1)\};$$

namely it is the average of the function at the endpoints. Note that the trapezoid rule is exact for all polynomials of degree less than or equal to one. The Peano kernel for the trapezoid rule is given by

$$K(t) = -\tfrac{1}{2}t(1-t)$$

and so it is of constant sign. Further we have that

$$\int_0^1 K(t)\,dt = -\frac{1}{12}$$

and therefore for $f \in C^2[0, 1]$ we have

$$E_1 = -\frac{f^{(2)}(t_0)}{12}$$

for some $t_0 \in [0, 1]$.

When $n = 3$ we have *Simpson's rule* which is given by

$$\frac{1}{6}f(0) + \frac{2}{3}f(1/2) + \frac{1}{6}f(1).$$

This is a weighted average of the function at the endpoints and the midpoint of the interval. Simpson's rule exactly integrates all polynomials of degree three or less. This is easily seen by noting that, because it is an interpolatory rule based on three points, Simpson's rule integrates exactly all polynomials of degree two or less. Then observe that the formula for Simpson's rule gives the value 1/4 when $f(x) = x^3$. The Peano kernel for Simpson's rule is given by

$$K(t) = \begin{cases} \frac{1}{72}t^3(3t-2) & 0 \le t \le \frac{1}{2} \\ \frac{1}{72}(3t-1)(t-1)^3 & \frac{1}{2} \le t \le 1 \end{cases}$$

which is of constant sign. Further

$$\int_0^1 K(t)\,dt = -\frac{1}{2880}$$

and therefore for $f \in C^4[0,1]$ we have that

$$E_1 = -\frac{f^{(4)}(t_0)}{2880}$$

for some $t_0 \in [0,1]$. Simpson's rule is commonly used so it is convenient to record the formula for this rule after compounding m times in the interval $[a,b]$; namely

$$\sum_{i=1}^{m}\left(\frac{b-a}{6m}\right)\left[f\left(a+\frac{(i-1)(b-a)}{m}\right)+4f\left(a+\frac{(2i-1)(b-a)}{2m}\right)\right.$$
$$\left.+f\left(a+\frac{i(b-a)}{m}\right)\right]$$
$$=\left(\frac{b-a}{6m}\right)\left[f(a)+f(b)+4\sum_{i=1}^{m}f\left(a+\frac{(2i-1)(b-a)}{2m}\right)\right.$$
$$\left.+2\sum_{i=2}^{m}f\left(a+\frac{(i-1)(b-a)}{m}\right)\right].$$

Table 5.1 gives some higher order closed Newton–Cotes rules. Higher order Newton–Cotes rules have faster rates of convergence under compounding for sufficiently smooth functions, so it may seem that the higher order rules are desirable. However, associated with higher order rules are weights of increasing magnitude. Also, all Newton–Cotes rules of order $n \ge 11$ have some negative weights which

Table 5.1. Weights for closed Newton–Cotes rules of order n

n	w_1	w_2	w_3	w_4	w_5	w_6	w_7
4	1/8	3/8	3/8	1/8			
5	7/90	32/90	32/90	32/90	7/90		
6	19/288	75/288	50/288	50/288	75/288	19/288	
7	41/840	216/840	27/840	272/840	27/840	216/840	41/840

are undesirable because of the possibility of cancellation error. These deficiencies sometimes lead to numerical problems when calculating higher order Newton–Cotes rules. It is also possible to prove that, as $n \to \infty$, the sequence of Newton–Cotes rules $\sum_{i=1}^{n} w_i f(x_i)$ does not converge to $\int_0^1 f(x)\,dx$ for general f.

We have restricted attention to Newton–Cotes rules where the weight function is the identity; i.e. $w(x) \equiv 1$. For more general weight functions on [0, 1] we obtain *weighted Newton–Cotes* rules. We consider some of these in the exercises.

5.3.4 GAUSS RULES

Gauss rules possess an optimality property in the class of interpolatory rules. We note that the points $\{x_1, \ldots, x_n\}$ can be freely chosen in developing an interpolatory rule. Note that once we have chosen the points for an interpolatory rule the weights are fixed. We look then for the set of points which calculates

$$\int_A p(x)w(x)\,dx$$

exactly for every polynomial of highest possible degree. We show that the highest degree is $2n-1$ when using n points. This highest degree is attained by the Gauss rule based on n points for the weight function w. The properties of Gauss rules depend heavily on the properties of orthogonal polynomials. We refer the reader to Hochstadt (1986) and Jackson (1941) for general discussions of orthogonal polynomials including many more interesting and useful results than we mention here.

Now suppose that w is a probability density on the set $A \subset R^1$ such that it has its first $2n$ moments. Define a (pseudo) inner product on the real vector space of functions

$$L_w^2 = \left\{ f\colon \int_A f^2(x)w(x)\,dx < \infty \right\}$$

by

$$\langle f, g \rangle = \int_A f(x)g(x)w(x)\,dx$$

where the qualifier "pseudo" is used because we may have $\langle f, f \rangle = 0$ when f is 0 almost everywhere w but not identically 0 on A. Note that the inner product leads to a norm on L^2_w given by $\|f\| = \sqrt{\langle f, f\rangle}$. Now using this inner product we apply the Gram–Schmidt process to the sequence of polynomials $1, x, x^2, \ldots$ to give a sequence of orthonormal polynomials $p_0, p_1, p_2, \ldots$ where deg $p_i = i$; i.e.

$$\langle p_i, p_j\rangle = \delta_{ij}.$$

We have that $p_0(x) \equiv 1$ and for $i > 1$

$$p_i(x) = \frac{x^i - \sum_{j=0}^{i-1}\langle x^i, p_j\rangle p_j(x)}{\left\|x^i - \sum_{j=0}^{i-1}\langle x^i, p_j\rangle p_j(x)\right\|}. \tag{5.11}$$

Sometimes we use the unnormalized versions of these polynomials, namely $P_i(x) = x^i - \sum_{j=0}^{i-1}\langle x^i, P_j\rangle P_j(x)/\langle P_j, P_j\rangle$.

Rather than use (5.11) to calculate these polynomials it is easier to use the *three-term recurrence relation* as given in the following result.

Theorem 5.8 The sequence of polynomials $P_0, P_1, P_2, \ldots$ satisfies

$$P_{i+1}(x) = (x - a_{i+1})P_i(x) - b_{i+1}P_{i-1}(x)$$

where

$$a_{i+1} = \frac{\langle xP_i(x), P_i(x)\rangle}{\|P_i(x)\|^2}$$

$$b_{i+1} = \frac{\langle xP_i(x), P_{i-1}(x)\rangle}{\|P_{i-1}(x)\|^2}.$$

Proof: Exercise.

In some cases the constants a_i and b_i can be worked out exactly and so the integrations are avoided.

As we will see the points $\{x_1, \ldots, x_n\}$ of the Gauss rule of degree n associated with w are the roots of the polynomial P_n. We prove the following basic fact about these roots.

Theorem 5.9 The roots $\{x_1, \ldots, x_n\}$ of P_n are distinct and they all lie in the interior of the support of w.

Proof: When $n > 1$ we have that

$$\langle P_n, 1\rangle = \int_A P_n(x)w(x)\,dx = 0$$

and so P_n must change sign in the interior of the support of w which we denote by $(\text{supp}(w))^o$. Suppose that $z_1, \ldots, z_m$ are the roots of P_n in $(\text{supp}(w))^o$. Then $P_n(x)\prod_{i=1}^m(x - z_i)$ is of constant sign in $(\text{supp}(w))^o$. But if $m < n$ then

$$0 = \int_A P_n(x)\prod_{i=1}^{m}(x - z_i)w(x)\,dx$$

because $\prod_{i=1}^m(x - z_i)$ can be expressed as a linear combination of $P_0, P_1, \ldots, P_m$. Therefore we must have that $m = n$.

Now define

$$w_i = \int_A \frac{P_n(x)}{(x - x_i)P_n^{(1)}(x_i)} w(x)\,dx. \tag{5.12}$$

We then define the *Gauss rule of degree n* with respect to the weight function w on A by $R_n = \{(x_1, w_1), \ldots, (x_n, w_n)\}$. We have the following result.

Theorem 5.10 The Gauss rule R_n is a polynomial interpolation rule and satisfies

$$\sum_{i=1}^{n} w_i p(x_i) = \int_A p(x)w(x)\,dx$$

for all polynomials p of degree $2n - 1$ or less.

Proof: Let p be a polynomial of degree less than or equal to $2n - 1$. Then

$$P(x) = \sum_{i=1}^{n} p(x_i) \frac{P_n(x)}{(x - x_i)P_n^{(1)}(x_i)}$$

interpolates p at $x_1, \ldots, x_n$ (see Exercise 17) and this establishes that R_n is a polynomial interpolation rule. Therefore $P(x) - p(x)$ is a polynomial of degree less than or equal to $2n - 1$ and $P(x_i) - p(x_i) = 0$ for $i = 1, \ldots, n$ which implies that

$$r(x) = \frac{P(x) - p(x)}{P_n(x)}$$

is a polynomial of degree less than or equal to $n - 1$. Therefore

$$0 = \int_A P_n(x)r(x)w(x)\,dx = \sum_{i=1}^{n} w_i P_n(x_i)r(x_i)$$

which implies that

$$\begin{aligned}
\sum_{i=1}^{n} w_i p(x_i) &= \sum_{i=1}^{n} w_i \{P(x_i) - P_n(x_i)r(x_i)\} \\
&= \sum_{i=1}^{n} w_i P(x_i) - \int_A P_n(x)r(x)w(x)\,dx \\
&= \sum_{i=1}^{n} \left\{ \int_A \frac{P_n(x)}{(x - x_i)P_n^{(1)}(x_i)} w(x)\,dx \right\} p(x_i) \\
&\quad - \int_A P_n(x)r(x)w(x)\,dx
\end{aligned}$$

$$= \int_A P(x)w(x)\,dx - \int_A P_n(x)r(x)w(x)\,dx$$
$$= \int_A \{P(x) - P_n(x)r(x)\}w(x)\,dx$$
$$= \int_A p(x)w(x)\,dx.$$

The optimality property of the Gauss rules follows from the following result.

Theorem 5.11 Amongst rules of the form $\sum_{i=1}^n w_i f(x_i)$ no rule can integrate exactly all polynomials of degree $2n$ or less.

Proof: Let $f(x) = \prod_{i=1}^n (x - x_i)^2$. Then

$$\int_A f(x)w(x)\,dx > 0$$

but $\sum_{i=1}^n w_i f(x_i) = 0$.

Some useful facts can be proved about the weights w_i which are sometimes referred to as *Christoffel numbers*. We have the following result.

Theorem 5.12 For the Gauss rule on n points with respect to the weight function w we have that $w_i > 0$ for $i = 1, \ldots, n$ and $\sum_{i=1}^n w_i = 1$.

Proof: It is immediate that $1 = \int_A w(x)\,dx = \sum_{i=1}^n w_i$. Now put

$$f_i(x) = \prod_{j \neq i} \left(\frac{x - x_j}{x_i - x_j} \right)^2$$

which is a polynomial of degree $2n - 2$. Then

$$0 < \int_A f_i(x)w(x)\,dx = \sum_{j=1}^n w_j f_i(x_j) = w_i f_i(x_i) = w_i.$$

Of some interest is what happens to the Gauss rule approximation as $n \to \infty$. The following provides a satisfying answer.

Theorem 5.13 For the Gauss rule on n points with respect to the weight function w we have that

$$\lim_{n\to\infty} \sum_{i=1}^n w_i f(x_i) = \int_A f(x)w(x)\,dx$$

whenever f is continuous and A is compact.

Proof: For $\epsilon > 0$, by the Weierstrass approximation theorem, there is a polynomial g such that $|f(x) - g(x)| < \epsilon/2$ for every $x \in A$. Choose n_0 such that for all $n \geq n_0$

$$\sum_{i=1}^{n} w_i g(x_i) = \int_A g(x)w(x)\,dx.$$

Then

$$\begin{aligned}\left|\int_A f(x)w(x)\,dx - \sum_{i=1}^{n} w_i f(x_i)\right| &\leq \left|\int_A f(x)w(x)\,dx - \int_A g(x)w(x)\,dx\right| \\ &\quad + \left|\int_A g(x)w(x)\,dx - \sum_{i=1}^{n} w_i g(x_i)\right| \\ &\quad + \left|\sum_{i=1}^{n} w_i g(x_i) - \sum_{i=1}^{n} w_i f(x_i)\right| \\ &\leq \frac{\epsilon}{2} + 0 + \frac{\epsilon}{2} = \epsilon.\end{aligned}$$

Since ϵ is arbitrary this establishes the result.

In fact, versions of Theorem 5.13 also hold for special weight functions when A is not compact; for example, see Szegö (1985).

In the noncompact case Theorem 5.13 is more significant as compounding is unavailable. In the case when $A = [a, b]$ and w is constant a typical application chooses n and then compounds the rule. Accordingly, the following results give the rate of convergence under compounding via Theorem 5.7 under these circumstances.

Theorem 5.14 If $f \in C^{2n}[0, 1]$ then the error in the n-point Gauss rule based on the weight function w is given by

$$E_1 = \frac{f^{(2n)}(t_0)}{(2n)!} \int_0^1 \left\{\prod_{i=1}^{n}(x - x_i)\right\}^2 w(x)\,dx$$

for some $t_0 \in [0, 1]$.

Proof: Let p denote the polynomial of degree $2n - 1$ which solves the Hermite interpolation problem $p(x_i) = f(x_i)$, $p'(x_i) = f'(x_i)$ for $i = 1, \ldots, n$ as given in (5.8). Therefore $\int_0^1 p(x)w(x)\,dx = \sum_{i=1}^{n} w_i p(x_i) = \sum_{i=1}^{n} w_i f(x_i)$ and this implies that

$$\begin{aligned}\int_0^1 f(x)w(x)\,dx - \sum_{i=1}^{n} w_i f(x_i) &= \int_0^1 f(x)w(x)\,dx - \int_0^1 p(x)w(x)\,dx \\ &= \int_0^1 \{f(x) - p(x)\}w(x)\,dx.\end{aligned}$$

Now the Lagrange interpolation error based on the points $\{x_1, \ldots, x_n, x_{n+1}, \ldots, x_{2n}\}$ satisfies, see the proof of Theorem 5.3,

$$f(x) - \sum_{i=1}^{2n} f(x_i) l_i(x) = \frac{f^{(2n)}(\zeta(x))}{(2n)!} \prod_{i=1}^{2n} (x - x_i) \tag{5.13}$$

where $f^{(2n)}(\zeta(x))$ is a continuous function of $x \in [0, 1]$. Now let $x_{n+i} \to x_i$ for $i = 1, \ldots, n$. Then the left-hand side of (5.13) goes to $f(x) - p(x)$ (Exercise 5),

$$\frac{1}{(2n)!} \prod_{i=1}^{2n} (x - x_i) \to \frac{1}{(2n)!} \prod_{i=1}^{n} (x - x_i)^2$$

and so $f^{(2n)}(\zeta(x))$ goes to the ratio of these quantities. The continuity of $f^{(2n)}$ implies the existence of $\zeta^*(x) \in [0, 1]$ such that

$$f(x) - p(x) = \frac{f^{(2n)}(\zeta^*(x))}{(2n)!} \prod_{i=1}^{n} (x - x_i)^2$$

for all $x \in [0, 1]$. Furthermore it is easy to show, as in the proof of Theorem 5.3, that $f^{(2n)}(\zeta^*(x))$ is a continuous function of x. Then since

$$\prod_{i=1}^{n} (x - x_i)^2 w(x)$$

is of constant sign the result follows.

A disadvantage of using Gauss rules is that the calculation of the points and weights for arbitrary weight functions is not generally straightforward. Software exists, however, for the calculation of Gauss points and weights corresponding to some special and important weight functions. We provide some important (unnormalized) weight functions in Table 5.2. Observe that the Legendre rule is a special case of the Jacobi rule ($\alpha = \beta = 0$). If we transform the Jacobi rule to the interval $(0, 1)$ via the transformation $y = (1 + x)/2$ then the weight function is proportional to $y^{\beta}(1 - y)^{\alpha}$ which is proportional to a beta$(\beta + 1, \alpha + 1)$ density function. The weight functions of the Laguerre rules and the Hermite rules are defined on infinite

Table 5.2. Classical weight functions for Gauss rules

Rule	Weight function w	Parameter restrictions	Domain
Legendre	1		$[-1, 1]$
Jacobi	$(1 - x)^{\alpha}(1 + x)^{\beta}$	$\alpha > -1, \beta > -1$	$[-1, 1]$
Laguerre	$x^{\alpha} e^{-x}$	$\alpha > -1$	$[0, \infty)$
Hermite	$e^{-x^2/2}$		$(-\infty, \infty)$

intervals. This produces a very convenient choice when we have to integrate over these domains. For compact intervals it is better to use a lower order Legendre rule and compound this rather than use a very high order Gauss rule. This is because the accuracy of points and weights becomes a numerical issue as n gets large. Further, we note that the Laguerre, Hermite and (transformed) Jacobi rules have natural applications in statistics as the corresponding weight functions are proportional to gamma$(\alpha + 1, 1)$, $N(0, 1)$ and beta$(\beta + 1, \alpha + 1)$ densities respectively. The points and weights for these classical weight functions are available from many sources and in particular from the NAG and IMSL packages. Also see Stroud and Secrest (1966) for tables and algorithms for computing these rules.

If a weight function w commonly arises in applications, like those in Table 5.2, then it makes sense to construct the Gauss rules; i.e. the weights and points, associated with it. Conceptually this is straightforward as we need only determine the sequence of polynomials $P_0, P_1, P_2, \ldots$ associated with w, determine the roots $(x_1, \ldots, x_n)$ of P_n and then use (5.12) to determine the weights $(w_1, \ldots, w_n)$. This may involve some integrations but in principle these computations need only be done once and so a substantial effort can be made to ensure high accuracy. In practice, however, these computations can be very unstable. In particular, in the iterative determination of the orthogonal polynomials, small round-off errors can propagate, leading to big inaccuracies in relatively low degree polynomials. A partial solution to this is increased precision. We refer the reader to Davis and Rabinowitz (1984) and Ma, Rokhlin and Wandzura (1996) for discussions of algorithms used to construct Gauss rules.

5.3.5 ERROR IN INTERPOLATORY RULES

The expressions determined for the errors in applications of interpolatory rules perhaps have more theoretical than practical value as they all involve the need to maximize a higher order derivative. In practice these are rarely used. Typically one assesses the error in an approximation by computing approximations based on successively higher values of n, or via increasing the number of times m we compound a rule, and observing how the approximations change. When the changes become small enough then we feel "confident" that we have achieved a desired accuracy.

Of course this is not fool-proof. For example, consider the approximation of

$$\int_0^{10000} \frac{1}{1+x^2}\,dx$$

using the trapezoid rule and suppose that our stopping criterion is to stop when two approximations differ by less than 0.01. If we compare the values of the approximations obtained via the trapezoid rule for $n = 800$ and $n = 801$, we obtain 6.381 and 6.373 respectively and stop. To two decimal places this integral equals 1.57 and so our approximation is substantially in error. Of course one can argue that our stopping rule is rather simplistic and that a more sophisticated approach would result in a more accurate approximation in this example. This is true but it is also true that one can always construct examples where a particular stopping rule will perform badly.

For many rules, e.g. Gauss rules, increasing the order of the rule or compounding requires a completely new set of function evaluations. When the integrand is expensive to evaluate this is very inefficient. Various rules have been developed that take advantage of previous function evaluations. For example, in the context of Gauss rules we have the Gauss–Kronrod rules; e.g. see Davis and Rabinowitz (1984).

5.4 Envelope rules and exact error analysis

A natural approach to the approximation of $\int_A f(x)\,dx$ is to first approximate f by a function h such that the approximating integral $\int_A h(x)\,dx$ can be evaluated exactly. This is the motivation behind interpolatory rules where h is an interpolating function for f. Other obvious choices for h are truncated expansions of f; e.g. Taylor series. As we have seen, however, one of the difficulties with this approach is that is not clear what the error in the approximation is and so this is not entirely satisfactory.

As mentioned in Section 2.6, the highest level of error analysis, and perhaps the goal of all our discussions, is to be able to state an approximation with a definite bound on the error. At this point in time, this is rarely achievable except in certain contexts. There has been some work on this problem in the context of evaluating some of the special functions that arise in applied work. For example, many of the special functions of applied mathematics can be evaluated to essentially machine accuracy at least over certain ranges of the argument. In Wang and Kennedy (1994) methods are described for obtaining approximations with guaranteed error bounds for some of the common one-dimensional distribution functions that arise in applications of probability theory.

The question arises as to how we can construct an approximation to a general integral in such a way that a definite bound on the error can be given as well. Here we describe a method to do this that is applicable for many one-dimensional integrals at least. While it is also applicable to multiple integrals it suffers from the curse of dimensionality as one might expect. This approach is discussed in Evans and Swartz (1997, 1998a, 1998b) and is closely related to the methods for constructing generators for probability distributions discussed in Section 3.6. In general, we construct upper and lower envelopes U and L to the integrand f and such that $\int_A U(x)\,dx$ and $\int_A L(x)\,dx$ are easily evaluated. Then because

$$\int_A L(x)\,dx \le \int_A f(x)\,dx \le \int_A U(x)\,dx \tag{5.14}$$

we have that

$$0 \le \int_A U(x)\,dx - \int_A f(x)\,dx \le \int_A U(x)\,dx - \int_A L(x)\,dx \tag{5.15}$$

and

$$\left|\frac{\int_A U(x)\,dx - \int_A f(x)\,dx}{\int_A U(x)\,dx}\right| \leq \left|1 - \frac{\int_A L(x)\,dx}{\int_A U(x)\,dx}\right| \tag{5.16}$$

so that (5.15) and (5.16) give computable exact bounds on the absolute and relative error in the approximation $\int_A U(x)\,dx$. Similar bounds are obtained for the approximation $\int_A L(x)\,dx$. We refer to $\int_A L(x)\,dx$ as a *lower envelope rule* and $\int_A U(x)\,dx$ as an *upper envelope rule*.

In Section 3.6, envelope methods are developed and used for variate generation via adaptive rejection sampling. We remark that in adaptive rejection sampling, a density does not need to be extraordinarily well approximated by the bounding functions since the adaptation component of the algorithm quickly improves the approximation. Moreover, there is no meaningful efficiency gain in an algorithm which, for example, accepts variates 99.9% of the time over an algorithm which accepts variates only 90% of the time. Accordingly, piecewise linear envelopes are fine for random variate generation but it turns out that they are not adequate for integral approximations. For this reason, we consider the construction of piecewise polynomial approximations to f.

For this suppose that the domain A can be decomposed into a finite set of m intervals $[a_i, b_i]$ such that there exists a transformation T_i on $[a_i, b_i]$ with $(T_i(f))^{(n)}$ either concave or convex there. Now dropping the subscript i, suppose that $(T(f))^{(n)}$ is concave on $[a, b]$. Then the chord

$$c(x) = (T(f))^{(n)}(a) + \frac{(T(f))^{(n)}(b) - (T(f))^{(n)}(a)}{b-a}(x-a) \tag{5.17}$$

bounds $(T(f))^{(n)}$ from below and the tangent

$$t(x) = (T(f))^{(n)}(a) + (T(f))^{(n+1)}(a)(x-a) \tag{5.18}$$

bounds $(T(f))^{(n)}$ from above. If $(T(f))^{(n)}$ is convex on $[a, b]$, then the role of the chord and tangent expressions is interchanged in (5.17) and (5.18). The bounding chords and tangents allow for the construction of upper and lower envelopes to $f(x)$ on the interval $[a, b]$. For this we use the obvious result given in the following lemma.

Lemma 5.15 If $f^{(1)}(x) \leq g^{(1)}(x)$ for all $x \in [a, b]$ then $f(x) \leq g(x) - g(a) + f(a)$ for all $x \in [a, b]$

Then since

$$c(x) \leq (T(f))^{(n)}(x) \leq t(x) \tag{5.19}$$

on $[a, b]$, repeated application of Lemma 5.15 to (5.19) leads immediately to the upper and lower envelopes U and L to f via the following result.

Theorem 5.16 Suppose that T is a strictly increasing function and that $(T(f))^{(n)}$ is concave on the interval $[a, b]$. Then for every $x \in [a, b]$, we have that $L(x) \leq f(x) \leq U(x)$ where

$$L(x) = T^{-1}\left[\sum_{k=0}^{n} \frac{(T(f))^{(k)}(a)}{k!}(x-a)^k + \frac{(T(f))^{(n)}(b) - (T(f))^{(n)}(a)}{b-a}\frac{(x-a)^{n+1}}{(n+1)!}\right]$$

and

$$U(x) = T^{-1}\left[\sum_{k=0}^{n+1} \frac{(T(f))^{(k)}(a)}{k!}(x-a)^k\right].$$

If T is strictly decreasing then the expressions for L and U in Theorem 5.16 are interchanged. Also if $(T(f))^{(n)}$ is convex on the interval $[a, b]$ with T strictly decreasing then the expressions of Theorem 5.16 apply while they are reversed if T is strictly increasing. Notice that it is clear from these expressions that the lower and upper envelope rules are exact for certain classes of functions.

Therefore, provided the expressions for L and U are easy to integrate exactly, we obtain (5.14) at least for $A = [a, b]$. When A is subdivided into a set of bounded intervals then (5.14) is obtained by summing the piecewise inequalities. When A is unbounded then we can apply the envelope inequality in Theorem 3.7 to the tails to obtain (5.14). We remark again that the transformations T can vary from interval to interval in the decomposition of A. Note that we have available all the transformations T discussed in Section 3.6. Also note that when $n = 0$ the approximations to $T(f)$ are piecewise linear.

The aim in applying this approach is to achieve some prescribed error bound ϵ for (5.15) or (5.16). Therefore we require an iterative procedure to construct the envelopes so that the error criterion can be met. There are various ways that the approximating envelopes $L(x)$ and $U(x)$ can be improved. First, the envelopes can generally be made tighter by increasing n. Second, it is a simple matter to improve the envelopes on any interval $[a, b]$ by compounding. By introducing a new point x^* where $a < x^* < b$, the concavity structure of $T(f)$ on $[a, x^*]$ and $[x^*, b]$ is the same as on $[a, b]$. Therefore improved lower and upper envelopes can be defined on each of the subintervals. We then proceed as follows. We use the envelope inequality to make the contributions from the tails less than $\epsilon/2$ and then we compound on the central part until the error on this part is also less than $\epsilon/2$.

Of course we have to prove that, when compounding, the integrals of the envelopes converge to the integral of f over the bounded interval in question. This is established in the following result when T is the identity transformation. The result for general T is left as an exercise (Exercise 17). Assume then that $f^{(n)}$ is concave on the bounded interval $[a, b]$. Let L_m and U_m denote the lower and upper envelopes to f on $[a, b]$ determined by compounding m times on $[a, b]$ by dividing this into m equal-length subintervals; i.e. the overall envelopes determined by applying Theorem 5.16 in each subinterval. The following result establishes the convergence of the approximations $\int_a^b U_m(x)\,dx$ and $\int_a^b L_m(x)\,dx$ and the rate of convergence as $m \to \infty$.

Theorem 5.17 Suppose that $f^{(n)}$ is concave on the bounded interval $[a, b]$ and $f^{(n+2)}$ is continuous. If E^m and E_m denote the errors in $\int_a^b U_m(x)\,dx$ and $\int_a^b L_m(x)\,dx$ respectively as approximations of $\int_a^b f(x)\,dx$, then as $m \to \infty$ we have that $E^m = O(1/m^{n+2})$ with

$$E^m \sim \frac{1}{m^{n+2}} \left\{ f^{(n+1)}(b) - f^{(n+1)}(a) \right\} \frac{(b-a)^{n+2}}{(n+3)!}$$

while $E_m = o(1/m^{n+1})$.

Proof: From Taylor's theorem we have that

$$f(x) = U(x) + f^{(n+2)}(\zeta(x)) \frac{(x-a)^{n+2}}{(n+2)!}$$

where $\zeta(x) \in (a, b)$. Therefore

$$\begin{aligned}
\int_a^b f(x)\,dx &= \int_a^b U(x)\,dx + \int_a^b f^{(n+2)}(\zeta(x)) \frac{(x-a)^{n+2}}{(n+2)!}\,dx \\
&= \int_a^b U(x)\,dx + f^{(n+2)}(\tau) \int_a^b \frac{(x-a)^{n+2}}{(n+2)!}\,dx \\
&= \int_a^b U(x)\,dx + f^{(n+2)}(\tau) \frac{(b-a)^{n+3}}{(n+3)!} \qquad (5.20)
\end{aligned}$$

where the second equality is justified by Exercise 1 for some $\tau \in (a, b)$. We can then directly apply Theorem 5.7 to get the results for E^m.

Also by Taylor's theorem

$$f(x) = L(x) + \left\{ f^{(n+1)}(\eta(x)) - \frac{f^{(n)}(b) - f^{(n)}(a)}{b-a} \right\} \frac{(x-a)^{n+1}}{(n+1)!}$$

where $\eta(x) \in (a, b)$, which gives

$$\begin{aligned}
&\int_a^b f(x)\,dx \\
&= \int_a^b L(x)\,dx + \int_a^b \left\{ f^{(n+1)}(\eta(x)) - \frac{f^{(n)}(b) - f^{(n)}(a)}{b-a} \right\} \frac{(x-a)^{n+1}}{(n+1)!}\,dx \\
&= \int_a^b L(x)\,dx + \left\{ f^{(n+1)}(\xi) - \frac{f^{(n)}(b) - f^{(n)}(a)}{b-a} \right\} \int_a^b \frac{(x-a)^{n+1}}{(n+1)!}\,dx \\
&= \int_a^b L(x)\,dx + \left\{ f^{(n+1)}(\xi) - f^{(n+1)}(\xi^*) \right\} \frac{(b-a)^{n+2}}{(n+2)!} \qquad (5.21)
\end{aligned}$$

where the second equality is justified by Exercise 1 for some $\xi \in (a, b)$ and the last equality is justified by the mean value theorem for some $\xi^* \in (a, b)$. The conclusion for E_m then follow immediately from the same argument used in the proof of Theorem 5.7.

Obviously there is a similar result when $f^{(n)}$ is convex on $[a, b]$. Note that Theorem 5.17 establishes that using a higher order polynomial improves the rate of convergence under compounding of the envelope method whether we use the lower or upper envelope. Notice, however, that the theorem seems to establish that it is better to use the upper envelope than the lower envelope for the approximation as $O(1/m^{n+2})$ convergence implies $o(1/m^{n+1})$ convergence. Also, the upper envelope rule is exact for all polynomials of degree $n + 1$ or less from (5.20) while, from (5.21), the lower envelope rule is exact for all polynomials of degree n or less.

It is clear that the effectiveness of this approach to approximating integrals is dependent upon knowing the intervals where $f^{(n)}$, or $(Tf)^{(n)}$, for some convenient transformation T, is either convex or concave. Also we require appropriate transformations for the tails as in Section 3.6. Actually, knowledge of the tail behavior of an integrand is a necessary part of any successful integration strategy. In fact, the envelope inequality provides a convenient approach for dealing with this problem for any integration problem. Also, in the center where we compound there seems to be very little value in the use of transformations, so our problem reduces to determining the inflection points of $f^{(n)}$ if we want to implement the envelope approach. Although this seems like a significant limitation, it turns out that the need to find these inflection points can often be completely avoided. For example, suppose that the integrand f factors according to $f = f_1 f_2$ on $[a, b]$ and that we can obtain envelopes $L_1 < f_1 \le U_1$ and $L_2 \le f_2 \le U_2$ giving rise to $L_1 L_2 \le f \le U_1 U_2$ such that it is easy to calculate the bounds

$$\int_a^b f_1(x) L_2(x)\, dx \le \int_a^b f(x)\, dx \le \int_a^b f_1(x) U_2(x)\, dx.$$

Then we take $L(x) = f_1(x)L_2(x)$ and $U(x) = f_1(x)U_2(x)$. Another useful strategy arises when we can write $f(x) = f_1(f_2(x))$ and we have $l(x) \le f_1(x) < u(x)$ for some l and u such that $\int_a^b l(f_2(x))\, dx$ and $\int_a^b u(f_2(x))\, dx$ can be easily evaluated. Then we take $L(x) = l(f_2(x))$ and $U(x) = u(f_2(x))$. Analogs of Theorem 5.17 are available for these situations. Sometimes we can determine the concavity structure of $f^{(n)}$ in $[a, b]$ but, in our experience, this is often completely unnecessary.

We consider some examples.

Example 5.1 *Envelopes for the exponential function*
Consider the integral

$$\int_a^b e^{-x}\, dx = e^{-b} - e^{-a}.$$

Since we can evaluate this in closed form we certainly don't need envelope rules but it turns out that the envelopes for e^{-x} are easy to evaluate and they are useful in other more complex problems as we illustrate.

Note that $(e^{-x})^{(n)} = (-1)^{(n)}e^{-x}$ and so the n-th derivative is concave when n is odd and convex when n is even in any interval $[a, b]$. Therefore, on the interval $[a, b]$, using the n-th derivative, we obtain the upper envelope for e^{-x} given by

$$U^n_{[a,b]}(x) = \begin{cases} e^{-a} \sum_{k=0}^{n+1} \frac{(-1)^k}{k!}(x-a)^k & n \text{ odd} \\ e^{-a} \sum_{k=0}^{n} \frac{(-1)^k}{k!}(x-a)^k + \frac{(-1)^n}{(n+1)!} \frac{e^{-b} - e^{-a}}{b-a}(x-a)^{n+1} & n \text{ even} \end{cases}$$

and the lower envelope $L^n_{[a,b]}(x)$ for e^{-x} is obtained by reversing the expressions for n odd and n even.

Example 5.2 *Makeham's distribution*
Makeham's distribution is a three-parameter family of distributions that has applications in actuarial science. The density function is given by

$$f(z) = (\alpha + \beta\gamma^z)\exp\{-\alpha z - \beta(\gamma^z - 1)/\ln(\gamma)\}$$

where $\beta > 0, \gamma > 1, \alpha > -\beta$ and $z \geq 0$. No closed form expression exists for the probabilities

$$\int_a^b f(z)\,dz \tag{5.22}$$

for this family of distributions.

Let $\alpha^* = \alpha/\ln(\gamma)$ and $\beta^* = \beta/\ln(\gamma)$ and make the transformation $y = \beta^*\gamma^z$ so that (5.22) equals

$$e^{\beta^*}(\beta^*)^{\alpha^*} \int_{\beta^*\gamma^a}^{\beta^*\gamma^b} (\alpha^* + y)y^{-(\alpha^*+1)} \exp(-y)\,dy = \int_{\beta^*\gamma^a}^{\beta^*\gamma^b} K(y)\exp(-y)\,dy$$

for $y > \beta^*$. Therefore

$$\int_{\beta^*\gamma^a}^{\beta^*\gamma^b} K(y)L^n_{(a,b)}(y)\,dy \leq \int_{\beta^*\gamma^a}^{\beta^*\gamma^b} K(y)\exp(-y)\,dy \leq \int_{\beta^*\gamma^a}^{\beta^*\gamma^b} K(y)U^n_{(a,b)}(y)\,dy \tag{5.23}$$

where $L^n_{[a,b]}$ and $U^n_{[a,b]}$ are given in Example 5.1. The left-hand and right-hand sides of (5.23) are easily evaluated in closed form as each integrand is a sum of powers of y. Therefore we have the right-hand side of (5.23) as an approximation to (5.22) with exact error bound given by the difference of the right and left-hand sides of (5.23).

Table 5.3. Upper and lower bounds for Makeham's distribution function at $x = 1$ when $\alpha = 1$, $\beta = 2$ and $\gamma = e^{0.5}$.

n	m	Lower bound	Upper bound	Relative error
2	1	0.587143	1.302045	1.217594
2	10	0.972498	0.972590	0.000946
2	100	0.972536	0.972536	0.000000
4	1	0.908503	1.059043	0.165701
4	10	0.972536	0.972536	0.000000
4	100	0.972536	0.972536	0.000000

For example, suppose that $\alpha = 1, \beta = 2, \gamma = \exp(0.5) = 1.648721$ and we want to evaluate the distribution function of this Makeham's distribution at 1; i.e. $a = 0$ and $b = 1$. Then Table 5.3 gives the upper and lower bounds for this integral for different values of the order of the polynomial n, and the order of the compounding m. We see that the approximations converge quickly to the exact value 0.972536. As suspected, as we increase n the accuracy of the approximation improves.

A problem arises in the evaluation of (5.22) if the interval $[a, b]$ is such that the integrand f is very close to 0 over great parts of the right part of the interval. In such circumstances it may happen that the upper and lower envelopes provide very poor approximations unless the intervals are very short; i.e. m is large. To avoid this problem we advise using the envelope inequality of Theorem 3.7 to first determine b^* such that $\int_{b^*}^{\infty} f(z)\,dz \le \epsilon$ for some prescribed value ϵ and then replace b by $\min\{b, b^*\}$. We demonstrate now how to do this. Putting $g(z) = \ln(f(z))$ we obtain

$$g^{(1)}(z) = -\left\{\beta^2(\gamma^z)^2 - \beta(\ln\gamma)\gamma^z + 2\alpha\beta\gamma^z + \alpha^2\right\}(\alpha + \beta\gamma^z)^{-1}$$

$$g^{(2)}(z) = -\beta(\ln\gamma)\left\{\beta^2(\gamma^z)^3 + 2\alpha\beta(\gamma^z)^2 - (\ln\gamma)\alpha\gamma^z + \alpha^2\gamma^z\right\}(\alpha + \beta\gamma^z)^{-2}.$$

Then making the transformation to $y = \gamma^z$ the third factor of $g^{(2)}(z)$ becomes $l(y) = y\{\beta^2 y^2 + 2\alpha\beta y - (\ln\gamma)\alpha + \alpha^2\}$. The roots of $l(y)$ are 0 and $(1/\beta)(-\alpha \pm \sqrt{\alpha\ln\gamma})$ and the roots greater than 1 give the inflection points of the log-density. Observe that $l(y) \ge 0$ for large y and this implies that $g^{(2)}(z) \le 0$ for large z so that the tail of f is inevitably log-concave. Therefore the right tail of f is log-concave whenever $\alpha < 0$ or, if $\alpha \ge 0$, when

$$z \ge \frac{\ln\left\{\max\left(\dfrac{-\alpha + \sqrt{\alpha\ln\gamma}}{\beta}, 1\right)\right\}}{\ln\gamma}.$$

Further, the tangent to $\ln f$ at z has a negative slope whenever $\ln(\gamma) < 4\alpha$ or, if this inequality doesn't hold, whenever

$$\beta^2 y^2 - \beta(\ln\gamma)y + 2\alpha\beta y + \alpha^2, \text{ roots: } \begin{cases} \dfrac{\frac{1}{2}\ln\gamma - \alpha + \frac{1}{2}\sqrt{(\ln^2\gamma - 4(\ln\gamma)\alpha)}}{\beta} \\ \dfrac{\frac{1}{2}\ln\gamma - \alpha - \frac{1}{2}\sqrt{(\ln^2\gamma - 4(\ln\gamma)\alpha)}}{\beta} \end{cases}$$

$$z > \frac{\ln\left[\frac{1}{2\beta}\left\{(\ln\gamma - 2\alpha) + \sqrt{\ln(\gamma)}\sqrt{\ln\gamma - 4\alpha}\right\}\right]}{\ln\gamma}$$

as is readily determined by calculating the roots of the polynomial in y given by

$$\beta^2 y^2 - \beta(\ln\gamma)y + 2\alpha\beta y + \alpha^2$$

and noting that $g^{(1)}(z)$ is negative whenever γ^z is bigger than the largest root. Therefore if z is in the log-concave right tail and such that the slope of $g^{(1)}(z)$ is negative then the envelope inequality for this distribution with $z_0 = z_r = z$ is given by

$$1 - F(z) \le f(z)\left\{\alpha + \beta\gamma^z - (\ln\gamma)\frac{\beta\gamma^z}{\alpha + \beta\gamma^z}\right\}^{-1}$$

and we can use this to determine an appropriate upper limit of integration for any integral we approximate.

5.5 Product rules

The simplest method of constructing multiple quadrature rules for dimension $k > 1$ is to form *product rules*. Suppose that we have univariate rules

$$R_{jn_j} = \{(x_{ji}, w_{ji}) \colon i = 1, \ldots, n_j\}$$

to approximate integrals on the interval (a_j, b_j) for $j = 1, \ldots, k$. Then the product rule $\prod_{j=1}^{k} R_{jn_j}$ approximates the integral

$$\int_{a_k}^{b_k} \cdots \int_{a_1}^{b_1} f(x_1, \ldots x_k)\, dx_1 \cdots dx_k \tag{5.24}$$

by the sum

$$\sum_{i_1=1}^{n_1} \cdots \sum_{i_k=1}^{n_k} w_{1i_1} \cdots w_{ki_k} \frac{f(x_{1i_1}, \ldots, x_{ki_k})}{w_1(x_{1i_1}) \cdots w_k(x_{ki_k})} \tag{5.25}$$

where w_i corresponds to the weight function associated with the i-th component rule. Note that Theorem 5.1 applies in the context where all the weight functions are constant so that under compounding into subrectangles we have the convergence of (5.25) to (5.24).

Properties of product rules follow directly from the properties of the component univariate rules. For example, suppose that $f/(w_1 \cdots w_k)$ is a polynomial in $x_1, \ldots, x_k$ of degree l; i.e. $f/(w_1 \cdots w_k)$ is the sum of monomials $a_{j_1 \ldots j_k} x_1^{j_1} \cdots x_k^{j_k}$ where $0 \leq j_1 + \cdots + j_k \leq l$ and there exists $(j_1, \ldots, j_k)$ such that $j_1 + \cdots + j_k = l$ and $a_{j_1 \ldots j_k} \neq 0$. Then if each R_{jn_j} exactly integrates polynomials of degree l over (a_j, b_j) the rule $\prod_{j=1}^k R_{jn_j}$ exactly integrates $f/(w_1 \cdots w_k)$ over $\prod_{j=1}^k (a_j, b_j)$.

We consider an example of the use of product rules.

Example 5.3

Consider the $N_2(\mu, \Sigma)$ distribution discussed in Example 1.3 and suppose that we want to compute integrals for the marginal distribution of the ratio $Y = X_1/X_2$. Therefore we want to approximate integrals of the form

$$I(m) = \int_{-\infty}^{\infty} \int_{-\infty}^{\infty} m(x_1/x_2) \frac{1}{2\pi |\Sigma|^{1/2}} \exp\left\{ -\tfrac{1}{2}(x-\mu)'\Sigma^{-1}(x-\mu) \right\} dx_1\, dx_2$$

for various choices of $m : R^2 \to R^1$. Here we consider the choices $\mu = (5, 4)'$, $\sigma_{11} = 4$, $\sigma_{22} = 1$ and $\sigma_{12} = \sigma_{21} = 0.5$.

We begin by truncating R^2 into $(-15, 15) \times (-15, 15)$ which contains all but negligible probability for this distribution. The truncation allows us to use simple interpolatory rules based on the weight function $w(x) = 1$. In practice, we should try several truncations to check that our integral approximations do not change. Alternatively, we could try to avoid the truncation by using rules that are applicable to infinite intervals but we note that the only way to improve the approximation is to then increase the order of the rule. In this case there are restrictions that apply; e.g. the software will only compute Hermite or Laguerre rules up to a certain order. Therefore, it is generally better to truncate and use rules for compact intervals as then compounding can be used. The truncation process, however, does add another component to the error analysis.

We note that some care must also be taken due to the singularity at $x_2 = 0$. We handle this by breaking the region of integration for the x_2 variable into $(-15, 0) \cup (0, 15)$ and applying open rules in each interval. In particular, we use the midpoint rule along the x_2-axis. Along the x_1-axis, we use Simpson's rule. With 30 levels of compounding in each coordinate direction, the computation time is instantaneous and we obtain 6 digits of accuracy in the calculation of $E[X_1/X_2] = 1.31119$ (i.e. $m(x_1/x_2) = x_1/x_2$).

We note that product rules are seemingly restrictive as they only apply to the approximation of multiple integrals over rectangles. In fact, we can generally transform an integral so that the domain of integration is a rectangle. For example, many two-dimensional integrals can be written as

$$\int_{a_2}^{b_2} \int_{u_1(x_2)}^{v_1(x_2)} f(x_1, x_2)\, dx_1\, dx_2 \tag{5.26}$$

for some functions $u_1, v_1 : (a_2, b_2) \to R^1$ satisfying $0 < v_1(x_2) - u_1(x_2) < \infty$ for $x_2 \in (a_2, b_2)$. Then the transformation

$$u = \frac{x_1 - u_1(x_2)}{v_1(x_2) - u_1(x_2)}$$

applied to the inner integral in (5.26) leads to the integral

$$\int_{a_2}^{b_2} \int_0^1 f((v_1(x_2) - u_1(x_2))u + u_1(x_2), x_2)(v_1(x_2) - u_1(x_2))\, du\, dx_2.$$

If only $v_1(x_2) > 0$ and $u_1(x_2) \equiv -\infty$ then we can use the transformation

$$u = \frac{x_1}{v_1(x_2)}$$

to transform (5.26) to an integral over an infinite rectangle, etc. When (5.26) doesn't hold we can typically decompose the domain of integration into a finite partition where the conditions do apply on each piece. Effectively, most two-dimensional integrals can be approximated by multiple quadrature rules for rectangles. Notice, however, that this may result in a much more complicated integrand than the one we started with. This technique can be generalized to higher dimensions (see Exercise 24).

There are several standard transformations that can be used to transform particular regions into rectangles. These lead to transformed product rules. We discuss several of the most important of these in the following examples.

Example 5.4 *Integrals over simplices*
Suppose that we want to compute

$$\int_{S_k} f(x)\, dx \tag{5.27}$$

where $S_k = \{x : x_i \geq 0, 0 \leq x_1 + \cdots + x_k \leq 1\}$ is the k-simplex in R^k. For this we make the transformation $x \to y$ discussed in Exercise 2.8.8. Then (5.27) transforms to

$$\int_0^1 \cdots \int_0^1 f(y_1, (1 - y_1)y_2, \ldots, (1 - y_1) \cdots (1 - y_{k-1})y_k) \\ \times (1 - y_1)^{k-1}(1 - y_2)^{k-2} \cdots (1 - y_{k-1})\, dy_1 \cdots dy_k \tag{5.28}$$

which is an integral over the unit cube $[0, 1]^k$. Notice that if we use a transformed product on $[0, 1]^k$, where the i-th rule is based on the weight function $w_i(y) = (1 - y_i)^{k-i}$, i.e. a transformed Jacobi rule on $(0, 1)$ and has n points then this product rule computes (5.27) exactly whenever f is a polynomial of degree less than or equal to l provided $2n - 1 \geq l$. Note that we can transform such a rule on $[0, 1]^k$ back to S_k using the discussion in Section 5.2 and so obtain a *simplicial rule*. Such a rule requires at least n^k function evaluations.

Example 5.5 *Integrals over balls and spheres*
Suppose that we want to compute

$$\int_{B^k} f(x)\,dx \tag{5.29}$$

where $B^k = \{x : x_1^2 + \cdots + x_k^2 \le 1\}$ is the unit ball in k dimensions. Using the transformation of Exercise 2.8.7 this integral becomes

$$\int_0^1 \int_{-\pi/2}^{\pi/2} \cdots \int_{-\pi/2}^{\pi/2} \int_0^{2\pi} f(r\cos\theta_{k-1}\cos\theta_{k-2}\cdots\cos\theta_1, \ldots, r\sin\theta_{k-1})$$
$$\times r^{k-1}(\cos\theta_{k-1})^{k-2}(\cos\theta_{k-2})^{k-3}\cdots(\cos\theta_2)\,d\theta_1\cdots d\theta_{k-2}d\theta_{k-1}dr \tag{5.30}$$

which is an integral over a rectangle. In fact we can also use a transformed product Jacobi rule for (5.30). For this let

$$f_S(x) = f(x_1, x_2, \ldots, x_k) + f(-x_1, x_2, \ldots, x_k)$$

and note that (5.30) equals

$$\int_0^1 \int_{-\pi/2}^{\pi/2} \cdots \int_{-\pi/2}^{\pi/2} \int_{-\pi/2}^{\pi/2} f_S(r\cos\theta_{k-1}\cos\theta_{k-2}\cdots\cos\theta_1, \ldots, r\sin\theta_{k-1})$$
$$\times r^{k-1}(\cos\theta_{k-1})^{k-2}(\cos\theta_{k-2})^{k-3}\cdots(\cos\theta_2)\,d\theta_1\cdots d\theta_{k-2}d\theta_{k-1}dr \tag{5.31}$$

since $(\cos(\theta_1), \sin(\theta_1))$ for $\theta_1 \subset (-\pi/2, \pi/2)$ gives every point on the unit circle with positive first coordinate and $(-\cos(\theta_1), \sin(\theta_1))$ gives every point on the unit circle with negative first coordinate. Then in (5.31) make the transformation

$$y_i = \frac{1 + \sin\theta_i}{2}$$

for $i = 1, \ldots, k-1$ and $y_k = r$ so that $\sin\theta_i = 2y_i - 1$ and

$$\cos\theta_i = 2y_i^{1/2}(1 - y_i)^{1/2}$$

for $i = 1, \ldots, k-1$. This transformation has Jacobian $\prod_{i=1}^{k-1} y_i^{-1/2}(1 - y_i)^{-1/2}$ so (5.31) becomes

$$2^{(k-1)(k-2)/2} \int_0^1 \int_0^1 \cdots \int_0^1$$
$$f_S\left(2^{k-1} y_k y_{k-1}^{1/2}(1 - y_{k-1})^{1/2} \cdots y_1^{1/2}(1 - y_1)^{1/2}, \ldots, y_k(2y_{k-1} - 1)\right)$$
$$\times y_k^{k-1}\left[y_{k-1}^{k-1/2-1}(1 - y_{k-1})^{k-1/2-1}\right]\cdots\left[y_1^{1/2-1}(1 - y_1)^{1/2-1}\right] dy_1 \cdots dy_k. \tag{5.32}$$

Therefore it makes sense to use a transformed product Jacobi rule on $[0, 1]^k$ for (5.32) where the i-th rule is with respect to the weight function

$$w_i(y) = y_i^{i/2-1}(1 - y_i)^{i/2-1}$$

for $i = 1, \ldots, k - 1$ and

$$w_k(y) = y^{k-1}$$

all on $(0, 1)$.

Now observe that if f is a monomial $a_{j_1 \ldots j_k} x_1^{j_1} \cdots x_k^{j_k}$ then (5.29) equals 0 whenever at least one of the j_i is odd because the function f is antisymmetric with respect to the plane specified by $x_i = 0$. Note that

$$f_S(x) = a_{j_1 \ldots j_k} x_1^{j_1} \cdots x_k^{j_k} \{1 - (-1)^{j_1}\}.$$

Now assume that j_i is odd for some i. If $i = 1$ then $f_S \equiv 0$ and the above product Jacobi rule gives the value 0 as well. Suppose that $i \in \{2, \ldots, k\}$. Then observe that since $i \neq 1$ the term $\sin\theta_{i-1}$ appears in x_i when we express it in polar coordinates. Therefore the term $(2y_{i-1} - 1)^{j_i}$ appears when we substitute the y variables into f_S. The reflection of a point $y \in [0, 1]$ through $1/2$ is $1/2 + (1/2 - y) = 1 - y$ and note

$$\{2(1 - y_{i-1}) - 1\}^{j_i} = (-1)^{j_i}\{2y_{i-1} - 1\}^{j_i}$$

while $y_{i-1}^{1/2}(1 - y_{i-1})^{1/2}$ is invariant under this reflection. Therefore in this case the function

$$f_S\big(2^{k-1} y_k y_{k-1}^{1/2}(1 - y_{k-1})^{1/2} \cdots y_1^{1/2}(1 - y_1)^{1/2}, \ldots, y_k(2y_{k-1} - 1)\big) \quad (5.33)$$

is antisymmetric with respect to the plane given by $y_{i-1} = 1/2$. Since the weight function w_{i-1} is symmetric about 1/2 this implies that the points in this rule are symmetric about 1/2 and symmetric points have the same weights. Finally this implies that the above product Jacobi rule produces the value 0 when applied to such an f. When all the j_i are even then (5.33) is a monomial in $y_1, \ldots, y_k$. Therefore we can choose the orders of the Jacobi rules in the above product Jacobi rule to exactly integrate a polynomial f.

Notice that if instead we want to calculate the surface integral

$$\int_{S^{k-1}} f(x)\, dx \quad (5.34)$$

where $S^{k-1} = \{x : x_1^2 + \cdots + x_k^2 = 1\}$ is the unit sphere in R^k then this is given by

$$\int_{-\pi/2}^{\pi/2} \cdots \int_{-\pi/2}^{\pi/2} \int_0^{2\pi} f(\cos\theta_{k-1}\cos\theta_{k-2}\cdots\cos\theta_1, \ldots, \sin\theta_{k-1})$$
$$\times (\cos\theta_{k-1})^{k-2}(\cos\theta_{k-2})^{k-3}\cdots(\cos\theta_2)\, d\theta_1 \cdots d\theta_{k-2}\, d\theta_{k-1}. \quad (5.35)$$

Therefore a product Jacobi rule is also appropriate for calculating integrals over spheres. Note that we can transform such a rule on $[0, 1]^{k-1}$ back to S^{k-1} using the discussion in Section 5.2 and so obtain a *spherical rule*.

Example 5.6 *Integrals over* R^k
Whenever we approximate an integral over an infinite set one possibility is to truncate the region so that we are integrating over a bounded set and then use product rules for rectangles perhaps after a further transformation. This is often appropriate but it introduces the necessity of assessing and controlling the error introduced due to the truncation. Sometimes we can avoid this by using product rules suitable for the original region but we recall that this has its limitations as well.

Suppose, however, that we are integrating over R^k and via a product Hermite rule; i.e. we use a Hermite rule for each variable. The product weight function for this is given by

$$w(x) = \exp\left(-\tfrac{1}{2}x^t x\right) = \exp\left(-\tfrac{1}{2}\sum_{i=1}^{k} x_i^2\right)$$

which is proportional to the density function of the $N_k(0, I)$ density discussed in Example 1.3. For such a rule to be appropriate then f/w should be well approximated by a polynomial. Preferably we want f/w to be well approximated by a low degree polynomial. So it is necessary that the tails of the integrand f be no longer than the tails of the $N_k(0, I)$ distribution so that $f(x)/w(x)$ does not go to 0 as x gets large.

Suppose then that f is nonnegative and integrable. If f is unimodal then it is recommended to transform the problem before applying the product Hermite approach. For this we calculate the mode μ of f and consider the first three terms of a Taylor expansion of $\log f$ about μ. This gives the approximation

$$\log f(x) \approx \log f(\mu) - \tfrac{1}{2}(x - \mu)^t \Sigma^{-1}(x - \mu), \tag{5.36}$$

where the linear term drops out because $\nabla \log f(\mu) = 0$, and $\Sigma^{-1} \in R^{k\times k}$ is given by

$$\Sigma^{-1} = \left(-\frac{\partial^2 \log f(\mu)}{\partial x_i \partial x_j}\right)$$

which is the negative of the Hessian matrix of $\log f$ at μ and which we note is positive definite. Then (5.36) leads to the approximation

$$f(x) \approx f(\mu) \exp\left\{-\tfrac{1}{2}(x - \mu)'\Sigma^{-1}(x - \mu)\right\}$$

which is proportional to a $N_k(\mu, \Sigma)$ density. Therefore it is customary to transform the integral

$$\int_{R^k} f(x)\,dx$$

by the transformation $y = \Sigma_\Delta^{-1}(x - \mu)$, where Σ_Δ is the *Cholesky factor* of Σ (the unique lower triangular matrix with positive diagonals satisfying $\Sigma = \Sigma_\Delta \Sigma'_\Delta$), to

$$\det \Sigma_\Delta \int_{R^k} f(\mu + \Sigma_\Delta y)\, dy$$

and then apply the product Hermite rule to the integrand $g(y) = f(\mu + \Sigma_\Delta y)$. The integrand g has its mode at 0. Also g should have more spherically shaped contours but there is no guarantee of this and significant skewness or significantly longer tails may result in a poor approximation. When the integrand takes negative as well as positive values we look to see if we can factor this as hf where f is nonnegative and integrable. We then compute μ and Σ as above, based on f, and then approximate the integral

$$\det \Sigma_\Delta \int_{R^k} h(\mu + \Sigma_\Delta y)(f(\mu + \Sigma_\Delta y)\, dy$$

using product Hermite rules. We discuss further refinements of this approach in Section 5.10.

Error analysis for product rules typically proceeds by evaluating approximations using successively more points and comparing the approximations. Bounds on the error can be obtained using the bounds derived for the univariate rules but these are rarely practically useful. We illustrate this via the following example.

Example 5.7 *Product trapezoid rule*
Suppose that $f \in C^2([0,1]^2)$ and that $f_i \in C^2([0,1])$ where $f_1(u_1) = \int_0^1 f(u_1, u_2)\, du_2$ with a similar definition for f_2. Then consider the product trapezoid rule with each component rule based on compounding the basic rule n times. Then applying Corollary 5.6 to the one-dimensional trapezoid rule and maximizing second derivatives to get an upper bound on the error, we have that

$$\left| \int_0^1 \int_0^1 f(u_1, u_2)\, du_1\, du_2 - \sum_{i=1}^n \sum_{j=1}^n w_i w_j f(x_i, x_j) \right|$$

$$\leq \left| \int_0^1 f_1(u_1)\, du_1 - \sum_{i=1}^n w_i f_1(x_i) \right|$$

$$+ \left| \sum_{i=1}^n w_i f_1(x_i) - \sum_{i=1}^n \sum_{j=1}^n w_i w_j f(x_i, x_j) \right|$$

$$\leq \frac{\sup\left\{ \left| f_1^{(2)}(u_1) \right| : u_1 \in [0,1] \right\}}{12n^2}$$

$$+ \sum_{i=1}^n w_i \left| \int_0^1 f(x_i, u_2)\, du_2 - \sum_{j=1}^n w_j f(x_i, x_j) \right|$$

$$\leq \frac{\sup\left\{\left|f_1^{(2)}(u_1)\right| : u_1 \in [0,1]\right\}}{12n^2} + \sum_{i=1}^{n} w_i \frac{\sup\left\{\left|\dfrac{\partial^2 f(x_i,u_2)}{\partial u_2^2}\right| : u_2 \in [0,1]\right\}}{12m^2}$$

$$\leq \frac{1}{12n^2}\left[\sup\left\{\left|f_1^{(2)}(u_1)\right| : u_1 \in [0,1]\right\} + \sup\left\{\left|\frac{\partial^2 f(u_1,u_2)}{\partial u_2^2}\right| : u_1, u_2 \in [0,1]\right\}\right].$$

This says that the error is $O(n^{-2})$ but this does not express the rate of convergence appropriately because n is not a true measure of the amount of computing resources expended in computing the approximation. Let $N = n^2$ equal the number of function evaluations so that $n = \sqrt{N}$. Now we see that the rate of convergence of the product rule, expressed in terms of the number of function evaluations, is $O(N^{-1})$.

Under similar conditions we can show that the product trapezoid rule when applied to an integral on $[0, 1]^k$ has rate of convergence given by $O(N^{-2/k})$. So as the dimension rises the performance decreases quickly. This is another illustration of the dimensional effect for integration problems. It is noteworthy that when $k > 4$ the performance is worse than crude Monte Carlo which samples points uniformly in $[0, 1]^k$ and averages the function at these points. A convenient measure of the accuracy of this average is given by its standard error which is a constant divided by $\sqrt{N}$. We discuss this at greater length in Chapter 6.

A significant drawback for product rules is that they suffer from the curse of dimensionality. We note that there are $n_1 \cdots n_k$ terms in the sum in (5.25) and this calculation can easily exceed reasonable computing time limits as the dimension k increases. This observation suggests that we need to be careful in choosing the component quadrature rules. In bivariate problems we may be able to use compounded versions of Simpson's rule but even here we note that a modest 10^3 function evaluations for each univariate rule entails 10^6 function evaluations for the full rule. For higher dimensions the use of such a rule is almost certainly prohibitive. This typically means we need to use more efficient rules, such as Gauss rules, in each dimension. We may also consider choosing n_j larger (smaller) according to the difficulty (ease) of integration in each of the component directions. Although the choice of $n_1, \ldots, n_k$ is difficult to determine by a visual inspection of the integrand f, we note that many of the adaptive routines discussed in Section 5.9 do this automatically. Oversimplifying the adaptive strategy, if the value of the product rule approximation changes significantly when going from n_j to $n_j + 1$

then this suggests that a higher order component rule should be used in the j-th coordinate direction.

As one potential solution to the dimensional problem with product rules Hammersley (1960) suggested sampling the terms in (5.25); i.e. use Monte Carlo methods to approximate the approximation. Typically this would involve sampling the terms in the sum uniformly. Recall, however, that Theorem 5.12 established that the weights for the Gauss rules form a probability distribution on the points. So in this context it is natural to use the product distribution probabilities $w_{1i_1} \cdots w_{ki_k}$ to sample and then average the values of $f(x_{1i_1}, \ldots, x_{ki_k})/(w_1(x_{1i_1}) \cdots w_k(x_{ki_k}))$. Evans and Swartz (1988b) provide an extensive analysis of this technique, and in particular, show that a variance reduction can sometimes be obtained when compared with the associated importance sampling (see Chapter 6) approach using the product weight function. Perhaps most interesting is the demonstration that sample sizes much smaller than $n_1 \cdots n_k$ can give accurate approximations to the sum. So the idea is to choose the values of the n_i so that (5.25) provides an accurate approximation of (5.24) and then sample the sum. In some contexts it is possible to choose the n_i appropriately but this is relatively rare.

5.6 Monomial rules

One approach to avoiding the dimensional effect associated with product rules is to construct rules that minimize the number of points in the rule while still retaining high accuracy. Here accuracy is typically measured by the ability of the rule to exactly integrate a large class of functions such as polynomials. While various classes of functions could be considered, polynomials seem quite natural and we restrict our attention to such rules. Since polynomials are linear combinations of monomials these are commonly called monomial rules.

Suppose then that we want to approximate the k-dimensional integral

$$\int_A f(x)\,dx = \int_A \frac{f(x)}{w(x)} w(x)\,dx$$

where w is a specified weight function such that $\text{supp}(f) \subset \text{supp}(w)$. A monomial of degree d takes the form $a_{i_1 \ldots i_k} x_1^{i_1} \cdots x_k^{i_k}$ where $i_1 + \cdots + i_k = d$. An n-point *monomial rule* of degree d is a rule $R_n = \{(x_i, w_i) \colon i = 1, \ldots, n\}$ which evaluates $\int_A f(x)\,dx$ exactly via

$$\sum_{i=1}^{n} w_i \frac{f(x_i)}{w(x_i)}$$

whenever f/w is a monomial of degree less than or equal to d. Note that there are

$$\dim(k, d) = \binom{k+d}{k}$$

distinct monomials of degree less than or equal to d in k variables and they are linearly independent (Exercise 29). For example, in R^2, a basis for the space of polynomials spanned by monomials of degree less than or equal to 2 is $\{1, x_1, x_2, x_1^2, x_1x_2, x_2^2\}$. Let us label this basis as $\{b_1, \ldots, b_{\dim(k,d)}\}$. The value of $\dim(k, d)$ grows very quickly and this causes some difficulty for this approach; e.g. $\dim(10, 4) = 1001$ and $\dim(20, 10) = 30, 045, 015$.

It is worth noting that the approach taken to constructing rules when $k = 1$, i.e. via polynomial interpolation, does not work when $k > 1$ because it is not possible in general to interpolate a function f at n arbitrary points by a linear combination of n linearly independent monomials. So a different approach to the construction of these rules must be taken.

If a rule $R_n = \{(x_1, w_1), \ldots, (x_n, w_n)\}$ integrates every polynomial of degree d or less then it must satisfy the polynomial system of equations

$$\sum_{j=1}^{n} w_j b_i(x_j) = \int_A w(x) b_i(x)\, dx \tag{5.37}$$

for $i = 1, \ldots, \dim(k, d)$. These are called the *moment equations* for the rule. To obtain the rule we must solve $\dim(k, d)$ polynomial equations in $nk+n$ unknowns where nk refers to the points $x_1, \ldots, x_n$ and n refers to the weights $w_1, \ldots, w_n$. Not every such system of equations has a solution and when one exists its derivation can be a considerable computational challenge. We expect a solution to exist, however, whenever $n \geq \dim(k, d)/(k+1)$. Of course we would like to find a rule satisfying (5.37) with the smallest value of n.

For given points $x_1, \ldots, x_n$ if there exists a unique solution $w_1, \ldots, w_n$ to the linear system of equations given by (5.37) then the rule is called *interpolatory*. In a context where more than one solution exists, then the evaluation linear functionals $x_1^*, \ldots, x_n^*$ defined on $L\{b_1, \ldots, b_{\dim(k,d)}\}$, i.e. $x_i^*(b) = b(x_i)$, are not linearly independent and we want to reduce to a subset of these points so that a unique solution exists; i.e. so that the corresponding evaluation functionals are linearly independent.

Many rules have been obtained by a variety of techniques. See Stroud (1971) and Cools and Rabinowitz (1993) for a discussion. Some of these suffer from defects. For example, sometimes some of the points lie outside the domain of integration A. This can be particularly inconvenient when the integrand is not defined in any natural way outside A. Further, sometimes the weights are not of constant sign. Given that the rule is a sum, and remembering our discussion of cancellation error in Chapter 2, we would like to avoid this if possible. In fact, the following result was established by Tchakaloff (1957).

Theorem 5.18 Suppose that A is compact and w is positive and integrable. Then there exists a monomial rule R_n of degree d such that $n \leq \dim(k, d)$, the points satisfy $\{x_1, \ldots, x_n\} \subset A$ and the weights satisfy $w_1, \ldots, w_n > 0$.

Proof: See Stroud (1971).

This is a very encouraging result but in fact it is only an existence proof and does not lead to the construction of such rules.

We consider one approach to the construction of monomial rules based on symmetry. Many of the standard domains of integration A and standard weight functions w possess symmetries. We assume hereafter that the weight function has support given by the region A. Symmetries are expressed mathematically by saying that the weight function is invariant under some subgroup of the group of orthogonal transformations $O(k)$; i.e. $w(Qx) = w(x)$ for every $Q \in G \subset O(k)$ where G is a group. Note that orthogonal transformations do not change volumes. For an integrand f and a finite group $G \subset O(k)$ define

$$f_G(x) = \frac{1}{\#(G)} \sum_{Q \in G} f(Qx); \tag{5.38}$$

i.e. $f_G(x)$ is the average value of f on the orbit $Gx = \{Qx\colon Q \in G\}$. Note that it may be that $\#(Gx) < \#(G)$ so that in fact we can compute (5.38) by averaging over far fewer points. Observe that f_G is now invariant under G as well and since G leaves the weight function w invariant so is f_G/w. Also we have that

$$\begin{aligned}
\int_A f(x)\,dx &= \int \frac{f(x)}{w(x)} w(x)\,dx \\
&= \int \frac{f_G(x)}{w(x)} w(x)\,dx \\
&= \int_A f_G(x)\,dx.
\end{aligned} \tag{5.39}$$

It might be wondered why we would want to replace f by a seemingly more complicated integrand f_G. Recall that in using monomial rules based on w we want the integrand divided by w to be as constant as possible as this presumably means that low order rules will give accurate approximations. By replacing f by its symmetrized version f_G we are making the integrand and the weight function more alike as they now share symmetries. So we can hope at least that lower order rules can be used with f_G than with f. Of course, if f is invariant or, nearly invariant, under G already then the additional computation caused by averaging will be wasted. We return to this topic when we discuss systematic sampling as a variance reduction technique in Chapter 6.

Given the invariance of the integrand f_G in (5.39) it is natural to look for rules that are also invariant with respect to G. By this we mean that $R_n = \{(x_1, w_1), \ldots, (x_n, w_n)\}$ is such that the orbits $Gx_i \subset \{x_1, \ldots, x_n\}$ for every i and if $x_j \in Gx_i$ then $w_j = w_i$. A subset $\{\hat{x}_1, \ldots, \hat{x}_m\} \subset \{x_1, \ldots, x_n\}$ is called a *generator set* if these points index the orbits; i.e. one of these points, and only one, is in each of the orbits of G acting on $\{x_1, \ldots, x_n\}$. Note that $f_G(x) = f_G(\hat{x}_i), w(x) = w(\hat{x}_i)$ for every $x \in G\hat{x}_i$. We suppose hereafter, without loss of generality, that $\{x_1, \ldots, x_n\}$ is a generator set. Then an invariant rule approximates $\int_A f(x)\,dx$ by

$$\sum_{i=1}^{n} w_i \frac{f_G(x_i)}{w(x_i)}. \tag{5.40}$$

Now suppose that we want the invariant rule to integrate exactly all polynomials of degree less than or equal to d. If f is a polynomial then it is immediate that f_G is an invariant polynomial with respect to G, So to achieve our goal we need only find a rule that exactly integrates every invariant polynomial of degree d or less. We need therefore only solve the moment equations using invariant monomials. This is a significant reduction in the complexity of the problem because the linear space of all invariant polynomials of degree d or less is of smaller dimension than the linear space of all polynomials of degree d or less. Again there is no guarantee that a solution exists.

Example 5.8 *Fully symmetric rules*
Consider the weight function $w(x) = I_{[-1,1]^k}(x)$; i.e. the indicator function for the cube $[-1, 1]^k$. This weight function is invariant under the permutation group which has $k!$ elements; i.e. $Q \in G$ is a full rank matrix with a single 1 in every row and all other elements 0. Also w is invariant under the group of reflections in coordinate hyperplanes which has 2^k elements; i.e. $Q \in G$ is a diagonal matrix with ± 1's along the diagonal. Denote the subgroup of $O(k)$ generated by these transformations as G and note that G has cardinality $2^k k!$. Rules that are invariant with respect to this group are called fully invariant rules and a number of these have been obtained. Any weight function which is invariant under these transformations is also referred to as being *fully symmetric*.

Consider the case with $w(x) = I_{[-1,1]^k}(x)$, $n = 1$, and suppose we look for x_1 of the form $x_1 = ae_1$ where $e_i \in R^k$ is the i-th standard basis vector. Note that to calculate $f_G(x_1)$ we need only average over the orbit Gx_1 which has $2k$ points in it. Also if the rule is to integrate constants exactly then we must have $w_1 = 2^k$ which is the volume of $[-1, 1]^k$. Therefore the rule takes the form

$$\frac{2^k}{2k} \sum_{i=1}^{k} \{f(ae_i) + f(-ae_i)\}$$

and we are left to determine a so that it has the maximal possible degree. Now note that any monomial f which has at least one odd power in it has integral 0 and also $f(ae_i) + f(-ae_i) = 0$ for every i so that the rule is exact for all such monomials. Therefore the rule has degree at least 1 and we see if a can be determined so it integrates the quadratic polynomials x_i^2 exactly. The integral of these polynomials is equal to $2^k/3$. So we require that

$$\frac{2^k}{2k}(a^2 + a^2) = \frac{2^k}{3}$$

or $a = \sqrt{k/3}$. With these choices this fully symmetric rule has degree 3 and it is obviously minimal. Notice, however, that when $k > 3$ the generator x_1 lies outside $[-1, 1]^k$.

Higher degree rules for the uniform weight function over $[-1, 1]^k$ and other weight functions can be found in Stroud (1971) and Cools and Rabinowitz (1993). See also Genz and Malik (1983) for more discussion of fully symmetric rules. Dellaportas and Wright (1991) discuss an approach to deriving embedded sequences of fully symmetric rules with positive weights based on product Gauss rules. This is another approach to avoiding the dimensional effect associated with product rules and further allows us to increase the number of points in a rule without discarding previous function evaluations.

We now consider the development of monomial rules for integrals over the unit sphere.

Example 5.9 *Spherical rules*
Suppose that we want to integrate f over R^k. Then putting $r = \|x\|$ and $u = x/\|x\|$ we can write

$$\begin{aligned}\int_{R^k} f(x)\,dx &= \int_0^\infty r^{k-1}\left\{\int_{S^{k-1}} f(ru)\,du\right\} dr \\ &= \int_0^\infty r^{k-1} F(r)\,dr\end{aligned}$$

and we can break the integration into two parts: integrating first over the unit sphere in R^k to obtain F and then integrating over $(0, \infty)$. If for each r the function $f(ru)$ does not vary much with u then we can expect low order rules on the sphere to give an accurate approximation to $F(r)$. Therefore it is useful to have rules for integrating over S^{k-1} that are exact for polynomials.

Accordingly we consider monomial rules for integrals of the form

$$\int_{S^{k-1}} f(u)\,du.$$

The best we can hope for is that f is spherically symmetric; i.e. $f(Qx) = f(x)$ for every $Q \in O(k)$. In this case $f(u)$ is constant as a function of u and the integral over S^{k-1} is just this constant times the surface area of S^{k-1} which is

$$A_k = \frac{2\pi^{k/2}}{\Gamma(k/2)}.$$

Symmetrizing f with respect to some finite subgroup G of $O(k)$ makes the integrand more spherical as f_G is invariant with respect to this subgroup. If $-Q \in G$ whenever $Q \in G$ then every monomial containing an odd power is integrated exactly over S^{k-1} and we restrict our attention to such rules. The simplest such group is given by $G = \{I, -I\}$ and we see immediately that with $n = 1$, x_1 any point in S^{k-1} and $w_1 = A_k$ then the rule

$$\frac{w_1}{2}\{f(x_1) + f(-x_1)\}$$

has degree 1.

Consider now the group $G = \{I, -I\}$ and the generators consisting of $\{e_1, \ldots, e_k\}$. Now since the integral of $1 = \sum_{i=1}^{k} u_i^2$ over S^{k-1} is $w_1 = A_k$ we must have

$$\int_{S^{k-1}} u_i^2 \, du = \frac{w_1}{k}$$

and so we see that the rule

$$\frac{A_k}{2k}\sum_{i=1}^{k}\{f(e_i) + f(-e_i)\}$$

is a monomial rule of degree 3 on the sphere. Higher degree rules for the surface of the sphere can be found in Stroud (1971) and Cools and Rabinowitz (1993).

5.7 Bayesian quadrature

Bayesian quadrature is a method for constructing quadrature rules that does not involve the exact integration of polynomials. Rather than supposing that the integrand f can be well approximated by polynomials it is supposed that f arises from some stochastic process indexed by points in the domain of integration A. We then have the freedom of choosing the points $\{x_1, \ldots, x_n\} \subset A$ where we observe the value of f. Using this information we construct an estimator of the integral of f. The stochastic process assumption leads to a natural optimality criterion for selecting the points—a so-called *design problem.*

The use of the word Bayesian to describe the methodology reflects the fact that we can choose the stochastic process based upon what we know about the integrand f and so it resembles a prior distribution as discussed in Example 1.2. It may seem strange that we conceive of being uncertain about f when in fact we know it to the extent that we can presumably compute any value $f(x)$. While this is true, it is perhaps fair to say that the major stumbling block in designing an effective multidimensional integration strategy is that it is typically difficult to determine the characteristics of f necessary to achieving an adequate approximation. Accordingly, the stochastic viewpoint seems appropriate. Also, we can conceive of real-world problems where the integrand is indeed the outcome of a stochastic process.

It has to be acknowledged, however, that the ability of a practitioner to effectively choose the stochastic process in a wide variety of practical problems has yet to be proven. Another concern is that, even with what seem to be fairly standard assumptions for the stochastic process, the design problem, i.e. determining the optimal points $\{x_1, \ldots, x_n\}$, has proven to be intractable except in fairly limited contexts. It is worth noting, however, that the design problem only has to be solved once and then the rule is available to be applied to many integrands. At this time the theory and practice of this approach is not fully developed but it seems to hold some promise for further development. We provide a brief introduction to the subject. Further discussion can be found in Eberlein (1962), Kimeldorf and Wahba (1970), O'Hagan (1991, 1992, 1996), Sacks and Ylvisaker (1970, 1985) and Diaconis (1988).

Consider then the evaluation of the integral

$$\int_{R^k} f(x)w(x)\,dx \tag{5.41}$$

where w is a weight function as previously discussed and which is chosen so that the factor f could conceivably be thought of as arising from a particular stochastic process. Denote this stochastic process by the probability model $(R^{\mathrm{supp}(w)}, \mathcal{B}^{\mathrm{supp}(w)}, P)$; i.e. it is a class of random variables indexed by $x \in \mathrm{supp}(w)$ and $\mathcal{B}$ is the class of Borel sets. A natural a priori estimate of (5.41), i.e. before we observe f, is then given by the a priori mean (assuming it exists) of (5.41) under P, i.e.

$$E\left[\int_{R^k} f(x)w(x)\,dx\right]. \tag{5.42}$$

If P is a second order process, i.e. each random variable has a finite second moment, with mean function

$$m(x) = E[f(x)]$$

and covariance function

$$v(x, y) = \mathrm{Cov}[f(x), f(y)]$$

then, under certain conditions, we have that (5.42) satisfies

$$E\left[\int_{R^k} f(x)w(x)\,dx\right] = \int_{R^k} m(x)w(x)\,dx$$

and

$$\mathrm{Var}\left[\int_{R^k} f(x)w(x)\,dx\right] = \int_{R^k}\int_{R^k} v(x, y)w(x)w(y)\,dx\,dy.$$

See Lamperti (1977) for a discussion of these results.

Suppose now that we have decided to observe the random function f at $\{x_1, \ldots, x_n\}$. We then use the information that f came from P together with the observed values $\{(x_1, f(x_1)), \ldots, (x_n, f(x_n))\}$ to estimate (5.41). Conditional probability prescribes that this be based on the conditional stochastic process; i.e. the posterior process

$$P(\cdot|\{(x_1, f(x_1)), \ldots, (x_n, f(x_n))\})$$

which assigns probabilities to the possible integrands given what we have observed. If P is a second order process then the posterior process will have mean function

$$m(x|x_1, \ldots, x_n) = E[f(x)|\{(x_1, f(x_1)), \ldots, (x_n, f(x_n))\}]$$

and covariance function

$$v(x, y|x_1, \ldots, x_n) = \text{Cov}[f(x), f(y)|\{(x_1, f(x_1)), \ldots, (x_n, f(x_n))\}].$$

Note that while we have suppressed the dependence on the values $\{f(x_1), \ldots, f(x_n)\}$ for convenience from both the notation for the conditional mean function and the notation for the conditional covariance function these will in general depend on these quantities. A natural estimator of (5.41) is then its posterior mean (provided it exists); i.e.

$$\hat{I} = E\left[\int_{R^k} f(x)w(x)\,dx \;\middle|\; \{(x_1, f(x_1)), \ldots, (x_n, f(x_n))\}\right]$$

which under the second order assumptions equals

$$\int_{R^k} m(x|x_1, \ldots, x_n)w(x)\,dx. \tag{5.43}$$

Of course if the integral (5.43) is intractable then nothing has been accomplished and so the ability to relatively easily evaluate (5.43) places a restriction on the stochastic processes that can be used. The posterior MSE (mean-squared error) of this estimator is given by

$$E\left[\left(\int_{R^k} f(x)w(x)\,dx - \hat{I}\right)^2 \;\middle|\; \{(x_1, f(x_1)), \ldots, (x_n, f(x_n))\}\right] \tag{5.44}$$

which under the second order assumptions equals

$$\int_{R^k}\int_{R^k} v(x, y|x_1, \ldots, x_n)w(x)w(y)\,dx\,dy.$$

The posterior MSE is an appropriate measure of accuracy for a given integral estimate but its dependence on the values $\{f(x_1), \ldots, f(x_n)\}$ makes it inappropriate for design purposes as these depend on a particular f. We can measure the "goodness" of a particular choice of $\{x_1, \ldots, x_n\}$ by

$$E\left[E\left[\left(\int_{R^k} f(x)w(x)\,dx - \hat{I}\right)^2 \;\middle|\; \{(x_1, f(x_1)), \ldots, (x_n, f(x_n))\}\right]\right] \tag{5.45}$$

where the outer expectation is with respect to the distribution of $(f(x_1), \ldots, f(x_n))$. The design problem is then to choose $\{x_1, \ldots, x_n\}$ to make

(5.44) as small as possible. If (5.44) does not depend on $\{f(x_1), \ldots, f(x_n)\}$ then (5.44) equals (5.45).

For example, it can be shown that the trapezoid rule arises when we take P to be Brownian motion and when P is integrated Brownian motion then we get the rule corresponding to integrating cubic splines. To illustrate the theory we consider an example.

Example 5.10 *Gaussian processes*
Suppose that f follows a Gaussian process on R^k with mean function $m : R^k \to R^1$, such that $\int_{R^k} m(x)w(x)\,dx$ exists finitely, and covariance function $v : R^k \times R^k \to R^1$, such that $\int_{R^k}\int_{R^k} v(x,y)w(x)w(y)\,dx\,dy$ exists finitely. Therefore for any $(x_1, \ldots, x_n)$ and n we have that

$$\begin{aligned}\mathbf{f}_n(x_1, \ldots, x_n) &= \begin{pmatrix} f(x_1) \\ f(x_2) \\ \vdots \\ f(x_n) \end{pmatrix} \\ &\sim N_n\left(\begin{pmatrix} m(x_1) \\ m(x_2) \\ \vdots \\ m(x_n) \end{pmatrix}, \begin{pmatrix} v(x_1,x_1) & \cdots & v(x_1,x_n) \\ v(x_2,x_1) & \cdots & v(x_2,x_n) \\ \vdots & & \vdots \\ v(x_n,x_1) & \cdots & v(x_n,x_n) \end{pmatrix}\right) \\ &= N_n(\mathbf{m}_n(x_1, \ldots, x_n), \mathbf{V}_n(x_1, \ldots, x_n))\end{aligned}$$

which is an equivalent way to specify the process. It can be shown (Exercise 32) that this implies that

$$\int_{R^k} f(x)w(x)\,dx \sim N\left(\int_{R^k} m(x)w(x)\,dx, \int_{R^k}\int_{R^k} v(x,y)w(x)w(y)\,dx\,dy\right)$$

which is the prior distribution of the integral. The a priori estimate of (5.41) is then given by

$$\int_{R^k} m(x)w(x)\,dx \tag{5.46}$$

which is the expected value of the mean function with respect to the distribution given by w.

Given $\{(x_1, f(x_1)), \ldots, (x_n, f(x_n))\}$ the conditional distribution of f is again a Gaussian process and for any $\{y_1, \ldots, y_l\}$ and l

$$\begin{aligned}&\mathbf{f}_l(y_1, \ldots, y_l)|\{(x_1, f(x_1)), \ldots, (x_n, f(x_n))\} \\ &\quad\sim N_l(\mathbf{m}_l(y_1, \ldots, y_l|x_1, \ldots, x_n), \mathbf{V}_{l,n}(y_1, \ldots, y_l|x_1, \ldots, x_n))\end{aligned}$$

where

$$\begin{aligned}
&\mathbf{m}_l(y_1, \dots, y_l | x_1, \dots, x_n)\\
&\quad = \mathbf{m}_l(y_1, \dots, y_l)\\
&\quad\quad + \mathbf{V}_{l+n,12}(y_1, \dots, y_l, x_1, \dots, x_n)\mathbf{V}_{l+n,22}^{-1}(y_1, \dots, y_l, x_1, \dots, x_n)\\
&\quad\quad \times (\mathbf{f}_n(x_1, \dots, x_n) - \mathbf{m}_n(x_1, \dots, x_n)),\\
&\mathbf{V}_{l+n}(y_1, \dots, y_l, x_1, \dots, x_n)\\
&\quad = \begin{pmatrix} \mathbf{V}_{l+n,11}(y_1, \dots, y_l, x_1, \dots, x_n) & \mathbf{V}_{l+n,12}(y_1, \dots, y_l, x_1, \dots, x_n) \\ \mathbf{V}_{l+n,21}(y_1, \dots, y_l, x_1, \dots, x_n) & \mathbf{V}_{l+n,22}(y_1, \dots, y_l, x_1, \dots, x_n) \end{pmatrix}\\
&\quad = \begin{pmatrix} \mathbf{V}_l(y_1, \dots, y_l) & \mathbf{V}_{l+n,12}(y_1, \dots, y_l, x_1, \dots, x_n) \\ \mathbf{V}_{l+n,21}(y_1, \dots, y_l, x_1, \dots, x_n) & \mathbf{V}_n(x_1, \dots, x_n) \end{pmatrix}
\end{aligned}$$

and

$$\begin{aligned}
&\mathbf{V}_{l,n}(y_1, \dots, y_l | x_1, \dots, x_n)\\
&\quad = \mathbf{V}_l(y_1, \dots, y_l) - \mathbf{V}_{l+n,12}(y_1, \dots, y_l, x_1, \dots, x_n)\mathbf{V}_n^{-1}(x_1, \dots, x_n)\\
&\quad\quad \times \mathbf{V}_{l+n,12}^t(y_1, \dots, y_l, x_1, \dots, x_n).
\end{aligned}$$

This in turn implies that the conditional distribution of (5.41) given $\{(x_1, f(x_1)), \dots, (x_n, f(x_n))\}$ is

$$\begin{aligned}
&\int_{R^k} f(x)w(x)\,dx | \{(x_1, f(x_1)), \dots, (x_n, f(x_n))\}\\
&\quad \sim N\left(\int_{R^k} m(x|x_1, \dots, x_n)w(x)\,dx,\right.\\
&\quad\quad \left.\int_{R^k}\int_{R^k} v(x, y|x_1, \dots, x_n)w(x)w(y)\,dx\,dy\right)
\end{aligned}$$

where

$$\begin{aligned}
&\int_{R^k} m(x|x_1, \dots, x_n)w(x)\,dx\\
&\quad = \int_{R^k} m(x)w(x)\,dx\\
&\quad\quad + \left(\int_{R^k} \mathbf{V}_{1+n,12}(x, x_1, \dots, x_n)\mathbf{V}_n^{-1}(x_1, \dots, x_n)w(x)\,dx\right)\\
&\quad\quad \times (\mathbf{f}_n(x_1, \dots, x_n) - \mathbf{m}_n(x_1, \dots, x_n))\\
&\quad = \int_{R^k} m(x)w(x)\,dx + \mathbf{w}^t(\mathbf{f}_n(x_1, \dots, x_n) - \mathbf{m}_n(x_1, \dots, x_n))\\
&\quad = \left[\int_{R^k} m(x)w(x)\,dx - \sum_{i=1}^n w_i m(x_i)\right] + \sum_{i=1}^n w_i f(x_i) \qquad (5.47)
\end{aligned}$$

with

$$\mathbf{w}^t = \int_{R^k} \mathbf{V}_{1+n,12}(x, x_1, \ldots, x_n)\mathbf{V}_n^{-1}(x_1, \ldots, x_n)w(x)\,dx$$

and

$$\begin{aligned}
&\int_{R^k}\int_{R^k} v(x, y|x_1, \ldots, x_n)w(x)w(y)\,dx\,dy \\
&\quad = \int_{R^k}\int_{R^k} v(x, y)w(x)w(y)\,dx\,dy \\
&\quad - \int_{R^k}\int_{R^k} \mathbf{V}_{1+n,12}(x, x_1, \ldots, x_n)\mathbf{V}_n^{-1}(x_1, \ldots, x_n) \\
&\quad \times \mathbf{V}_{1+n,12}^t(y, x_1, \ldots, x_n)w(x)w(y)\,dx\,dy. \qquad (5.48)
\end{aligned}$$

Notice that in this case the posterior variance of the random integral does not depend on the values $\{f(x_1), \ldots, f(x_n)\}$.

We note several things about this development. First, the estimate of (5.41) is given by (5.47) and this takes the form of a constant plus a quadrature rule. Also notice that (5.47) is the a priori estimate (5.46) plus a linear combination of the differences between f and m at the points $\{x_1, \ldots, x_n\}$. It is certainly preferable if (5.47) can be evaluated in closed form but this is not essential as it need only be evaluated once and can then be applied to many integration problems.

Second the design problem takes the form of finding $\{x_1, \ldots, x_n\}$ which minimizes (5.48) or equivalently which minimizes the expected value of the form

$$(v(x, x_1), \ldots, v(x, x_n))\mathbf{V}_n^{-1}(x_1, \ldots, x_n)((v(y, x_1), \ldots, v(y, x_n)))^t$$

where the expectation is taken with respect to the joint density $w(x)w(y)$. Such a choice of points together with the weights gives a Gaussian Bayesian quadrature rule. This optimization problem is very difficult and at this point there aren't good approaches to its general solution.

The key ingredients in determining the success of Gaussian Bayesian quadrature rules, when applied to a particular f, are the mean function m, the covariance function v and the weight function w. The choice of w seems relatively simple. For example, we might take w to be a multivariate normal distribution or a multivariate Student distribution. The function m would then represent something we know about the behavior of f. Similarly, the choice of v should reflect this as well; e.g. that the function is relatively smooth can be expressed by the existence of correlations amongst values $f(x)$ and $f(y)$ when x and y are far apart. O'Hagan (1991, 1992, 1996) discusses such issues and generalizes the Bayesian nature of the problem to allow prior distributions for the mean and covariance functions as well. This results in a somewhat different theory than that presented above. Some new quadrature rules are also developed there but these are at present only for low dimensions.

5.8 Quasi-random rules

Consider the integration problem

$$\int_{[0,1]^k} f(x)\,dx;$$

i.e. integration over the unit cube. The methods of this section require that a transformation to the unit cube has been made. As noted in Section 5.5 it is possible to transform many integration domains to rectangles (whether bounded or unbounded). It is then easy to transform such a domain to the unit cube; e.g. consider a probability transform such as that discussed in Exercise 3.9.9 where F is an absolutely continuous distribution function on this rectangle. It should be noted, however, that there is no guarantee that we are making the integration problem easier by employing such a transformation. If we can choose the transformation, however, so that the integrand becomes more constant then this generally makes the problem easier. This suggests choosing F so that the probability mass of the distribution lies in the region where the greatest contribution to the initial integral occurs.

The motivation for quasi-random methods follows from the following somewhat surprising result. We make use of the fractional part function $\{x\}$ which equals the fractional part of the real number x.

Theorem 5.19 Suppose that f is Riemann integrable on [0, 1] and z_0 is an irrational number. Then

$$\lim_{n\to\infty} \frac{1}{n+1} \sum_{i=0}^{n} f(\{iz_0\}) = \int_0^1 f(x)\,dx. \tag{5.49}$$

Proof: Consider $f(x) = \exp(2\pi ihx)$ where h is an integer. If $h = 0$ then (5.49) holds. If $h \neq 0$ then $\int_0^1 f(x)\,dx$ equals 0. Now since $\exp(2\pi ihk) = 1$ for any integer k we have that

$$\begin{aligned}\frac{1}{n+1}\sum_{j=0}^{n}\exp(2\pi ih\{jz_0\}) &= \frac{1}{n+1}\sum_{j=0}^{n}\exp(2\pi ihjz_0)\\ &= \frac{1}{n+1}\,\frac{1-\exp(2\pi ih(n+1)z_0)}{1-\exp(2\pi ihz_0)}\\ &\to 0\end{aligned}$$

as $n \to \infty$ and so (5.49) holds for this f. Note that the irrationality of z_0 is necessary for the convergence to 0.

Now consider continuous, periodic f; i.e. $f(0) = f(1)$. By the Weierstrass approximation theorem (thinking of f as being defined now on the unit circle) for a given $\epsilon > 0$ there is a *trigonometric polynomial* $h(x) = \sum_{j=1}^{l} c_j \exp(2\pi i j x)$ such that $|f(x) - h(x)| \leq \epsilon/3$ for all $x \in [0, 1]$. Therefore

$$\left|\int_0^1 f(x)\,dx - \frac{1}{n+1}\sum_{i=0}^{n} f(\{iz_0\})\right|$$

$$\leq \left|\int_0^1 f(x)\,dx - \int_0^1 h(x)\,dx\right| + \left|\int_0^1 h(x)\,dx - \frac{1}{n+1}\sum_{i=0}^{n} h(\{iz_0\})\right|$$

$$+ \left|\frac{1}{n+1}\sum_{i=0}^{n} f(\{iz_0\}) - \frac{1}{n+1}\sum_{i=0}^{n} h(\{iz_0\})\right|$$

$$\leq \epsilon/3 + \left|\int_0^1 h(x)\,dx - \frac{1}{n+1}\sum_{i=0}^{n} h(\{iz_0\})\right| + \epsilon/3$$

and by the first part of the proof we can choose n so large that

$$\left|\int_0^1 h(x)\,dx - \frac{1}{n+1}\sum_{i=0}^{n} h(\{iz_0\})\right| \leq \epsilon/3$$

which establishes (5.49) for f.

Now suppose that f is the indicator function for an interval $[a, b] \subset [0, 1]$. Then there exist continuous, periodic function g_1, g_2 such that $g_1 \leq f \leq g_2$ and for given $\epsilon > 0$

$$\left|\int_0^1 g_2(x)\,dx - \int_0^1 g_1(x)\,dx\right| \leq \epsilon.$$

Then

$$\begin{aligned}
\int_0^1 g_1(x)\,dx &= \lim_{n\to\infty} \frac{1}{n+1}\sum_{i=0}^{n} g_1(\{iz_0\}) \\
&\leq \liminf_{n\to\infty} \frac{1}{n+1}\sum_{i=0}^{n} f(\{iz_0\}) \\
&\leq \limsup_{n\to\infty} \frac{1}{n+1}\sum_{i=0}^{n} f(\{iz_0\}) \\
&\leq \lim_{n\to\infty} \frac{1}{n+1}\sum_{i=0}^{n} g_2(\{iz_0\}) \\
&= \int_0^1 g_2(x)\,dx
\end{aligned}$$

and

$$\int_0^1 g_1(x)\,dx \le \int_0^1 f(x)\,dx \le \int_0^1 g_2(x)\,dx.$$

Since we can choose g_1 and g_2 to make the lower and upper bounds on these inequalities arbitrarily close we must have that

$$\liminf_{n\to\infty} \frac{1}{n+1}\sum_{i=0}^{n} f(\{iz_0\}) = \limsup_{n\to\infty} \frac{1}{n+1}\sum_{i=0}^{n} f(\{iz_0\}) = \int_0^1 f(x)\,dx$$

and (5.49) holds for f. It immediately follows that (5.49) holds for any step function as well.

Now suppose that f is any Riemann integrable function. Then we can find step functions g_1, g_2 such that $g_1 \le f \le g_2$ and for given $\epsilon > 0$

$$\left|\int_0^1 g_2(x)\,dx - \int_0^1 g_1(x)\,dx\right| \le \epsilon.$$

The proof then concludes just as in the case for an indicator function of an interval.

Corollary 5.20 For any interval $[a, b) \subset [0, 1]$

$$\lim_{n\to\infty} \frac{1}{n+1}\sum_{i=0}^{n} I_{[a,b)}(\{iz_0\}) = b - a.$$

The content of Corollary 5.20 is that the limiting proportion of values in the sequence $\{nz_0\}$ is equal to the length of the interval. Note that it is immediate from the proof of Theorem 5.19 that if f is a trigonometric polynomial then the error in the approximation satisfies

$$\left|\int_0^1 f(x)\,dx - \frac{1}{n+1}\sum_{i=0}^{n} f(\{iz_0\})\right| = O\left(\frac{1}{n}\right).$$

This result can be extended to a wider class of functions. For example, in Davis and Rabinowitz (1984) it is shown that if $f \in C^3[0, 1]$ is periodic and z_0 is the root of a quadratic polynomial then again the error is $O(1/n)$. This is better than Monte Carlo where, as we see in Chapter 6, the error declines like $1/\sqrt{n}$. This improvement in rate of convergence over Monte Carlo is the primary reason quasi-random rules are advocated.

It can be shown (Exercise 33) that any sequence $x_0, x_1, \ldots$ in $[0, 1]$ that satisfies

$$\lim_{n\to\infty} \frac{1}{n+1}\sum_{i=0}^{n} I_{[a,b)}(x_i) = b - a \tag{5.50}$$

for every $[a, b) \subset [0, 1]$ also satisfies

$$\lim_{n\to\infty} \frac{1}{n+1} \sum_{i=0}^{n} f(x_i) = \int_0^1 f(x)\, dx \tag{5.51}$$

for every Riemann integrable f. A sequence with the property (5.50) is said to be *equidistributed* in [0, 1] and (5.51) establishes the importance of this property for integration. Theorem 5.19 exhibits an example of an equidistributed sequence.

In general, we say that a sequence $x_0, x_1, \ldots$ is equidistributed in $[0, 1]^k$ if for any rectangle $[a, b) = \prod_{i=1}^{k} [a, b)$

$$\lim_{n\to\infty} \frac{1}{n+1} \sum_{i=0}^{n} I_{[a,b)}(x_i) = \prod_{i=1}^{k} (b_i - a_i).$$

And for such a sequence we again have that

$$\lim_{n\to\infty} \frac{1}{n+1} \sum_{i=0}^{n} f(x_i) = \int_{[0,1]^k} f(x)\, dx$$

for any Riemann integrable f. If $z_1, \ldots, z_k$ are irrational numbers such that $1, z_1, \ldots, z_k$ are linearly independent over Q then it can be shown that the sequence given by $x_i = (\{iz_1\}, \ldots, \{iz_k\})$ is equidistributed in $[0, 1]^k$, see Kuipers and Niederreiter (1974). For computational purposes, however, this sequence is not very convenient as we cannot handle irrational numbers on the computer.

Many equidistributed sequences have been developed for practical use and we discuss some of these. How does one select amongst them? This is typically based on an asymptotic error analysis; i.e. the rate of convergence of

$$\left| \int_{[0,1]^k} f(x)\, dx - \frac{1}{n+1} \sum_{i=0}^{n} f(x_i) \right|$$

as $n \to \infty$. In practical circumstances error assessment is performed by computing

$$\frac{1}{n+1} \sum_{i=0}^{n} f(x_i)$$

for different values of n and comparing the results.

It is important to recognize that an equidistributed sequence of points does not behave like a random sequence and therefore error assessment cannot be based on probabilistic reasoning. To see this quite plainly suppose, without loss of generality, that $z_0 \in [0, 1/2]$ is irrational. Now consider $[a, b) \subset [0, z_0)$ and $[c, d) \subset [2z_0, 1]$. Then when $\{iz_0\} \in [a, b)$ we have that $\{(i + 1)z_0\} = \{iz_0\} + z_0$ and so

$$\begin{aligned}
&\lim_{n\to\infty} \frac{1}{n+1} \sum_{i=0}^{n} I_{[a,b)\times[c,d)}(\{iz_0\}, \{(i+1)z_0\}) \\
&= \lim_{n\to\infty} \frac{1}{n+1} \sum_{i=0}^{n} I_{[a,b)}(\{iz_0\}) I_{[c,d)}(\{iz_0\} + z_0) \\
&= 0
\end{aligned}$$

rather than equalling $(b - a)(d - c)$. Therefore the sequence $\{iz_0\}$ definitely does not have the independence properties that we would want a random sequence to possess. Because the equidistributed sequences have all the integration properties that a truly random sequence possesses, but do not have the independence or randomness properties of a random sequence, they are often referred to as quasi-random sequences.

To compare quasi-random sequences, there are various criteria used to measure effectiveness. One measure is the *star discrepancy* of a sequence $x_1, x_2, \ldots$ which is given by

$$D_n^* = \sup_{(t_1,\ldots,t_k)\in[0,1]^k} \left| \frac{1}{n+1} \sum_{i=1}^{n} I_{R(t_1,\ldots,t_k)}(x_i) - t_1 \cdots t_k \right|$$

where $R(t_1, \ldots, t_k) = \prod_{i=1}^{k} [0, t_i]$. This measures the worst absolute error of the quasi-random rule when approximating the distribution function of the uniform distribution on $[0, 1]^k$. The *discrepancy* D_n is defined similarly as the supremum error over all subrectangles of $[0, 1]^k$. The star discrepancy seems sufficient when comparing sequences for integration problems because we can easily show that $D_n^* \le D_n \le 2^k D_n^*$ (Exercise 36). Therefore if $D_n^* \to 0$ as $n \to \infty$ then $D_n \to 0$ as well and this implies that the sequence is equidistributed.

The *Vitali variation* of a function $f \in C^k([0, 1]^k)$ is given by

$$V^{(k)}(f) = \int_{[0,1]^k} \left| \frac{\partial^k f(x_1, \ldots, x_k)}{\partial x_1 \cdots \partial x_k} \right| dx_1 \cdots dx_k$$

which of course is finite. This definition can be weakened to apply to a much broader class of functions but it becomes much more complicated; see Niederreiter (1992) for details. For a function f, for $1 \le l \le k$ and $1 \le i_1 < \cdots < i_l \le k$, define the function $f_{i_1,\ldots,i_l}$ to be f restricted to the set

$$\{(x_1, \ldots, x_k) \colon x_i = 0 \text{ whenever } i \ne i_1, \ldots, i_l \text{ and } 0 \le x_i \le 1 \text{ otherwise}\};$$

i.e. an l-dimensional face of $[0, 1]^k$. Then the *Hardy–Krause variation* of f is defined by

$$V(f) = \sum_{l=1}^{k} \sum_{1 \le i_1 < \cdots < i_l \le k} V^{(l)}(f_{i_1,\ldots,i_l}).$$

We have the following famous inequality.

Theorem 5.21 (*Koksma–Hlawka inequality*) For any sequence $x_0, x_1, \ldots$ in $[0, 1]^k$

$$\left| \int_{[0,1]^k} f(x)\,dx - \frac{1}{n+1} \sum_{i=0}^{n} f(x_i) \right| \le V(f) D_n^*.$$

Proof: See Hlawka (1961) or Keng and Yuan (1981).

This result implies that we can compare sequences with respect to their rate of convergence by looking at the asymptotic behavior of D_n^*.

Classes of sequences of practical importance for applications of this approach are discussed in the following examples.

Example 5.11 *Van der Corput sequences*
Let $r > 1$ be an integer and note that $m \in \mathcal{N}$ can be uniquely expressed in base (radix) r as $m = a_0 + a_1 r + a_2 r^2 + \cdots + a_l r^l$ where each $a_i \in \{0, 1, \ldots, r-1\}$, $a_l \ne 0$ and $r^l \le m < r^{l+1}$; i.e. the values $l, a_1, \ldots, a_l$ are unique. The base r *radical inverse function* is $\phi_r : \mathcal{N} \to [0, 1]$ given by

$$\phi_r(m) = a_0 r^{-1} + a_1 r^{-2} + a_2 r^{-3} + \cdots + a_l r^{-l-1}.$$

Therefore if the base r representation of m is $a_l a_{l-1} \cdots a_1 a_0$ the base r representation of $\phi_r(m)$ is $0.a_0 a_1 \cdots a_l$; i.e. the radical inverse function reflects the number through the decimal point. The sequence with $x_i = \phi_r(i)$ is called a *Van der Corput sequence*. It can be shown that $D_n^* = O(n^{-1} \log n)$ so this is indeed an equidistributed sequence. This is in fact the best rate of convergence that can be obtained by an equidistributed sequence when $k = 1$. Also

$$\limsup_{n\to\infty} \frac{n D_n^*}{\log n} = \begin{cases} \dfrac{r^2}{4(r+1)\log r} & r \text{ even} \\ \dfrac{r-1}{4\log r} & r \text{ odd} \end{cases}$$

which implies that the optimal choice is $r = 3$; see Niederreiter (1992) for details on these results. Note that Van der Corput sequences can be computed relatively easily on a computer; see Fox (1986) for details.

Example 5.12 *Halton sequences*
For integrals over $[0, 1]^k$ we can use a generalization of the Van der Corput sequences known as *Halton sequences*. For this suppose that we have the bases

$r_1, \ldots, r_k$ which are pairwise relatively prime (so as to avoid correlations existing amongst the coordinates). Then the i-th element of the Halton sequence corresponding to this sequence of bases is given by

$$x_i = (\phi_{r_1}(i), \ldots, \phi_{r_k}(i)).$$

It is common to take $r_1, \ldots, r_k$ to be the first k primes. It can be shown that

$$D_n^* \leq \left[\prod_{i=1}^{k} \frac{r_i - 1}{2 \log r_i}\right] \frac{(\log n)^k}{n} + \frac{(\log n)^{k-1}}{n}$$

which implies that $D_n^* = O(n^{-1}(\log n)^k)$ and it is conjectured that this is the best asymptotic error rate for a sequence. Note, however, the dimensional effect implicit in the asymptotic constant. See Niederreiter (1992) for details on these results. Note that if we take the asymptotic error rate for Monte Carlo to be $n^{-1/2}$ then $n^{-1}(\log n)^k / n^{-1/2} = n^{-1/2}(\log n)^k \to 0$ as $n \to \infty$. Therefore integration based on the Halton sequence has a much better asymptotic error rate for all dimensions. Note, however, as we will see, the constant for Monte Carlo does not exhibit a dimensional effect at least explicitly. We will discuss this in greater detail in Chapter 6.

Example 5.13 *Hammersley sets*
Each Halton sequence leads to a sequence of Hammersley sets depending on n. For the Halton sequence in $[0, 1]^{k-1}$ using bases $r_1, \ldots, r_{k-1}$ the Hammersley set in $[0, 1]^k$ with n elements is given by

$$H_n = \left\{\left(\frac{i}{n}, \phi_{r_1}(i), \ldots, \phi_{r_{k-1}}(i)\right) : i = 0, \ldots, n-1\right\}$$

and note that $H_n \not\subseteq H_{n+1}$. It can be shown that for H_n

$$D_n^* \leq \left[\prod_{i=1}^{k-1} \frac{r_i - 1}{2 \log r_i}\right] \frac{(\log n)^{k-1}}{n} + \frac{(\log n)^{k-2}}{n}$$

which implies that $D_n^* = O(n^{-1}(\log n)^{k-1})$ for the sequence of sets. It is conjectured that this is the best asymptotic error rate for a sequence of finite sets and note that it is better than the rate for the infinite Halton sequence. Of course to increase the number of points in the approximation requires that we compute f at every point in the Hammersley set while with the Halton sequence we can reuse previous computations.

In Wozniakowski (1991) an optimality result, within the context of *information-based complexity* theory, is established for a small modification of the Hammersley sets (the first coordinate is modified to $(i + t^*)/n$ for a fixed constant t^*). For this, suppose that f is an occurrence from a Gaussian process with mean function identically 0 and

$$\mathrm{Cov}[f(s), f(t)] = \prod_{i=1}^{k} \min\{s_i, t_i\}.$$

The error in a set $\{x_1, \ldots, x_n\}$ is measured by

$$E\left[\left(\int_{[0,1]^k} f(x)\, dx - \frac{1}{n}\sum_{i=1}^{n} f(x_i)\right)^2\right]. \tag{5.52}$$

The *computational complexity* of the integration problem is a function Comp : $(0, \infty) \to (0, \infty)$ defined by setting Comp(ϵ) equal to the smallest n such that (5.52) is no greater than ϵ for some set of n points. Therefore Comp(ϵ) is a measure of how much computation is needed to achieve a given accuracy on average for an integrand coming from the specified stochastic process. It is shown that a sequence of sets exists that is optimal in the following sense: there exist positive constants c_1, c_2, c_3 such that for all $\epsilon \le c_3$ there is a set, in fact a slight modification of a Hammersley set, with $n(\epsilon)$ points whereby (5.52) is no greater than ϵ and

$$c_1 \le \frac{n(\epsilon)}{\mathrm{Comp}(\epsilon)} \le c_2.$$

So the modified Hammersley set is about as good as such rules get at least in the sense prescribed. For further discussion of this result and the theory of information-based complexity see Traub and Wozniakowski (1991).

Example 5.14 (t, m, k)-*nets* and (t, k)-*sequences*
Rules based on (t, m, k)-nets and (t, k)-sequences were introduced by Niederreiter to achieve the (presumed) optimal rates of convergence for sequences and sequences of sets while avoiding the extreme growth in the asymptotic constant that we observed in Examples 5.12 and 5.13. In fact, the growth in these constants has led to the recommendation that Halton sequences and Hammersley sets not be used in high dimensions; see Niederreiter (1992).

The definitions are motivated by the following property of the Van der Corput sequence based on radix r. First we note that for any integers $l \ge 0, m \ge 1$ we have r^m points x_n with $lr^m \le n < (l+1)r^m$. From the construction of the x_n it follows that each interval $[br^{-m}, (b+1)r^{-m})$ for $b = 0, 1, \ldots, r^m - 1$ contains exactly one of the points x_n with $lr^m \le n < (l+1)r^m$. Therefore these r^m points distribute in a very regular way. For example, if $l = 0$ then each point in

$$\{\phi_r(n)\colon 0 \le n < r^m\} = \left\{\sum_{i=0}^{m-1} a_i r^{-i-1}\colon 0 \le a_i < r, i = 0, \ldots, m-1\right\}$$

is the left-hand endpoint of an interval of the form $[br^{-m}, (b+1)r^{-m})$, etc.

Now consider the situation when $k \geq 1$. An *elementary interval* base r is defined to be a set of the form

$$E = \prod_{i=1}^{k} \left[\frac{b_i}{r^{m_i}}, \frac{b_i + 1}{r^{m_i}} \right)$$

where $b_i, m_i \in \mathcal{N}_0$ and $0 \leq b_i < r^{m_i}$ for $1 \leq i \leq k$. For $t, m \in \mathcal{N}_0$ with $0 \leq t \leq m$, a (t, m, k)-*net* in base r is a set $S \subset [0, 1]^k$ of r^m points such that

$$\frac{\#(E \cap S)}{r^m} = r^{t-m}$$

for every elementary interval E base r for which $\text{Vol}(E) = r^{t-m}$. Therefore for every $l, m \in \mathcal{N}_0$ the set $\{x_n\colon lr^m \leq n < (l+1)r^m\}$ of points in the Van der Corput sequence is a $(0, m, 1)$-net in base r. For $t \in \mathcal{N}_0$ a sequence $x_0, x_1, \ldots$ in $[0, 1]^k$ is a (t, k)-*sequence* in base r if for every $l, m \in \mathcal{N}_0$ with $m > t$ the set $S_{l,m} = \{x_n\colon lr^m \leq n < (l+1)\, r^m\}$ is a (t, m, k)-net in base r. Note that the Van der Corput sequence in base r is a $(0, 1)$-sequence.

The importance of these concepts derives from the fact that Niederreiter (1992) has established that for a (t, m, k)-net in base r containing $n = r^m$ points

$$D_n^* \leq B(k, r) r^t O\big(n^{-1} (\log n)^{k-1}\big) + O\big(n^{-1} (\log n)^{k-2}\big)$$

with

$$B(k, r) = \begin{cases} \left(\dfrac{r-1}{2 \log r} \right)^{k-1} & \text{when } k = 2 \text{ or } r = 2, k = 3, 4 \\ \dfrac{1}{(k-1)!} \left(\dfrac{\lfloor r/2 \rfloor}{\log r} \right)^{k-1} & \text{otherwise.} \end{cases}$$

Therefore a sequence of (t, m, k)-nets in base r satisfies $D_n^* = O(n^{-1} (\log n)^{k-1})$ and it can be shown that the asymptotic constant is growing more slowly with k than that for the Hammersley set. Similarly, Niederreiter (1992) shows that for a (t, k)-sequence in base r

$$D_n^* \leq C(k, r) r^t O\big(n^{-1} (\log n)^k\big) + O\big(n^{-1} (\log n)^{k-1}\big)$$

where

$$C(k, r) = \begin{cases} \dfrac{1}{k} \left(\dfrac{r-1}{2 \log r} \right)^{k} & \text{when } k = 2 \text{ or } r = 2, k = 3, 4 \\ \dfrac{1}{k!} \dfrac{r-1}{2 \lfloor r/2 \rfloor} \left(\dfrac{\lfloor r/2 \rfloor}{\log r} \right)^{k} & \text{otherwise.} \end{cases}$$

Therefore a (t, k)-sequence in base r satisfies $D_n^* = O\big(n^{-1} (\log n)^k\big)$ and it can be shown that the asymptotic constant is growing more slowly with k than that for the Halton sequence. Niederreiter (1992) contains many results concerning the construction of (t, m, k)-nets and (t, k)-sequences. At this point in time these are the preferred quasi-random rules to use.

Niederreiter (1992) contains a discussion of other quasi-random sequences and an in-depth treatment of the theory of quasi-random and pseudo-random sequences. Also see the text by Kuipers and Niederreiter (1974) for a discussion of the equidistribution property. Shaw (1988), Johnson (1992), Fang, Wang and Bentler (1994) and Fang and Wang (1994) discuss aspects of applying these rules in statistical contexts. The current state of quasi-random theory is covered in a review paper by Spanier and Maize (1994). Of particular interest is a discussion in this paper of the analogs of various Monte Carlo variance reduction techniques.

At this point we see no categorical reason to abandon pseudo-random sequences for quasi-random sequences even though there are examples that indicate substantially improved accuracy with the quasi-random approach. The powerful analytical tools that probability theory provides for the analysis and design of algorithms is too much to give up. These topics are extensively discussed in Chapters 6 and 7. In practical integration problems, the most important aspect of any algorithm is the appropriate placement of the points with respect to the particular integrand under consideration and this has nothing to do with the uniformity of the underlying sequence. No matter which method we choose, quasi-random sequences or pseudo-random sequences, if many of the points are placed outside the region which is significant for the evaluation of the integral then a poor approximation will result. Avoiding this is what makes the integration problem difficult. From this point of view, the debate over the use of quasi-random versus pseudo-random sequences does not seem all that material.

5.9 Lattice rules

Again we consider the integration problem

$$\int_{[0,1]^k} f(x)\,dx.$$

Lattice rules are equal weight quadrature rules based on averaging f over a lattice of points in $[0, 1]^k$. It is required for the application of these rules that f be periodic in each coordinate; i.e. $f(x_1, \ldots, x_{i-1}, 0, x_{i+1}, \ldots, x_k) = f(x_1, \ldots, x_{i-1}, 1, x_{i+1}, \ldots, x_k)$ for $i = 1, \ldots, k$. Accordingly, to apply such a rule to an f defined on $[0, 1]^k$ we typically first have to apply a periodizing transformation. There are several possibilities for this. Perhaps the most direct is to symmetrize f with respect to the group G of reflections through the hyperplanes defined by $x_i = 1/2$ for $i = 1, \ldots, k$; i.e. replace f by

$$f_G(x) = \frac{1}{\#(G)} \sum_{g \in G} f(gx).$$

The original integral is invariant under this transformation. Since $\#(G) = 2^k$ this is only feasible for small k. To avoid this we look for a relatively simple nonlinear transformation to accomplish the periodicity. As discussed in Sloan and Joe (1994) the simplest polynomial periodizing transformation is given by $x = \phi(y) = \left(3y_1^2 - 2y_1^3, \ldots, 3y_k^2 - 2y_k^3\right)$ since this replaces f in the integral by the integrand

$$f\left(3x_1^2 - 2x_1^3, \ldots, 3x_k^2 - 2x_k^3\right)6^k \prod_{i=1}^{k} x_i(1 - x_i)$$

and note that this function takes the value 0 whenever a coordinate equals 0 or 1. Another simple periodizing transformation is $x_i = y_i - (\sin 2\pi y_i)/2\pi$. The trigonometric nature of this transformation makes it particularly appropriate when using lattice rules. Smoother periodizing transformations are also available in Sloan and Joe (1994); i.e. transformations that make higher derivatives of the integrand periodic as well. We suppose hereafter that f is periodic.

A *lattice* in R^k is a discrete set L which is a group under addition. If $A \in R^{k \times l}$ is of rank l then $L_A = \{Ag : g \in \mathcal{Z}^l\}$ is called the lattice determined by A. If $A = I$ then $L_I = \mathcal{Z}^k$ and is known as the *cubic lattice*. A *basis* B for a lattice L is a set of linearly independent vectors such that every element of L can be written uniquely as a linear combination of elements of B when the coefficients are elements of $\mathcal{Z}$. The columns of the matrix A comprise a basis for L_A. In fact, every lattice has a basis and these basis vectors are also linearly independent over R^1. If the cardinality of the basis for L is k, and the basis vectors comprise the columns of $A \in R^{k \times k}$, then A has the smallest determinant amongst all matrices whose columns are linearly independent (over R^1) elements of L. The value of $\det A$ for a basis is called the *unit cell volume* of the lattice and note that it is independent of the particular basis used so we denote it as $\det L$ hereafter. If we consider each cell of the lattice to be closed only along those edges connecting to the point in the cell closest to the origin then the average number of lattice points in a unit volume is given by $1/\det L$. Obviously the standard basis of R^k serves as a basis for $\mathcal{Z}^k$ and the unit cell volume of the cubic lattice is 1. These and other facts about lattices can be found in Siegel (1980).

An *integration lattice* L is a lattice which contains $\mathcal{Z}^k$. Note that if L is an integration lattice and $m \in \mathcal{N}$ then $m^{-1}L$ is also an integration lattice and the unit cell volume of $m^{-1}L$ is $m^{-k} \det L$. For an integration lattice L, the *lattice rule* associated with L is given by

$$\frac{1}{\#\left\{L \cap [0,1)^k\right\}} \sum_{x \in L \cap [0,1)^k} f(x). \tag{5.53}$$

Obviously we want the unit cell volume of L to be small to increase the number of points in $L \cap [0, 1)^k$. We can do this by replacing L by $m^{-1}L$ as this has approximately $m^k/\det L$ points in $[0, 1)^k$ and $m^{-1}L$ is still an integration lattice. Note that we sum over the lattice points in $[0, 1)^k$ rather than $[0, 1]^k$ because of the periodicity of f.

If $x, y \in [0, 1]$ we define addition modulo 1 by $x \oplus y = \{x + y\}$ where $\{x\}$ denotes the fractional part of x. This turns $[0, 1]$ into a group. Similarly, if we define the group operation to be addition modulo 1 coordinatewise then $[0, 1]^k$ is a group. It is an easy exercise to show that $L \cap [0, 1)^k$ is a subgroup of $[0, 1]^k$. We note, however, that this will be an infinite subgroup whenever there is a point in L that has an irrational coordinate. To see this consider the sequence $\{nz_0\}$, when z_0 is irrational, discussed in Section 5.8 and also Exercise 35. Under these circumstances it would be more appropriate to define (5.53) as a limit. To ensure a finite sum we restrict attention to lattices L where all elements have coordinates in $\mathbb{Q}$, the set of rational numbers. Under these circumstances $L \cap [0, 1)^k$ is a subgroup of $[0, 1]^k$ using addition modulo 1 and, because of the periodicity of f, (5.53) can be thought of as the average of f over the entire lattice L. Note that in the spirit of Section 5.6 the average (5.53) equals $f_G(0)$ where $G = L \cap [0, 1)^k$; i.e. applying a lattice rule can be thought of as a symmetrization of an integrand to make it more like a weight function, which in this case is the constant 1.

The following result is proved in Sloan and Joe (1994).

Theorem 5.22 Let L be an integration lattice contained in $\mathbb{Q}^k$ and $\#\{L \cap [0, 1)^k\} \geq 2$. Then there exists a unique integer l (the *rank* of the lattice rule) with $1 \leq l \leq k$ and unique integers (the *invariants* of the rule) $m_1, \ldots, m_l$ with $m_l > 1$ such that m_{i+1} divides m_i for $i = 1, \ldots, l - 1$ and linearly independent $z_1, \ldots, z_l \in \mathcal{Z}^k$ such that the lattice rule (5.53) can be expressed as

$$\frac{1}{m_1 \cdots m_l} \sum_{i_1=1}^{m_1} \cdots \sum_{i_l=1}^{m_l} f\left(\left\{\frac{i_1 - 1}{m_1} z_1 + \cdots + \frac{i_l - 1}{m_l} z_l\right\}\right).$$

The number of points in the rule satisfies $\#\left\{L \cap [0, 1)^k\right\} = m_1 \cdots m_l$.

This representation gives a compact method for computing a lattice rule and avoids any redundancies in the sum.

Sloan and Joe (1994) present many additional results concerning lattice rules but we specialize for the remainder of the section to the simplest case where $l = 1$, so-called *good lattice point rules*. These show that lattice point rules have some interesting and useful properties. By Theorem 5.22 a good lattice point rule is specified by choosing $z \in \mathcal{Z}^k$ and $m > 1$. The rule is then given by

$$\frac{1}{m} \sum_{i=1}^{m} f\left(\left\{\frac{i - 1}{m} z\right\}\right) = \frac{1}{m} \sum_{i=1}^{m} f\left(\frac{i - 1}{m} z\right) \tag{5.54}$$

where the equality follows from the periodicity of f. Therefore a good lattice point rule can be thought of as simply an average of f on a finite set of points lying on the line containing the vector z.

Now suppose that f has Fourier expansion

$$f(x) = \sum_{h \in \mathcal{Z}^k} \hat{f}(h) \exp(2\pi i h' x)$$

where

$$\hat{f}(h) = \int_{[0,1]^k} \exp(-2\pi i h^t x) f(x)\, dx$$

and

$$\sum_{h \in \mathcal{Z}^k} |\hat{f}(h)| < \infty.$$

Notice that

$$\hat{f}(0) = \int_{[0,1]^k} f(x)\, dx.$$

Then applying the good lattice point rule to the Fourier expansion of f we obtain

$$\begin{aligned} \frac{1}{m}\sum_{j=1}^{m} f\left(\frac{j-1}{m}z\right) &= \sum_{h \in \mathcal{Z}^k} \hat{f}(h) \frac{1}{m}\sum_{j=1}^{m} \exp\left(2\pi i h^t \frac{j-1}{m}z\right) \\ &= \sum_{h \in \mathcal{Z}^k} \hat{f}(h) \frac{1}{m}\sum_{j=1}^{m} \left[\exp\left(2\pi i h^t \frac{1}{m}z\right)\right]^{j-1} \\ &= \sum_{h \in \mathcal{Z}^k, h^t z \equiv 0 \bmod m} \hat{f}(h) \end{aligned}$$

since

$$\frac{1}{m}\sum_{j=1}^{m} \left[\exp\left(2\pi i h^t \frac{1}{m}z\right)\right]^{j-1} = \begin{cases} 1 & \text{if } h \in \mathcal{Z}^k,\ h^t z \equiv 0 \bmod m \\ 0 & \text{otherwise.} \end{cases}$$

Now usually$|\hat{f}(h)|$ decreases rapidly as $|h_1| + \cdots + |h_k|$ increases. Therefore we want to choose z so that any $h \in \mathcal{Z}^k$satisfying $h'z \equiv 0 \bmod m$ is such that $|h_1| + \cdots + |h_k|$ is large. This will imply that the absolute error in the good lattice point approximation, namely

$$\left| \sum_{h \in \mathcal{Z}^k, h^t z \equiv 0 \bmod m, h \neq 0} \hat{f}(h) \right| \leq \sum_{h \in \mathcal{Z}^k, h^t z \equiv 0 \bmod m, h \neq 0} \left|\hat{f}(h)\right|,$$

is small, at least if $|\hat{f}(h)|$ decreases rapidly. Therefore a *good* lattice point is a $z \in \mathcal{Z}^k$ where the smallest value of $|h_1| + \cdots + |h_k|$, for any nonzero $h \in \mathcal{Z}^k$ satisfying $h^t z \equiv 0 \bmod m$, is large. Sloan and Joe (1994) discuss techniques for finding good lattice points. These are basically search methods.

We consider a simple example of a not necessarily "good" lattice point rule.

Example 5.15
Suppose we take $z = (1, \ldots, 1)^t$. To see how good this point is we have to determine the smallest value of $|h_1| + \cdots + |h_k|$ amongst all nonzero $h \in \mathcal{Z}^k$ satisfying $h^t z \equiv 0 \bmod m$. Therefore we are looking amongst those h which satisfy $h_1 + \cdots + h_k = m$ and we must find the point on this hyperplane with integer coordinates which minimizes $|h_1|+\cdots+|h_k|$. This can be accomplished by searching. This value then tells us how to rank z relative to other possible choices. See Sloan and Joe (1994) for specific recommendations.

Note that if f is a trigonometric polynomial of degree $n < m$, i.e.

$$f(x) = \sum_{h \in \mathcal{Z}^k, |h_1|+\cdots+|h_k| \leq n} c_h \exp(2\pi i h^t x),$$

then

$$\frac{1}{m} \sum_{j=1}^{m} f\left(\frac{j-1}{m} z\right) = \int_{[0,1]^k} f(x)\, dx;$$

i.e. this rule integrates such functions exactly. Of course, a similar fact is true for any good lattice point rule, with the maximum degree of the trigonometric polynomials integrated exactly determined by the good lattice point.

5.10 Adaptive quadrature

We repeatedly emphasize in the text that the success of an integration strategy depends upon using as much information about the specific integrand as possible. For example, one should locate the modes of the integrand, understand the shape of the contours of the integrand, determine the tail behavior of the integrand, identify singularities, etc. It is our view that general purpose algorithms that avoid taking such information into account are doomed to failure even for some problems in low dimensions. Of course, in practical problems it is often exceedingly difficult or even impossible to obtain this information and make use of it or at the very least exceedingly time consuming to do this. Accordingly, we would like to use the computer to learn the essential facts and design an appropriate strategy. In a general sense this is what is meant by adaptive algorithms.

Adaptation is no guarantee of success either as integrands can always be constructed that defeat whatever adaptive strategy is devised. If one takes such a worst-case perspective then there seems to be little use for adaptation. Further, there is a dearth of mathematical theorems establishing the value of adaptation and even some that indicate otherwise from the point of view of efficiency; e.g. see Wozniakowski (1991). Still experience teaches that in many cases adaptation is a useful technique. The overriding caution, however, is that we are a long way from obtaining black-boxes for integration. In this section, we briefly outline some of the ideas behind adaptive integration. More information is provided in Krommer and Ueberhuber (1998); see also Bernsten (1989) and Lyness (1983).

For univariate integrals, there are a number of adaptive programs and many of these work well. Adaptive programs in one dimension are often written for particular classes of problems; e.g. finite intervals, infinite intervals, oscillatory functions, functions with singularities, etc., and they are available in many numerical libraries. For example, QUADPACK, NAG and IMSL all contain univariate adaptive integration programs and we discuss some of the adaptive strategies used for one dimensional integrals in Sections 5.10.1 and 5.10.2. There are fewer adaptive routines available for multivariate integration. The use of adaptive integration with product Hermite rules was pioneered by Naylor and Smith (1982) and we discuss this in Section 5.10.3. The program ADBAYS written by A. Genz can be considered the current state-of-the-art in adaptive integration and we discuss this in Section 5.10.4.

5.10.1 ADAPTIVE ROMBERG INTEGRATION

Suppose that a function T satisfies $T(h) = c_0 + c_1 h^p + O(h^q)$ as $h \to 0$ for $q > p$ and we want to evaluate $T(0) = c_0$ but, for whatever reason, we can only evaluate $T(h)$ for nonzero h. Then for $d < 1$ we have that $T(dh) = c_0 + c_1 d^p h^p + O(h^q)$ as $h \to 0$ and combining these expressions gives

$$\begin{aligned} c_0 &= T(h) - c_1 h^p + O(h^q) \\ &= T(h) - \frac{T(h) - T(dh)}{1 - d^p} + O(h^q). \end{aligned}$$

Therefore we can approximate c_0 using $T(h) - (T(h) - T(dh))/(1 - d^p)$ rather than $T(h)$ as the former gives a more accurate approximation for small h. If we can write $T(h) = c_0 + c_1 h^p + c_2 h^q + O(h^r)$ as $h \to 0$ for $r > q > p$ then we can apply this procedure to $T(h)$, $T(dh)$, $T(d^2h)$ obtaining three equations in the three unknowns c_0, c_1, c_2. Solving these for c_0 gives

$$\begin{aligned} c_0 = T(h) &- \frac{d^q(T(h) - T(dh)) - (T(dh) - T(d^2h))}{(1 - d^p)(d^q - d^p)} \\ &- \frac{d^p(T(h) - T(dh)) - (T(dh) - T(d^2h))}{(1 - d^q)(d^p - d^q)} + O(h^r) \end{aligned}$$

and we obtain an even more accurate approximation for small h. This process is known as *repeated Richardson extrapolation* and it can obviously be continued provided the necessary expansion holds. We stop the iteration when the difference between two approximations is less than some prescribed tolerance. A simple computational scheme for carrying this out can be found, for example, in Gautschi (1997).

Now let

$$T\left(\frac{1}{n}\right) = \frac{1}{n}\sum_{i=0}^{n} f\left(\frac{i}{n}\right) - \frac{1}{2n}(f(0) + f(1))$$

be the approximation to $\int_0^1 f(x)\,dx$ given by the compounded trapezoid rule. The *Euler–Maclaurin formula*, see Davis and Rabinowitz (1984) for a proof, gives that, whenever $f \in C^{2l+1}([0,1])$, then

$$T\left(\frac{1}{n}\right) - \int_0^1 f(x)\,dx = \sum_{i=1}^{l} \frac{B_{2i}}{(2i)!}\left[f^{(2i-1)}(1) - f^{(2i-1)}(0)\right]\frac{1}{n^{2i}} + O\left(\frac{1}{n^{2l+1}}\right) \tag{5.55}$$

where B_i is the i-th Bernoulli number which can be obtained from the generating function

$$\frac{t}{e^t - 1} = \sum_{i=0}^{\infty} B_i \frac{t^i}{i!}.$$

Therefore repeated Richardson application can be applied to the trapezoid rule and this is known as *Romberg integration*. It is common in such contexts to take $d = 1/2$. The difference between successive approximations gives an estimate of the error in an approximation. Various adaptive schemes have been devised that make use of Romberg integration.

An interesting observation concerning the accuracy of the trapezoid rule approximation can be made from (5.55). Recall that earlier in this chapter we established that the error in the trapezoid rule is $O(1/n^2)$ whenever $f \in C^2([0,1])$. Now suppose that $f \in C^3([0,1])$ and $f^{(1)}(1) = f^{(1)}(0)$; then in fact the error is $O(1/n^3)$. If $f \in C^5([0,1])$ and $f^{(1)}(1) = f^{(1)}(0)$, $f^{(3)}(1) = f^{(3)}(0)$ then the error is $O(1/n^5)$, etc. As we expect this kind of behavior with highly periodic smooth functions it is sometimes said that the trapezoid rule is appropriate in such contexts.

5.10.2 GAUSS–KRONROD RULES

A difficulty with the use of Gauss rules for integration when we want to compute approximations for varying numbers of function evaluations; e.g. for the estimation of error, is that function evaluations for lower order rules are not used for higher orders. This seems wasteful. The Gauss–Kronrod scheme is an attempt to remedy this.

Suppose we start with the Gauss rule $R_n = \{(x_1, w_1), \ldots, (x_n, w_n)\}$ which is exact for the integrals of all polynomials of degree $2n - 1$ or less with respect to the weight function w. The *Gauss–Kronrod rule*, when it exists, replaces R_n by

$$R^*_{2n+1} = \left\{\left(x_1, w^*_1\right), \ldots, \left(x_n, w^*_n\right), \left(x^*_{n+1}, w^*_{n+1}\right), \ldots, \left(x^*_{2n+1}, w^*_{2n+1}\right)\right\}$$

by adding the $n + 1$ points $x^*_{n+1}, \ldots, x^*_{2n+1}$ and computing the weights

$$w^*_1, \ldots, w^*_n, w^*_{n+1}, \ldots, w^*_{2n+1}$$

in such a way that R^*_{2n+1} is exact for all polynomials of degree $3n + 1$ or less. The Gauss–Kronrod rule is therefore not optimal for a rule based on $2n + 1$ points, but it has the advantage of retaining the function evaluations from the application of R_n.

The Gauss–Kronrod rule has been shown to exist for the important case of the Legendre rules; see Davis and Rabinowitz (1984) for details. In this case it is known that the additional points lie in the interval of integration $[-1, 1]$ and that the weights are positive. Therefore an adaptive strategy for an integral over $[a, b]$ involves subdividing $[a, b]$ into subintervals, applying R_n and R^*_{2n+1} in each subinterval and estimating the error contribution from that subinterval by computing the absolute value of the difference of the approximations. Something like this is done in some of the routines in QUADPACK and the reader is referred to Piessens, de Doncker-Kapenga, Ueberhuber and Kahaner (1983) for a discussion of this and also for a listing of Gauss–Kronrod rules. In certain circumstances it is possible to iterate this procedure; i.e. add still more points, and we refer the reader to Davis and Rabinowitz (1984) for a discussion.

5.10.3 ADAPTIVE PRODUCT HERMITE RULES

In Example 5.6 we discussed the use of product Hermite rules to approximate

$$\int_{R^k} f(x)\,dx.$$

For this it was recommended, when f is nonnegative, to first transform the integral via the transformation $y = \mu + \Sigma_\Delta x$ where μ is the mode of f and Σ_Δ is the Cholesky factor of the matrix Σ specified by

$$\Sigma^{-1} = \left(-\frac{\partial^2 \log f(\mu)}{\partial x_i \partial x_j} \right).$$

In a sense we are, in these circumstances, considering μ as being an approximation to the mean and Σ as an approximation to the variance matrix of the density proportional to f. Presumably improvements could be made in the transformed Hermite rule approximation if we replace these estimates by the actual mean and variance matrix of this density. For example, if f is unimodal and highly skewed such an improvement seems likely as the initial approximations will tend to concentrate the transformed product Hermite points around the mode.

We briefly describe an adaptive procedure designed to correct for this and compute more accurate approximations. For more details and many applications see Naylor and Smith (1982, 1988), Smith, Skene, Shaw, and Naylor (1987), Smith, Skene, Shaw, Naylor and Dransfield (1985). For this we start with (μ_0, Σ_0) as specified above. Then using the method specified in Example 5.6 the integrals

$$\int_{R^k} x_i^{\epsilon_i} x_j^{\epsilon_j} f(x)\,dx$$

for $\epsilon_i, \epsilon_j \in \{0, 1\}, i, j \in \{1, \ldots, k\}$ are estimated. From these new estimates (μ_1, Σ_1) of the mean and variance are computed. This process is iterated until these estimates stabilize. The above papers give many examples where this strategy is very successful. Shaw (1999) develops diagnostics that assess the convergence of this iterative procedure.

5.10.4 SUBREGION ADAPTIVE INTEGRATION

Subregion adaptive algorithms are described in van Dooren and de Ridder (1976), Genz and Malik (1980), Bernsten, Espelid and Genz (1991) and Genz (1991, 1992a, 1992b). For low dimensional contexts they are greatly recommended as they produce accurate answers over a wide class of problems. Related algorithms are discussed in Lepage (1978) and Friedman and Wright (1981). For these algorithms the problem is transformed so that the domain of integration is $[0, 1]^k$.

The basic algorithm then proceeds as follows. Let $\epsilon > 0$ be the global absolute error desired and let $n_{\max}$ be the maximum number of subregions where we start with one region $R_{11} = [0, 1]^k$. At the n-th step $[0, 1]^k$ has been partitioned into n subregions $R_{n1}, \ldots, R_{nn}$. Multiple quadrature rules are applied in each subregion to get estimates $I_{n1}, \ldots, I_{nn}$ of the corresponding integrals and also absolute error estimates $E_{n1}, \ldots, E_{nn}$. The error estimate E_{ni} is obtained by computing I_{ni} using rules of different orders and comparing the results. If $E_{n1} + \cdots + E_{nn} < \epsilon$ or if $n \geq n_{\max}$, the algorithm stops. Otherwise, the region R_{ni} is split if $E_{ni} = \max\{E_{nj} \mid 1 \leq j \leq n\}$. The partitioning algorithm takes the regions to be rectangles with sides parallel to the coordinate axes and uses a criterion to choose which side of the chosen rectangle to split to form the subrectangles. Different criteria can be used but they are all based on selecting the coordinate direction, in the chosen subregion, where the integrand is most variable. Variants of what we have described are also used. For example, relative error can be substituted for absolute error and we can base the stopping rule on the total number of function evaluations as well. Subregion adaptive integration has been applied to a number of problems in statistics. See, for example, Genz and Kass (1991, 1997) and Genz (1992b).

Example 5.16 *Calculating an orthant probability*
We consider the computation of an orthant probability in R^6 with high accuracy; i.e.

$$\int_{R^6} m(x) f(x)\, dx$$

where $m(\theta) = I_{[0,\infty)^6}(\theta)$ is the indicator function of $[0, \infty)^6$ and f is the density of the $N_6(0, \Sigma)$ distribution. As a specific example we take $\Sigma^{-1/2} = \operatorname{diag}(0, 1, 2, 3, 4, 5) + ee'$ where $e = (1, 1, 1, 1, 1, 1)'$. The exact value of the integral to 10 decimal places is 0.166625×10^{-4}. We make use of a sequence of reparameterizations discussed in Genz (1992a). The first transformation is from $x \in [0, \infty)^6$ to $u = C^{-1}x$ where C is the lower triangular Cholesky factor of Σ. The second transformation is from $u \in R^6$ to $v = \Phi(u)$ where Φ is the $N(0, 1)$ distribution function applied coordinatewise. The interval of integration for v_i is

$(a_i, 1)$ where $a_1 = 1/2$ and

$$a_i = \Phi\left(\{-c_{i1}\Phi^{-1}(v_1) - \cdots - c_{ii-1}\Phi^{-1}(v_{i-1})\}/c_{ii}\right)$$

for $i > 1$. The third transformation is given by $w_i = (v_i - a_i)/(1 - a_i)$ for $i = 1, \ldots, 6$. Note that the integration with respect to w_6 can be carried out exactly.

For these computations we used the routine ADBAYS developed by A. Genz. This algorithm uses fully symmetric rules of orders 5, 7 and 9 for subregion estimates and for error estimates; see Genz and Malik (1983) for a discussion of the particular rules. Further, a criterion based on fourth differences is used to decide how to split a chosen subregion. The program runs until a specified relative error is attained or too many function evaluations have been carried out.

Based on $N = 10^5$ function evaluations the approximation and its relative error are given by $\hat{I}(m) = 0.166626 \times 10^{-4}(0.6 \times 10^{-6})$ and required several minutes of CPU time on a Sparc workstation. For $N = 10^6$ the approximation is exact to 10 decimal places and required more, although not 10 times as much, CPU time. Overall the computations were a bit slow but highly accurate. For more discussion of this example see Evans and Swartz (1995a) and Genz *et al.* (1996).

5.11 Exercises

(1) Prove the following mean-value theorem for integrals: if w is of constant sign in $[a, b]$ and f is continuous in $[a, b]$ then there exists $x_0 \in [a, b]$ such that $\int_a^b f(x)w(x)\,dx = f(x_0)\int_a^b w(x)\,dx$.

(2) Obtain the interpolatory rule of order 4 corresponding to $x_1 = a = 0$, $x_2 = 2$, $x_3 = 6$, $x_4 = b = 10$ and $w(x) = 1$.

(3) Compute the weights for the spline interpolation rule of order 1 for $A = [0, 1]$, points $\{0, 1/3, 2/3, 1\}$ and $w(x) \equiv 1$.

(4) Prove that the n-th order Hermite rule integrates exactly all polynomials of degree $2n - 1$ or less.

(5) Show that Hermite interpolation at n points $x_1, \ldots, x_n$ arises from Lagrange interpolation at $2n$ points $x_1, \ldots, x_n, x_{n+1}, \ldots, x_{2n}$ by taking limits as $x_{n+i} \to x_i$ for $i = 1, \ldots, n$. Hint: do it first for $n = 2$ and then proceed by induction.

(6) Establish the rate of convergence under compounding for a rule satisfying Corollary 5.6.

(7) Calculate the Peano kernels K for the rectangle rules and $\int_0^1 K(t)\,dt$.

(8) Calculate the Peano kernel for the midpoint rule and $\int_0^1 K(t)\,dt$.

(9) Calculate the Peano kernel for the trapezoid rule and $\int_0^1 K(t)\,dt$.

(10) Calculate the Peano kernel for Simpson's rule and $\int_0^1 K(t)\,dt$.

(11) Let Simpson's rule be compounded m_1 times and the trapezoid rule be compounded m_2 times. What is the relationship between m_1 and m_2 so that the compounded version of Simpson's rule and the compounded version of the trapezoid rule have the same number of function evaluations?

(12) Determine the weighted trapezoid rule using the weight function $w(x) = 6x(1-x)$ on $[0, 1]$. Determine its Peano kernel.

(13) Determine the weighted trapezoid rule using the weight function

$$\frac{\Gamma\left(\frac{3}{2}\right)}{\Gamma\left(\frac{1}{2}\right)}(1-x)^{-1/2}$$

on $[0, 1]$.

(14) Show that $L_w^2 = \left\{f\colon \int_A f^2(x)w(x)\,dx < \infty\right\}$ is a real vector space when w is a density on A and prove that $\langle f, g\rangle = \int_A f(x)g(x)w(x)\,dx$ is a (pseudo) inner product defined on L_w^2.

(15) Establish $\langle p_i, p_j\rangle = \delta_{ij}$ for the sequence of orthonormal polynomials with respect to the density w on A.

(16) Prove Theorem 5.8.

(17) In Theorem 5.10 establish that the polynomial P satisfies $P(x_i) = p(x_i)$ for $i = 1, \ldots, n$.

(18) In the proof of Theorem 5.14 establish the continuity of $f^{(2n)}(\zeta^*(x))$.

(19) Establish a version of Theorem 5.17 for general strictly increasing T when T^{-1} is continuously differentiable.

(20) Prove an analog of Theorem 5.17 for the situation when $f(x) = f_1(x)f_2(x)$, $f_2^{(n)}$ is concave on (a, b) and we put $L(x) = f_1(x)L_2(x)$, $U(x) = f_1(x)U_2(x)$.

(21) Prove an analog of Theorem 5.17 for the situation when $f(x) = f_1(f_2(x))$, $f_1^{(n)}$ is concave on (a, b) and we put $L(x) = L_1(f_1(x)), U(x) = U_1(f_1(x))$.

(22) Establish the envelope inequality for the right-tail of Makeham's distribution in Example 5.2.

(23) Give an example of a two-dimensional integral which cannot be written in the form (5.26).

(24) Consider the integral

$$\int_{a_k}^{b_k} \cdots \int_{u_2(x_3,\ldots,x_k)}^{v_2(x_3,\ldots,x_k)} \int_{u_1(x_2,\ldots,x_k)}^{v_1(x_2,\ldots,x_k)} f(x_1, \ldots, x_k)\,dx_1 \cdots dx_k.$$

Provide a transformation so that the integration is over the rectangle $\prod_{j=1}^{k}(a_j, b_j)$. Be sure to state the conditions necessary for its applicability.

(25) In Example 5.3 compute $P(Y > 5/4)$ using product rules.

(26) In Example 5.4 determine the exact form of the product rule on the unit cube; i.e. determine the points and weights in terms of transformed Jacobi rules, and the exact form of the rule when transformed to the simplex.

(27) Obtain a product spherical rule on S^3 that will exactly integrate all polynomials of degree 3 or less.

(28) Obtain the rate of convergence of a product trapezoid rule on $[0, 1]^3$ for an integrand of appropriate smoothness where each component trapezoid rule is compounded n times.

(29) Prove that there are $\binom{k+d}{k}$ distinct monomials of degree less than or equal to d and that they are linearly independent.

(30) Find an example of three points in R^2 and a function f such that no linear combination of 1, x_1 and x_2 interpolates f at these points.

(31) Prove that the set of all polynomials invariant under a group is a linear space. Consider the space of all polynomials on R^2 of degree less than or equal to 2. Determine the dimension of this space. Let G be the group consisting of the identity and reflection through the origin. Determine the dimension of the space of all G-invariant polynomials of degree less than or equal to 2.

(32) Prove that if f follows a Gaussian process as described in Example 5.10 then

$$\int_{R^k} f(x)w(x)\,dx \sim N\left(\int_{R^k} m(x)w(x)\,dx, \int_{R^k}\int_{R^k} v(x, y)w(x)w(y)\,dx\,dy\right)$$

holds when w is the indicator of a bounded rectangle. Hint: consider the distribution of a Riemann sum and what happens to this distribution as the mesh goes to 0.

(33) Prove that an equidistributed sequence $x_0, x_1, \ldots$ in $[0, 1]$ also satisfies

$$\lim_{n\to\infty} \frac{1}{n+1}\sum_{i=0}^{n} f(x_i) = \int_0^1 f(x)\,dx$$

for any Riemann integrable f.

(34) Prove that the error in the approximation of Theorem 5.19 is $O(1/n)$ for any trigonometric polynomial.

(35) Prove that for any irrational z_0 the sequence $\{nz_0\}$ is dense in $[0, 1]$.

(36) For any sequence $x_1, x_2, \ldots \in [0, 1]^k$ prove that $D_n^* \le D_n \le 2^k D_n^*$.

(37) Show that the transformation $x_i = y_i - (\sin 2\pi y_i)/2\pi$ for $i = 1, \ldots, k$ periodizes the integrand in $\int_{[0,1]^k} f(x)\, dx$.

(38) Suppose that L is a lattice in R^k. Show that $L \cap [0, 1]^k$ is a subgroup of $[0, 1]^k$ using addition modulo 1. Show that it is a finite subgroup whenever $L \subset \mathbb{Q}^k$.

(39) Establish the expressions given for c_0 in Section 5.10.1.

6
Independent importance sampling

6.1 Introduction

We use the term *independent importance sampling* to refer to what might be considered the most straightforward Monte Carlo integration technique; namely generating independent values $x_1, \ldots, x_n$ from a density w and then taking a weighted average of the integrand evaluated at the sampled points with the weights given by the reciprocal of w evaluated at the sample values. With some minor restrictions there is great freedom in the choice of w and the term "importance sampling" refers to choosing w so that the sampled values lie in the region that is important for the value of the integral; i.e. not in a region where the integrand is negligible. *Crude sampling* usually refers to the use of some standard density w, such as a uniform density on a compact domain of integration. As we will see, a well-chosen *importance sampler* w can lead to greatly improved estimates of an integral when compared to the nominal choice. We discuss the importance sampling algorithm in Section 6.2 and in Section 6.3 we discuss the construction of importance samplers.

Importance sampling is sometimes viewed as a variance reduction technique. However, we view the selection of the importance sampler as only the first step in the construction of a Monte Carlo integration algorithm. After it has been selected according to criteria that we describe, the next step is to use a variety of techniques, which we refer to as *variance reduction methods*, to make the estimates as accurate as possible. We detail a fairly complete catalogue of variance reduction methods in Section 6.4. In Section 6.5 we discuss adaptive importance sampling; i.e. the iterative determination of an importance sampler.

Other sources containing discussion of importance sampling include Hammersley and Handscomb (1964), Rubinstein (1981), Bratley, Fox and Schrage (1983), Ripley (1987), Geweke (1989), Wolpert (1990) and Fishman (1996). This chapter discusses importance sampling in the context of generating independent and identically distributed variates from the importance sampler. More general situations where importance sampling is appropriate, such as approximating the expectation of a real-valued function of a Markov process, are described in Glynn and Iglehart (1989). Also see the discussion in Chapter 7 where importance sampling is considered in the context of Markov chain Monte Carlo. For historical interest, one of the earliest papers on Monte Carlo methods is Metropolis and Ulam (1949).

6.2 The independent importance sampling algorithm

Consider the approximation of the integral

$$I(f) = \int_{R^k} f(x)\,dx \tag{6.1}$$

where f is integrable. The method of importance sampling requires the specification of a density function w known as an importance sampler that satisfies $\text{supp}(f) \subset \text{supp}(w)$. We can then write (6.1) as

$$E_w\left[\frac{f}{w}\right] = \int_{R^k} \frac{f(x)}{w(x)} w(x)\,dx;$$

i.e. the original integral is written as an expectation with respect to the density w. As an additional requirement, we suppose that we have a reasonably efficient algorithm for generating from w. Then based on a sample $x_1, \ldots, x_n$ from w, the importance sampling estimator of (6.1) is given by

$$\hat{I}_{w,n} = \frac{1}{n}\sum_{i=1}^{n} \frac{f(x_i)}{w(x_i)}. \tag{6.2}$$

Note that $\hat{I}_{w,n}$ takes the form of a multiple quadrature rule where $R_n = \{(x_i, w_i) : i = 1, \ldots, n\}$ and $w_i = (nw\,(x_i))^{-1}$. Of course, a significant difference between this rule and those of Chapter 5 is that the points and weights are randomly determined.

The basic justification for (6.2) as an estimator is the strong law of large numbers which implies that

$$P\left(\lim_{n\to\infty} \hat{I}_{w,n} = I(f)\right) = 1;$$

e.g. see Durrett (1991). Therefore for large values of n we expect $\hat{I}_{w,n}$ to be a good approximation to I. To assess how large n has to be for the approximation to be adequate we have to examine the probability distribution of the error

$$\left|\hat{I}_{w,n} - I(f)\right|$$

which is in general very difficult to do. This distribution can be approximated, however, via the central limit theorem when $\hat{I}_{w,n}$ has finite variance; i.e.

$$\text{Var}_w\left[\hat{I}_{w,n}\right] = E_w\left[\left(\hat{I}_{w,n} - I(f)\right)^2\right] = \frac{\sigma_w^2}{n} < \infty$$

where

$$\sigma_w^2 = \text{Var}_w\left[\frac{f}{w}\right] = \int_{R^k}\left(\frac{f(x)}{w(x)} - I(f)\right)^2 w(x)\,dx = \int_{R^k} \frac{f^2(x)}{w(x)}\,dx - I^2(f). \tag{6.3}$$

We have the following result where $X_n \to^D X$ means that the sequence of random variables $\{X_n\}$ converges in distribution to the random variable X.

Theorem 6.1 If the importance sampler w is such that $\sigma_w^2 < \infty$ then

$$\frac{\hat{I}_{w,n} - I(f)}{\sigma_w/\sqrt{n}} \to^D Z$$

as $n \to \infty$ where $Z \sim N(0, 1)$.

Proof: This is just the central limit theorem; e.g. see Durrett (1991).

For sequences $\{X_n\}, \{Y_n\}$ of random variables we write $X_n = O_p(Y_n)$ if for any $\epsilon > 0$ there exists $M_\epsilon > 0$, $N_\epsilon > 0$ such that

$$P\left(\left|\frac{X_n}{Y_n}\right| > M_\epsilon\right) < \epsilon$$

for all $n \geq N_\epsilon$; i.e. the sequence X_n/Y_n is *bounded in probability*. Then Theorem 6.1 implies that

$$\hat{I}_{w,n} - I(f) = O_p\left(\frac{1}{\sqrt{n}}\right).$$

Since $1/\sqrt{n} \to 0$ as $n \to \infty$, we say that Monte Carlo estimates converge at the rate $1/\sqrt{n}$. Note that this requires the finiteness of the variance of the estimator. This is obviously a very slow rate of convergence but it has the apparent virtue of being independent of the dimension k.

While Theorem 6.1 tells us something about the rate of convergence it doesn't really tell us how good a particular estimate $\hat{I}_{w,n}$ is. Intuitively, we know that the interval $\hat{I}_{w,n} \pm 3\sigma_w/\sqrt{n}$ contains the value $I(f)$ with virtual certainty when n is large and so the half-length $3\sigma_w/\sqrt{n}$ indicates how accurate $\hat{I}_{w,n}$ is. Since we don't know σ_w, or even an upper bound on this quantity, we must estimate it. A natural estimator of σ_w^2 is

$$s_w^2 = \frac{1}{n}\sum_{i=1}^{n}\left(\frac{f(x_i)}{w(x_i)} - \hat{I}_{w,n}\right)^2.$$

We have the following result.

Theorem 6.2 If the importance sampler w is such that $\sigma_w^2 < \infty$ then

$$\frac{\hat{I}_{w,n} - I(f)}{s_w/\sqrt{n}} \to^D Z$$

as $n \to \infty$ where $Z \sim N(0, 1)$.

Proof: By the strong law of large numbers we have that $s_w^2 \to \sigma_w^2$ with probability 1 as $n \to \infty$. Then by Theorem 6.1 and Slutsky's theorem (see Serfling 1980) the result follows.

So we can quote the half-length $3s_w/\sqrt{n}$ as an estimate of how much error there is in the approximation $\hat{I}_{w,n}$. Again this has the interpretation that the interval $\hat{I}_{w,n} \pm 3s_w/\sqrt{n}$ contains the value $I(f)$ with virtual certainty when n is large.

It is important to note that these results tell us nothing about how large n has to be for the approximations to make sense. Particularly important is that s_w be a good approximation to σ_w. If a poor importance sampler is chosen then s_w can be very misleading, even with very large sample sizes. To illustrate the basic problem consider the following example.

Example 6.1
Suppose that f is the density of the $N(c, 1)$ density and w is the $N(0, 1)$ density. Obviously $I(f) = 1$ and

$$\sigma_w^2 = \int_{-\infty}^{\infty} \frac{f^2(x)}{w(x)}\,dx - 1 = \exp(c^2) - 1.$$

In Table 6.1 we provide some values of σ_w^2 and sample sizes n required for $3\sigma_w/\sqrt{n} \leq 0.01$. From this table we observe that even a modest deviation in the importance sampler from where the integrand f has most of its mass, e.g. $c = 1$, leads to a barely acceptable sample size in order to obtain two-decimal accuracy. If the deviation is more serious, e.g. $c = 5$, then the sample sizes required are completely unacceptable in terms of current technology.

This example also illustrates a more serious concern with importance sampling, however. In Table 6.2 we present the results of a simulation where we estimate $I(f)$, σ_w^2 and $3\sigma_w/\sqrt{n}$ for various choices of n and c. From this we see that the results of a simulation can be very misleading. Note that when $c = 5$, the error in the approximation, as measured by $3s_w/\sqrt{n}$, appears to be decreasing with increasing n and this is incorrect because the estimate of σ_w^2 is very wrong. Even when $c = 2$ and $n = 10^6$ there is a serious underestimation of the amount of error in the approximation $\hat{I}_{w,n}$.

This phenomenon occurs because the importance sampler virtually never puts points in the region necessary for the accurate evaluation of

$$\int_{-\infty}^{\infty} \frac{f^2(x)}{w(x)}\,dx.$$

Table 6.1. Values of variances and minimal sample sizes to achieve two decimal place accuracy in Example 6.1.

c	σ_w^2	n
1	1.718×10^0	1.5465×10^5
2	5.360×10^1	4.824×10^6
5	7.200×10^{10}	6.48×10^{15}

Table 6.2. Simulation results for Example 6.1.

c	n	$\hat{I}_{w,n}$	s_w^2	$3s_w/\sqrt{n}$
1	10^3	0.918	0.927	0.0913
1	10^4	1.013	1.919	0.042
1	10^5	1.004	1.734	0.012
1	10^6	1.002	1.733	0.004
2	10^3	0.651	3.152	0.168
2	10^4	1.083	64.097	0.240
2	10^5	1.006	47.618	0.065
2	10^6	1.007	46.701	0.021
5	10^3	0.008	0.020	0.014
5	10^4	0.591	1326.530	1.092
5	10^5	0.541	9887.605	0.943
5	10^6	0.492	6816.608	0.248

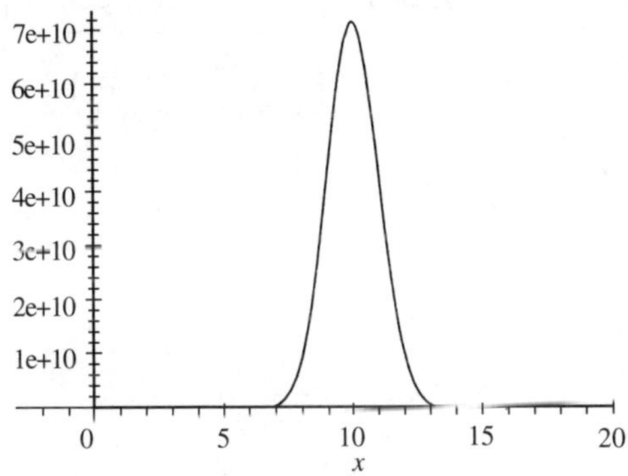

Fig. 6.1. Plot of f^2/w when $c = 5$ in Example 6.1.

To see this we have plotted the integrand f^2/w in Fig. 6.1 based on $c = 5$. Most of the mass for the integrand lies in (7, 13) while most of the mass for the importance sampler lies in $(-3, 3)$.

Of course we want to avoid the phenomenon demonstrated in Example 6.1. We discuss general strategies to ensure a reasonable choice of importance sampler in Section 6.3. We note, however, that the problem becomes increasingly difficult to avoid as the dimension grows. In Example 6.1 it is obvious that the importance sampler is inadequate and we can see this by simply graphing various functions. Still, it is not always easy to find an appropriate range to graph a particular function even when $k = 1$. Assessing the adequacy of an importance sampler becomes increasingly difficult as the dimension increases. Therefore, even though the expression for the variance does not reflect it, there is a dimensional effect involved with importance sampling.

Perhaps a more sensible method for estimating the error in $\hat{I}_{w,n}$, at least when $E_w[\hat{I}_{w,n}] = I(f) \neq 0$, is to use the coefficient of variation

$$\mathrm{CV}_w\left[\hat{I}_{w,n}\right] = \frac{\mathrm{SD}_w\left[\hat{I}_{w,n}\right]}{E_w\left[\hat{I}_{w,n}\right]} = \frac{1}{\sqrt{n}}\frac{\mathrm{SD}_w\left[\frac{f}{w}\right]}{E_w\left[\hat{I}_{w,n}\right]} = \frac{1}{\sqrt{n}}\mathrm{CV}_w\left[\frac{f}{w}\right]$$

because this is scale invariant. Naturally this is estimated by

$$\frac{1}{\sqrt{n}}\frac{s_w}{\hat{I}_{w,n}}.$$

The square of this quantity has an interesting interpretation whenever $f \geq 0$ as

$$\begin{aligned}\left(\frac{1}{\sqrt{n}}\frac{s_w}{\hat{I}_{w,n}}\right)^2 &= \frac{1}{n}\left[\frac{\frac{1}{n}\sum_{i=1}^{n}\left(\frac{f(x_i)}{w(x_i)}\right)^2 - \left(\frac{1}{n}\sum_{i=1}^{n}\frac{f(x_i)}{w(x_i)}\right)^2}{\left(\frac{1}{n}\sum_{i=1}^{n}\frac{f(x_i)}{w(x_i)}\right)^2}\right] \\ &= \frac{1}{n}\left[n\sum_{i=1}^{n}w_{*i}^2 - 1\right] \qquad (6.4)\end{aligned}$$

where

$$w_{*i} = \frac{\frac{f(x_i)}{w(x_i)}}{\sum_{i=1}^{n}\frac{f(x_i)}{w(x_i)}}$$

is the i-th normalized weight; i.e. $0 \leq w_{*i} \leq 1$ and $\sum_{i=1}^{n} w_{*i} = 1$. The value w_{*i} is a measure of the significance of the point x_i in terms of its effect on the estimate $\hat{I}_{w,n}$. Since (6.4) is greater than or equal to 0 we have that

$$\frac{1}{n} \leq \sum_{i=1}^{n} w_{*i}^2 \leq 1.$$

Further $\sum_{i=1}^{n} w_{*i}^2 = 1/n$ if and only if each $w_{*i} = 1/n$ and $\sum_{i=1}^{n} w_{*i}^2 = 1$ if and only of only one of the $w_{*i} = 1$ and the rest are 0 (Exercise 1). So a sensible diagnostic in this case is to compute $\sum_{i=1}^{n} w_{*i}^2$ and see if it is close to 1 as this indicates a few large weights and thus a failure of importance sampling. It is worth noting that (6.4) demonstrates that really nothing more can be learned about the success of the importance sampling by looking at the relative sizes of the weights beyond what is learned from the coefficient of variation of the estimate; e.g. a few big weights implies a high coefficient of variation. Recall also that $\sum_{i=1}^{n} w_{*i}^2 \approx 1/n$ does not say that the importance sampling has worked, as this can occur if we only sample on a subregion where f is relatively constant. When f is not of constant sign then it makes sense to consider the adequacy of w in estimating the integral of $|f|$ (see Theorem 6.5). Accordingly this discussion is relevant in the general case as well.

The above discussion points to the importance of ensuring that σ_w^2 is finite. Therefore in the case where supp(f) is unbounded we want to choose w so that the tails of $|f|$ go to 0 faster than those of w. In the case where $|f|$ is unbounded in a neighborhood of some point then we want $|f|$ to become infinite near such a point at a rate slower than w. Strictly speaking we should always prove that this is the case for any importance sampler we are using so that we guarantee $\sigma_w^2 < \infty$.

One method for guaranteeing that σ_w^2 is finite is to choose a w that satisfies $|f| \le cw$ for some constant c. Of course, this is very difficult to do in general. Suppose, however, that $|f| \le cw$. It is then possible to use an alternative integration algorithm to importance sampling, known as hit-or-miss integration, discussed in Section 3.5. We have the following result.

Theorem 6.3 If $|f| \le cw$ and we denote by $\hat{I}_n$ the hit-or-miss estimator of $I(f)$ based on a sample of n from w, then

$$\operatorname{Var}\left[\hat{I}_{w,n}\right] \le \operatorname{Var}\left[\hat{I}_n\right].$$

Proof: From (6.3) we have that

$$\begin{aligned}
\operatorname{Var}\left[\hat{I}_{w,n}\right] &= \frac{1}{n}\left[\int_{R^k} \frac{f^2(x)}{w(x)}\,dx - \left(\int_{R^k} f(x)\,dx\right)^2\right] \\
&\le \frac{1}{n}\left[c\int_{R^k} |f(x)|\,dx - \left(\int_{R^k} f(x)\,dx\right)^2\right] \\
&= \operatorname{Var}\left[\hat{I}_n\right]
\end{aligned}$$

by Theorem 3.6.

As a consequence of this result hit-or-miss integration should not be considered for integration problems.

In many applications, a practitioner is faced with approximating integrals where the integrand takes the form mf, with $f \ge 0$ and integrable, and there are a number of functions m of interest that are relatively simple with respect to f. By the "simplicity" of a function m with respect to f we mean that, if we could generate from the density proportional to f, i.e. using the importance sampler

$$w(x) = \frac{f(x)}{\int_{R^k} f(z)\,dz},$$

then the integral

$$\int_{R^k} m(x) f(x)\, dx$$

is adequately estimated by

$$\frac{1}{n}\sum_{i=1}^{n} \frac{m(x_i) f(x_i)}{w(x_i)} = \left(\int_{R^k} f(x)\, dx\right)\left(\frac{1}{n}\sum_{i=1}^{n} m(x_i)\right)$$

for computationally reasonable values of n. Specific examples of such functions m would be low degree polynomials or indicator functions for sets having probability contents that are not close to 0 with respect to the probability density given by the normalized f. This presupposes that we can evaluate $\int_{R^k} f(z)\, dz$, the inverse normalizing constant of f, which is not always the case. If in fact f is a density then $\int_{R^k} f(z)\, dz = 1$ and we are simply averaging m. Suppose then that we are in a context where m is simple with respect to f but we do not have an algorithm for generating from the density proportional to f. Since there are many m functions, it makes sense to find an importance sampler w that is good for approximating $\int_{R^k} f(z)\, dz$ and then using this w for all integrals for the functions of interest. If there is a function m of interest which is not relatively simple with respect to f then choosing an importance sampler in this way is not appropriate.

Actually, for the situation just described, we are often really interested in approximating the ratio of integrals

$$R = \frac{\int_{R^k} m(x) f(x)\, dx}{\int_{R^k} f(x)\, dx} = \frac{I(mf)}{I(f)}$$

which is just the expectation of m with respect to the density proportional to f. For example, this is the situation described in Example 1.3 where we discussed applications to Bayesian statistics. Suppose that we can neither generate from the density proportional to f nor evaluate the inverse normalizing constant. The natural estimator to use is then

$$\hat{R}_{w,n} = \frac{\sum_{i=1}^{n} \frac{m(x_i) f(x_i)}{w(x_i)}}{\sum_{i=1}^{n} \frac{f(x_i)}{w(x_i)}} \tag{6.5}$$

where $x_1, \ldots, x_n$ comprise a sample from w. Again, the strong law of large numbers implies that $\hat{R}_{w,n} \to R$ with probability 1 as $n \to \infty$. The following result establishes how we evaluate the error in (6.5).

Theorem 6.4 If $I(f) \neq 0$ and the importance sampler w is such that

$$\mathrm{Var}_w \left[\frac{f}{w}\right] < \infty$$

and

$$\mathrm{Var}_w\left[\frac{mf}{w}\right] < \infty$$

then

$$\frac{\hat{R}_{w,n} - R}{\sigma_{R,w}/\sqrt{n}} \to^D Z$$

as $n \to \infty$ where $Z \sim N(0, 1)$ and

$$\sigma^2_{R,w} = I^{-2}(f)\Big\{\mathrm{Var}_w\left[\frac{mf}{w}\right] + I^2(mf)I^{-2}(f)\mathrm{Var}_w\left[\frac{f}{w}\right] - 2I(mf)I^{-1}(f)\,\mathrm{Cov}_w\left[\frac{mf}{w}, \frac{f}{w}\right]\Big\}.$$

In addition, if $I(mf) \neq 0$ then we can write

$$\sigma^2_{R,w} = R^2\Big\{\mathrm{CV}^2_w\left[\frac{mf}{w}\right] + \mathrm{CV}^2_w\left[\frac{f}{w}\right] - 2\mathrm{CV}_w\left[\frac{mf}{w}\right]\mathrm{CV}_w\left[\frac{f}{w}\right]\mathrm{Corr}_w\left[\frac{mf}{w}, \frac{f}{w}\right]\Big\}.$$

Proof: Let N_n and D_n denote the numerator and denominator sums in (6.5). Then by the multivariate central limit theorem (see Durrett 1991) we have that

$$\frac{1}{n}\begin{pmatrix} N_n \\ D_n \end{pmatrix} \sim AN_k\left(\begin{pmatrix} \int_{R^k} m(x)f(x)\,dx \\ \int_{R^k} f(x)\,dx \end{pmatrix}, \frac{1}{n}\Sigma\right)$$

where

$$\Sigma = \begin{pmatrix} \sigma_{11} & \sigma_{12} \\ \sigma_{12} & \sigma_{22} \end{pmatrix}$$

$$\sigma_{11} = \mathrm{Var}_w\left[\frac{mf}{w}\right] = \int_{R^k} \frac{m^2(x)f^2(x)}{w(x)}\,dx - \left(\int_{R^k} m(x)f(x)\,dx\right)^2$$

$$\sigma_{12} = \mathrm{Cov}_w\left[\frac{mf}{w}, \frac{f}{w}\right] = \int_{R^k} \frac{m(x)f^2(x)}{w(x)}\,dx - \left(\int_{R^k} m(x)f(x)\,dx\right)\left(\int_{R^k} f(x)\,dx\right)$$

$$\sigma_{22} = \mathrm{Var}_w\left[\frac{f}{w}\right] = \int_{R^k} \frac{f^2(x)}{w(x)}\,dx - \left(\int_{R^k} f(x)\,dx\right)^2$$

and $X_n \sim AN_k(\mu_n, \Sigma_n)$ means $\Sigma_n^{-1/2}(X_n - \mu_n) \to^D N_k(0, I)$ as $n \to \infty$. Now let $h : R^1 \times \left(R^1\backslash\{0\}\right) \to R^1$ be given by $h(x, y) = x/y$. Then by the delta theorem (Serfling 1980, Theorem 3.3A) we have that

$$\hat{R}_{w,n} = h(n^{-1}N_n, n^{-1}D_n) \sim AN\left(R, \frac{1}{n}\sigma^2_{R,w}\right)$$

as $n \to \infty$ where

$$\sigma^2_{R,w} = \begin{pmatrix} \dfrac{\partial h\left(\int_{R^k} m(z)f(z)\,dz, \int_{R^k} f(z)\,dz\right)}{\partial x} \\ \dfrac{\partial h\left(\int_{R^k} m(z)f(z)\,dz, \int_{R^k} f(z)\,dz\right)}{\partial y} \end{pmatrix}^t$$
$$\times\, \Sigma \begin{pmatrix} \dfrac{\partial h\left(\int_{R^k} m(z)f(z)\,dz, \int_{R^k} f(z)\,dz\right)}{\partial x} \\ \dfrac{\partial h\left(\int_{R^k} m(z)f(z)\,dz, \int_{R^k} f(z)\,dz\right)}{\partial y} \end{pmatrix}$$

and this is easily verified (Exercise 2) to give the expression in the theorem for $\sigma^2_{R,w}$.

To estimate $\sigma^2_{R,w}$ we simply use the sample analogs of the quantities in its definition to obtain $\hat{\sigma}^2_{R,w}$. Then, as with Theorem 6.2, $\hat{R}_{w,n} \pm 3\hat{\sigma}_{R,w}$ will contain the exact value of R with virtual certainty provided n is large enough.

Now suppose that f/w dominates the behavior of mf/w in the sense that m is approximately constant in the region where most of the mass for w lies. For example, this can happen when w is a poor importance sampler for the problem, concentrating its mass in an inappropriate region for the integrand mf. Then $mf/w \approx bf/w$ in this region, $\text{Corr}_w[mf/w, f/w] \approx \pm 1$ with $\text{CV}_w[mf/w]$ of the same sign and therefore we have

$$\sigma^2_{R,w} \approx R^2 \left(\left| \text{CV}_w\left[\frac{mf}{w}\right] \right| - \left| \text{CV}_w\left[\frac{f}{w}\right] \right| \right)^2$$

which shows that the absolute coefficient of variation for estimating the ratio should be much smaller than $\text{CV}_w[mf/w]$ or $\text{CV}_w[f/w]$. So in a sense the ratio estimator partially compensates for a poor importance sampler.

The *sampling/importance resampling (SIR) algorithm* was introduced by Rubin (1988) as a method for approximately generating variates from the density proportional to $|f|$ using importance sampling. This proceeds by generating a sample $x_1, \ldots, x_n$ from w and then resampling with replacement from this sample, a sample $x_1^*, \ldots, x_{n^*}^*$ using the probability distribution on $\{x_1, \ldots, x_n\}$ with probabilities $p_1, \ldots, p_n$ where

$$p_i = \frac{\dfrac{|f(x_i)|}{w(x_i)}}{\sum_{j=1}^n \dfrac{|f(x_j)|}{w(x_j)}}.$$

Consider the case $k = 1$ and denote the distribution function of a resampled random variate by F_n^*. We then have that

$$F_n^*(x) = \sum_{\{j:\ x_j \le x\}} p_j$$

and by the strong law of large numbers

$$F_n^*(x) \to \frac{\int_{-\infty}^{x} |f(z)|\, dz}{\int_{-\infty}^{\infty} |f(z)|\, dz}$$

as $n \to \infty$. This implies that for large n the points $x_1^*, \ldots, x_{n^*}^*$ can be considered as an approximate sample of n^* from the density proportional to $|f|$. Obviously this result also holds when $k > 1$.

The SIR algorithm has also been proposed as a method of integration when we face a problem of the form (6.5) where $f \ge 0$; namely approximate R by

$$\hat{I}_{\text{SIR}}(m) = \frac{1}{n^*} \sum_{i=1}^{n^*} m\left(x_i^*\right).$$

Now

$$E\left[\hat{I}_{\text{SIR}}(m) \middle| x_1, \ldots, x_n\right] = \sum_{i=1}^{n} p_i m\left(x_i\right) = \hat{R}_{w,n}$$

and so

$$\begin{aligned}
\text{Var}\left[\hat{I}_{\text{SIR}}(m)\right] &- \text{Var}\left[E\left[\hat{I}_{\text{SIR}}(m) \middle| x_1, \ldots, x_n\right]\right] \\
&+ E\left[\text{Var}\left[\hat{I}_{\text{SIR}}(m) \middle| x_1, \ldots, x_n\right]\right] \\
&= \text{Var}\left[\hat{R}_{w,n}\right] + E\left[\text{Var}\left[\hat{I}_{\text{SIR}}(m) \middle| x_1, \ldots, x_n\right]\right] \\
&\ge \text{Var}\left[\hat{R}_{w,n}\right].
\end{aligned}$$

Therefore, while it is an interesting, and perhaps useful, point that $x_1^*, \ldots, x_{n^*}^*$ is an approximate sample from the density proportional to f, as an integration technique SIR is worse than importance sampling based on w. This was noted in Evans and Swartz (1995a) and further discussion can be found in Ross (1996).

6.3 Choosing an importance sampler

The choice of the importance sampler w for $I(f)$ in (6.1) is clearly the most significant consideration for a user of importance sampling. From Section 6.2 we see that we want to choose w to make σ_w^2 as small as possible. The best we can do is have $\sigma_w^2 = 0$ and this occurs if and only if $P_w(f(X)/w(X) = c) = 1$ for the constant $c = I(f)$ (Exercise 3). This can never happen if f is not of constant sign but we have the following result.

Theorem 6.5 If $0 < \int_{R^k} |f(z)|\, dz < \infty$ then the choice of w that minimizes the variance σ_w^2 is

$$q(x) = \frac{|f(x)|}{\int_{R^k} |f(z)|\, dz}$$

and the minimum variance is

$$\left(\int_{R^k} |f(x)|\, dx\right)^2 - \left(\int_{R^k} f(x)\, dx\right)^2.$$

Proof: From (6.3) we have that

$$\sigma_w^2 = \left(\int_{R^k} |f(x)|\, dx\right)^2 E_w\left[\left(\frac{q(x) - w(x)}{w(x)}\right)^2\right] + \left(\int_{R^k} |f(x)|\, dx\right)^2 - \left(\int_{R^k} f(x)\, dx\right)^2.$$

Hence minimizing σ_w^2 is equivalent to minimizing the *chi-square distance*

$$E_w\left[\left(\frac{q(x) - w(x)}{w(x)}\right)^2\right] = \int_{R^k} \frac{(q(x) - w(x))^2}{w(x)}\, dx$$

and this is clearly minimized by taking $w = q$.

As an immediate consequence of this result we have the following interesting formula for the coefficient of variation of the importance sampling estimator.

Corollary 6.6 When $I(f) \neq 0$ then

$$\mathrm{CV}_w^2\left[\frac{f}{w}\right] = \frac{\left(\int_{R^k} |f(x)|\, dx\right)^2}{\left(\int_{R^k} f(x)\, dx\right)^2}\left\{E_w\left[\left(\frac{q(x) - w(x)}{w(x)}\right)^2\right] + 1\right\} - 1.$$

Note that the chi-square distance is the expected squared relative error between the two densities. We can obtain similar expressions for the optimal w when we want to estimate R in (6.5); see Hesterberg (1991).

Theorem 6.5 is not really of much help in selecting w although we can carry away the qualitative goal of trying to choose w that is as similar to $|f|$ as possible in terms of relative error. In particular, it is the regions where q is large and where w is small that we have to worry about the most. So overall we try to make sure that w and q are roughly agreeable near the maxima of q and in the tails of w so that the relative error $(q(x) - w(x))/w(x)$ doesn't become too large.

What is needed are families of multivariate importance samplers that can accommodate a wide spectrum of shapes. We also need to be able to efficiently generate variates from these families and fit a particular family member to the problem at hand. A major limitation of the importance sampling approach is that there are few such families. We discuss several of these.

Example 6.2 *Dirichlet importance sampling on the simplex*
Suppose that the domain of integration is the k-simplex

$$S_k = \left\{ x \in R^k \colon x_i \geq 0, i = 1, \ldots, k \text{ and } \sum_{i=1}^{k} x_i \leq 1 \right\}.$$

We note that this can be transformed to the unit cell $[0, 1]^k$ via the transformation discussed in Exercise 2.8.8. Of course, the inverse of this transformation transforms $[0, 1]^k$ onto S_k. We note, however, that no families of distributions on the unit cell seem to exist that suggest themselves as natural candidates for importance sampling while the Dirichlet$(\alpha_1, \ldots, \alpha_{k+1})$ family of distributions is available on S_k. The density of the Dirichlet$(\alpha_1, \ldots, \alpha_{k+1})$ distribution on S_k for $\alpha_1 > 0, \ldots, \alpha_{k+1} > 0$ is given by

$$w(x) = \frac{\Gamma(\alpha_1 + \cdots + \alpha_{k+1})}{\Gamma(\alpha_1) \cdots \Gamma(\alpha_{k+1})} x_1^{\alpha_1 - 1} \cdots x_k^{\alpha_k - 1} \left(1 - \sum_{i=1}^{k} x_i \right)^{\alpha_{k+1} - 1}$$

So if faced with an integral on the unit cell we could transform to S_k and then use Dirichlet importance sampling. Example 3.9 discusses the generation of random values from Dirichlet distributions.

The question then is: how do we select $\alpha_1, \ldots, \alpha_{k+1}$ so that these choices give a good importance sampler for f? Suppose that f is unimodal and bounded on S_k with mode at $\hat{x}$. When $\alpha_i > 1$ for $i = 1, \ldots, k+1$ then the Dirichlet$(\alpha_1, \ldots, \alpha_{k+1})$ density is also bounded and unimodal. The mode of the Dirichlet$(\alpha_1, \ldots, \alpha_{k+1})$ is given by

$$x_i = \frac{\alpha_i - 1}{\sum_{j=1}^{k+1} \alpha_j - (k+1)}$$

provided $\alpha_i \geq 1$ for $i = 1, \ldots, k+1$. To match modes we then have that

$$\alpha_i = 1 + c\hat{x}_i$$

for $i = 1, \ldots, k$ and

$$\alpha_{k+1} = 1 + c \left(1 - \sum_{i=1}^{k} \hat{x}_i \right)$$

for any constant $c > 0$. Allowing $c = 0$ is convenient as this gives the uniform distribution on S_k. The parameter c determines the thickness of the tails of the Dirichlet distribution where small c gives thicker tails, but it also controls the spread of the distribution as well, as large c causes the distribution to concentrate at its mode. If we choose c too large the importance sampling distribution will be too concentrated and the estimates may have infinite variance. If f is bounded away from 0 then we will have to choose c so that $1 \le \alpha_i < 2$ to avoid an infinite variance estimator.

Note that when $\alpha_i < 1$ for some $i \le k$ then the density w becomes infinite as $x_i \to 0$. If $\alpha_{k+1} < 1$ then the density becomes infinite as $\sum_{i=1}^k x_i \to 1$. Accordingly, this family of densities could be used to integrate functions on the simplex which possess this kind of singularity.

Overall the Dirichlet family does not possess much flexibility. After matching modes we have only one degree of freedom left to adjust to the remaining behavior of the integrand. Examples of the use of this approach to approximate integrals on the simplex can be found in Evans and Swartz (1988a). In that paper the parameter c was set by trying different values and monitoring the variance of the importance sampling estimates. Heard and Swartz (1999) contains a high-dimensional example where the diagnostics were useful in demonstrating the failure of Dirichlet importance sampling even though the estimates had apparently "converged".

A more useful family of importance samplers is provided by the following example.

Example 6.3 *Multivariate Student importance sampling on R^k*
There are very few families of distributions in the multidimensional context that are in common use as importance samplers. When the problem is to integrate over R^k virtually the only family used is the multivariate Student family; see Example 3.11 for the definition of the $\text{Student}_k(\lambda, \mu, \Sigma)$ density and for an algorithm for generating from this distribution. Given the flexibility of this family it is generally good advice to try and transform your integration problem so that this family is applicable; i.e. transform so that the domain of integration is R^k.

The $\text{Student}_k(\lambda, \mu, \Sigma)$ density has a single mode at μ, ellipsoidal contours determined by Σ and tail-length determined by λ. The shortest tails are obtained as $\lambda \to \infty$ and a $\text{Student}_k(\lambda, \mu, \Sigma)$ density converges to the $N_k(\mu, \Sigma)$ density as $\lambda \to \infty$. So we think of the $N_k(\mu, \Sigma)$ densities as belonging to this family as well. The tails become progressively longer as $\lambda \to 0$. Note that along the ray $\{\mu + rx : r > 0, x \in R^k\}$ the $\text{Student}_k(\lambda, \mu, \Sigma)$ density goes to 0 like $(1 + cx^2)^{-(\lambda+k)/2}$ where $c = r^t \Sigma^{-1} r/\lambda$ and therefore the tails become shorter as the dimension k rises. This is some indication that the family may not be entirely suitable as dimension rises. Note that the mean of the $\text{Student}_k(\lambda, \mu, \Sigma)$ density exists if and only if $\lambda > 1$ and when it exists it equals μ (Exercise 5). The variance matrix of the $\text{Student}_k(\lambda, \mu, \Sigma)$ density exists if and only if $\lambda > 2$ and when it exists it equals $(\lambda/(\lambda - 2))\Sigma$ (Exercise 6).

For importance sampling we choose a density w in the family to mimic f. Of course f must exhibit roughly the same characteristics as a member of the Student family if this is to be successful and this is not possible if f varies in sign. As already mentioned, however, we are often able to factor the original integrand into mf where f is nonnegative and integrable and m is simple in the sense described in Section 6.2. Let us suppose then that we have $f \ge 0$ and f is unimodal with mode at $\hat{x}$ and has roughly ellipsoidal contours. Then one possibility is to take w as the density of

$$X = \hat{x} + \sqrt{\frac{\lambda - 2}{\lambda}} CZ \tag{6.6}$$

where C is the Cholesky factor of the negative inverse Hessian of $\log f$ at $\hat{x}$; i.e.

$$\hat{\Sigma} = -\left(\frac{\partial^2 \log f(\hat{x})}{\partial x_i \partial x_j}\right)^{-1},$$

and $Z \sim \text{Student}_k(\lambda, 0, I)$ for some $\lambda > 2$. With these choices $X \sim \text{Student}_k(\lambda, \hat{x}, \hat{\Sigma})$. The motivation for this choice is the same as with Laplace's approximation discussed in Chapter 4. This supposes of course that the density proportional to f has tails short enough so that variances exist. The parameter λ is then chosen low enough so that the estimate $\hat{I}_{w,n}$ has finite variance. Basically, the integrands mf have to be studied to determine an appropriate choice of λ and often this is very hard to determine. Alternatively, one can choose to take w as the density of

$$X = \hat{x} + \sqrt{\frac{\lambda + k}{\lambda}} CZ \tag{6.7}$$

with $Z \sim \text{Student}_k(\lambda, 0, I)$ as this ensures that $\log f$ and $\log w$ have the same Hessian matrix at $\hat{x}$. With this choice we can take $\lambda > 0$ as low as we like but note that (6.7) concentrates the probability much more closely about $\hat{x}$ than (6.6).

Attempts at providing more general multivariate importance samplers using variants of the multivariate Student distribution have been considered; for example, see Evans (1988) and Geweke (1989). These are sometimes known as split-t distributions as the multivariate Student is split at the origin providing tails of varying thicknesses.

We next consider an application of multivariate Student importance sampling to a real example.

Example 6.4 *Application of multivariate Student importance sampling*
For the integration problems discussed in Example 4.9 consider using (6.6) as an importance sampler with $\lambda = 5$. Partial results for importance sampling using this w are reported in Table 6.3 for a Monte Carlo sample size of $N = 10^5$. These computations require about 50 seconds of CPU time. The estimates are accurate but the estimated absolute coefficient of variation for $\hat{R}(\theta_4)$ is quite high. As it is difficult to determine a precise value for λ we tried importance sampling using different choices of $\lambda \in (2, 20)$ but the differences in the computational results were not large. Overall this importance sampler was quite successful in this example.

Table 6.3. Importance sampling estimates and estimated absolute coefficients of variation for Example 6.4 where R denotes the exact value and $\hat{R}$ denotes the importance sampling estimate.

i	$R(\theta_i)$	$\hat{R}(\theta_i)$	$R(\theta_i^2)$	$\hat{R}(\theta_i^2)$
1	2.043	2.043 (0.001)	4.263	4.265 (0.002)
2	0.095	0.096 (0.023)	0.081	0.080 (0.014)
4	0.018	0.017 (0.127)	0.069	0.070 (0.015)
10	−0.073	−0.073 (0.017)	0.033	0.033 (0.008)

Example 6.5 *Multivariate Student mixture importance sampling*
When the function f is multimodal on R^k, multivariate Student distributions do not adequately capture the shape of the integrand. Oh and Berger (1993) have modified the approach by considering mixtures of multivariate Student distributions. Suppose then that $f \geq 0$. If f has modes at $\hat{x}_1, \ldots, \hat{x}_l$, then a plausible importance sampler is given by the mixture distribution $p_1 w_1 + \cdots + p_l w_l$ where w_i is the $\text{Student}_k(\lambda_i, \hat{x}_i, ((\lambda_i - 2)/\lambda_i)\Sigma_i)$ density. In this formulation, Σ_i is the inverse Hessian of $-\ln f$ evaluated at $\hat{x}_i$ and the mixture probability p_i is such that $p_i \propto f(\hat{x}_i)/w_i(\hat{x}_i)$. Although the prescription based on Student mixtures is straightforward, the computational problem of finding the modes may prove overwhelming in high dimensional problems. See the response to the comments of Genz *et al.* (1996) for an example of the use of this methodology.

Going beyond the multimodal problem, mixture distributions allow for a variety of shapes and hold considerable promise for resolving some of the current limitations of importance sampling. In general, difficulties with the use of mixtures involve the determination of the number of elements in the mixture distribution and the determination of the parameters. In the approximation of $\int m(x) f(x)\,dx$, Hesterberg (1995) considers "defensive distributions" which are two-component mixtures $\lambda f + (1 - \lambda) g$ where f is a density from which we can sample and g is some other design density. The Hesterberg (1995) formulation is most appropriate in the context of a single integral with a single function m.

In some contexts we may know that a member of a family $\{w_\lambda : \lambda \in \Lambda\}$ performs adequately as an importance sampler but we do not know which one. It may then be possible to choose a mixture $w = p_1 w_{\lambda_1} + \cdots + p_l w_{\lambda_l}$ so that w reasonably covers the region where it is important to place mass. Such methods have been successfully used in a variety of contexts; see for example, Torrie and Valleau (1977) where this technique was introduced under the name *umbrella sampling*. In Section 6.5 we discuss adaptive methods for selecting an importance sampler based on a family $\{w_\lambda : \lambda \in \Lambda\}$.

It is sometimes said that as an integration technique, importance sampling is useful for low to moderate dimensional problems but is unsuitable for high dimensional problems. While there is an element of truth to this it is worth reflecting that this seems to be true of virtually all integration techniques. Further, it is not the case that importance sampling itself is a poor technique in high dimensions but rather that it becomes more difficult to find an appropriate importance sampler as the dimension rises. If an appropriate importance sampler is available, then importance sampling performs well. For example, consider the approximation of the first and second moments of the first coordinate of the $N_k(0, I)$ distribution using a multivariate Student$_k(\lambda, 0, ((\lambda - 2)/\lambda)I)$ importance sampler. We know that the first two moments are equal to 0.0 and 1.0 respectively. Table 6.4 gives the simulation results for various dimensions k. The results were obtained using $\lambda = 3$ and $N = 1000$ generated variates. It is clear from this that there is a deterioration in the results as the dimension rises but importance sampling still performs adequately even when $k = 1000$. In fact, a major problem with all integration methods is determining whether they have worked or failed and this is more difficult to ascertain in high dimensions.

Our rough criteria for choosing an importance sampler has been that w should mimic $|f|$. There are situations where this may not be ideal. For example, suppose that f is a density function and we wish to calculate the probability of an event A which has small probability with respect to this density. If we want an approximation with low relative error then it is clear that an importance sampler that mimics f will not perform adequately as we rarely generate values in A. Therefore, in such cases one should choose an importance sampler that mimics $m(x)f(x)$ rather than f. See Bucklew (1990) for a discussion of some situations where this is the case.

Another aspect to choosing w concerns the speed that variates can be generated from the importance sampler. We can combine these two optimality criteria by considering the efficiency of estimators. We say that the efficiency of importance sampler w_1 compared to importance sampler w_2 is

$$\frac{c_2\sigma_{w_2}^2}{c_1\sigma_{w_1}^2} \tag{6.8}$$

Table 6.4. Importance sampling estimates (with standard errors) based on Student importance sampling with $\lambda = 3$ and $N = 1000$ of the first two moments of a component of the $N_k(0, I)$ distribution.

k	First moment	Second moment	Computing time
10	0.012 (0.017)	1.006 (0.029)	0.7 secs
100	−0.023 (0.028)	0.926 (0.049)	5 secs
1000	0.004 (0.049)	0.921 (0.084)	53 secs

where c_i is the computational time to generate a variate using importance sampler w_i. We see that we would get approximately the same efficiency if we were to base our comparison on n generated values since increasing the number of variates by a factor of n reduces the variances by a factor of $1/n$ and increases the computational times by a factor of approximately n. For many problems the speed of generation is not a major issue; i.e. $c_1 \approx c_2$. However, in computationally intense problems speed of generation can be important and then (6.8) is the relevant quantity to use for comparison purposes. Note that (6.8) may be estimated using actual computational times and the estimated variances.

6.4 Variance reduction methods

Suppose that we want to approximate (6.1) and that we have an importance sampler w. There are many suggestions in the literature as to how one might improve the importance sampling algorithm through the reduction of the variance of the importance sampling estimator. However, very few of these improvements are guaranteed in the sense that they always lead to a variance reduction. Some of these methods, if applied under the wrong circumstances, could in fact increase the variance of the estimator. In this section we consider a number of variance reduction techniques that can be implemented with the importance sampling algorithm. We start with the variance reducers where the variance of the estimate is always guaranteed to be no greater than the variance of the estimator obtained by just using straight importance sampling from w. In such circumstances the extra work required to implement such a procedure could be wasted if there is no real reduction in variance.

6.4.1 DIMENSION REDUCTION

Intuitively, it seems reasonable that if we can integrate out exactly some of the variables of integration via exact methods, such that this doesn't involve us in extensive computation, then we should be able to improve the efficiency of the computation. This turns out to be correct under certain conditions but in general it is difficult to state that this is unequivocally true. In fact, in Chapter 7 we see that sometimes it pays to increase the dimension of the integration problem. The following result, however, gives one set of circumstances where this is indeed the correct approach.

Theorem 6.7 Suppose that we can integrate out the first coordinate of x in the integral, i.e. $\int_{R^k} f(x)\,dx = \int_{R^{k-1}} f_1(x)\,dx$ where $f_1(x_2,\ldots,x_k) = \int_{-\infty}^{\infty} f(x_1,\ldots,x_k)\,dx_1$, and that we can generate from

$$w_1(x_2,\ldots,x_k) = \int_{-\infty}^{\infty} w(x_1,\ldots,x_k)\,dx_1.$$

Then $\sigma_w^2 \geq \sigma_{w_1}^2$ where $\sigma_{w_1}^2$ is the variance associated with approximating $\int_{R^{k-1}} f_1(x)\,dx$ using importance sampling based on w_1.

Proof: We have that

$$\int_{R^k} f(x)\,dx = E_w\left[\frac{f}{w}\right] = E_w\left[E_w\left[f/w|x_2,\ldots,x_k\right]\right]$$

where

$$\begin{aligned} E_w\left[f/w|x_2,\ldots,x_k\right] &= \int_{R^1} \frac{f(x_1,x_2,\ldots,x_k)}{w(x_1,x_2,\ldots,x_k)} w(x_1|x_2,\ldots,x_k)\,dx_1 \\ &= \frac{f_1(x_2,\ldots,x_k)}{w_1(x_2,\ldots,x_k)}. \end{aligned}$$

Therefore

$$\begin{aligned} \sigma_w^2 &= \mathrm{Var}_w\left[\frac{f}{w}\right] = \mathrm{Var}_w\left[E_w\left[f/w|x_2,\ldots,x_k\right]\right] + E_w\left[\mathrm{Var}_w\left[f/w|x_2,\ldots,x_k\right]\right] \\ &\geq \mathrm{Var}_w\left[E_w\left[f/w|x_2,\ldots,x_k\right]\right] = \sigma_{w_1}^2 \end{aligned}$$

which establishes the result.

Exactly calculating some portion of an integral may first require a transformation of the integration problem. If we have an algorithm for generating from w there is no guarantee that we will have an algorithm for generating from the marginal density obtained after transforming and integrating out the variable(s) in question. It is this fact that prevents the method from being a generally useful variance reducer.

This technique is sometimes called *conditional Monte Carlo* (see, for example, Bratley, Fox and Schrage 1983), as it involves the explicit evaluation of a conditional expectation. On the other hand, this terminology is quite often used in the literature for a Monte Carlo technique concerned with evaluating conditional expectations; see Exercise 39. For this reason we have chosen to call the method of this section *dimension reduction* because that is its primary use in integration problems. It should be noted, however, that if we were able to evaluate $\int_D f(x)\,dx$ and $p = \int_D w(x)\,dx > 0$ for some $D \subset R^k$ and also generate from w conditioned to D^c then this gives a variance reduction although it is not a reduction in dimension (Exercise 12).

6.4.2 STRATIFIED SAMPLING

The technique of *stratified sampling* is based on a decomposition of the importance sampler w as

$$w = p_1 w_1 + \cdots + p_l w_l$$

where $w_1, \ldots, w_l$ are densities for which we have generating algorithms, with $\text{supp}(w) = \cup_{i=1}^{l} \text{supp}(w_i)$, $\text{supp}(w_i) \cap \text{supp}(w_j) = \emptyset$ when $i \neq j$ and

$$p_i = \int_{\text{supp}(w_i)} w(x)\,dx$$

for $i = 1, \ldots, l$. The sets supp (w_i) are the *strata* and of course $\sum_{i=1}^{l} p_i = 1$. Stratified sampling occurs when we sample a predetermined number n_i times from w_i and then estimate

$$\int_{\text{supp}(w_i)} f(x)\,dx$$

by $\hat{I}_{w_i,n_i}$. The estimate of $I(f)$ in (6.1) is then given by

$$\hat{I}_{\text{strat}} = \sum_{i=1}^{l} \hat{I}_{w_i,n_i}.$$

By the strong law of large numbers $\hat{I}_{\text{strat}}$ converges almost surely to $\int_{R^k} f(x)\,dx$ as $\min(n_1, \ldots, n_l) \to \infty$. Clearly we have that

$$\begin{aligned}\sigma^2_{\text{strat}}(n_1, \ldots, n_l) &= \sum_{i=1}^{l} \frac{1}{n_i} \text{Var}_{w_i}\left[\frac{f}{w_i}\right] = \sum_{i=1}^{l} \frac{\sigma^2_{w_i}}{n_i} \\ &= \sum_{i=1}^{l} \frac{1}{n_i}\left[\int_{\text{supp}(w_i)} \frac{f^2(x)}{w_i(x)}\,dx - \left(\int_{\text{supp}(w_i)} f(x)\,dx\right)^2\right]\end{aligned}$$

and we estimate this using

$$s^2_{\text{strat}} = \sum_{i=1}^{l} \frac{s^2_{w_i}}{n_i}$$

where $s^2_{w_i}$ is the sample estimate of $\sigma^2_{w_i}$. Provided that each $\sigma^2_{w_i} < \infty$ we have that

$$\frac{\hat{I}_{\text{strat}} - I(f)}{s_{\text{strat}}} \to^{D} Z$$

where $Z \sim N(0, 1)$ as $\min(n_1, \ldots, n_l) \to \infty$ (Exercise 13).

The idea behind stratified sampling is that we can allocate sample resources more carefully by deciding ahead of time to place more sample values in regions where f/w is highly variable; i.e. where the importance sampler w is not a good approximation to f. Stratified sampling may be viewed as a "divide and conquer" strategy. The following result shows that it is possible for a variance reduction to be achieved, and definitely to do no worse than importance sampling based on w, if we allocate the $n = \sum_{i=1}^{l} n_i$ sample values appropriately amongst the strata.

Theorem 6.8 (*Tschuprow–Neyman allocation*) For a given overall sample size n the variance $\sigma^2_{\text{strat}}(n_1, \ldots, n_l)$ is minimized by taking

$$n_i = n \frac{\sigma_{w_i}}{\sum_{j=1}^{l} \sigma_{w_j}}$$

and the minimized variance is

$$\frac{\left(\sum_{j=1}^{l} \sigma_{w_j}\right)^2}{n} \leq \frac{\sigma_w^2}{n}.$$

Proof: Using the Cauchy–Schwarz inequality together with the vectors $\left(\sqrt{n_1/n}, \ldots, \sqrt{n_l/n}\right)$ and $\left(\sigma_{w_1}/\sqrt{n_1}, \ldots, \sigma_{w_l}/\sqrt{n_l}\right)$ we have that

$$\sum_{i=1}^{l} \frac{\sigma_{w_i}^2}{n_i} \geq \frac{\left(\sum_{j=1}^{l} \sigma_{w_j}\right)^2}{n}$$

with equality if and only if for some constant c we have $\sqrt{n_i/n} = c\sigma_{w_i}/\sqrt{n_i}$ or $n_i = c\sqrt{n}\sigma_{w_i}$ for $i = 1, \ldots, l$ and we see that this implies that $c = \sqrt{n}/\sum_{j=1}^{l} \sigma_{w_j}$.

Finally generating $X \sim w$ is equivalent to generating j with probability p_j for $j = 1, \ldots, l$ and then generating $X \sim w_j$. Therefore

$$\begin{aligned}
\sigma_w^2 &= E[\text{Var}[f(X)/w(X)|j]] + \text{Var}[E[f(X)/w(X)|j]] \\
&= \sum_{i=1}^{l} \frac{\sigma_{w_i}^2}{p_i} + \text{Var}[E[f(X)/w(X)|j]] \\
&\geq \left(\sum_{j=1}^{l} \sigma_{w_j}\right)^2 \qquad (6.9)
\end{aligned}$$

where the final inequality follows from the Cauchy–Schwarz inequality applied to the vectors $\left(\sqrt{p_1}, \ldots, \sqrt{p_l}\right)$, $\left(\sigma_{w_1}/\sqrt{p_1}, \ldots, \sigma_{w_m}/\sqrt{p_l}\right)$ (to bound the first sum) and $\text{Var}[E[f(X)/w(X)|j]] \geq 0$.

Note that the inequality (6.9) establishes that the variance of stratified sampling with the Tschuprow–Neymann allocation is at least as small as the variance of importance sampling based on w. Now of course the n_i of Theorem 6.8 are not necessarily integers so the result is not strictly applicable. The idea, however, is to choose the n_i to be integers as close as possible to the values prescribed in Theorem 6.8. We note that this still does not lead to a practical method as the σ_{w_i} are unknown. There are several approaches to getting around this problem. One possibility is to estimate the σ_{w_i} in a pilot study before running importance sampling. Alternatively, and this is what is usually done, we use the following result.

Corollary 6.9 (*Proportional allocation*) For a given overall sample size n if we take $n_i = np_i$ then

$$\sigma^2_{\text{strat}}(n_1, \ldots, n_l) = \frac{1}{n}\sum_{i=1}^{l}\frac{\sigma^2_{w_i}}{p_i} \leq \frac{\sigma^2_w}{n}.$$

Proof: Immediate from (6.9).

Note that Corollary 6.9 establishes that stratification with proportional allocation is never worse than importance sampling using w. Observe from (6.9) that, with proportional allocation,

$$\sigma^2_w/n = \sigma^2_{\text{strat}}(n_1, \ldots, n_l) + \text{Var}[E[f(X)/w(X)|i]]/n.$$

Therefore to get the maximum benefit of stratification with proportional allocation we want to decompose w so that $\text{Var}[E[f(X)/w(X)|i]]$ is as large as possible. This is just the variance of the l numbers

$$\frac{1}{p_i}\int_{\text{supp}(w_i)} f(x)\,dx$$

with respect to the probability distribution given by the p_i. In particular, this means that we should stratify so that there are a few regions with small p_i but large values of

$$\int_{\text{supp}(w_i)} f(x)\,dx.$$

Also while the Tschuprow–Neymann allocation is not feasible to start with we can always start with proportional sampling for a preliminary sample and then switch to Tschuprow–Neymann allocation using the estimates $s^2_{w_i}$ obtained from proportional sampling.

It is also possible to estimate a ratio of integrals R as in (6.5) using stratification. Using the same stratification for the numerator and denominator let

$$I_i(mf) = \int_{\text{supp}(w_i)} m(x)f(x)\,dx$$

with estimate $\hat{I}_{w_i,n_i}(mf)$. Then we estimate R by

$$\hat{R}_{\text{strat}} = \frac{\sum_{i=1}^{l}\hat{I}_{w_i,n_i}(mf)}{\sum_{i=1}^{l}\hat{I}_{w_i,n_i}(f)}.$$

If $I(f) \neq 0$ then, under certain conditions, $\hat{R}_{\text{strat}}$ has asymptotic variance

$$I^{-2}(f)\sum_{i=1}^{l}\frac{1}{n_i}\left\{\text{Var}_{w_i}\left[\frac{mf}{w_i}\right]+I_i^2(mf)I_i^{-2}(f)\text{Var}_{w_i}\left[\frac{f}{w_i}\right]\right.$$
$$\left.-\,2I_i(mf)I_i^{-1}(f)\,\text{Cov}_{w_i}\left[\frac{f}{w_i},\frac{mf}{w_i}\right]\right\} \tag{6.10}$$

and if $I(mf) \neq 0$, then the asymptotic variance can be written as

$$R^2\sum_{i=1}^{l}\frac{1}{n_i}\left\{a_i^2\text{CV}_{w_i}^2\left[\frac{f}{w_i}\right]+b_i^2\text{CV}_{w_i}^2\left[\frac{mf}{w_i}\right]\right.$$
$$\left.-\,2a_ib_i\text{CV}_{w_i}\left[\frac{f}{w_i}\right]\text{CV}_{w_i}\left[\frac{mf}{w_i}\right]\text{Corr}_{w_i}\left[\frac{f}{w_i},\frac{mf}{w_i}\right]\right\} \tag{6.11}$$

where

$$a_i = \frac{I_i(f)}{\sum_{j=1}^{l} I_j(f)}$$
$$b_i = \frac{I_i(mf)}{\sum_{j=1}^{l} I_j(mf)}$$

(Exercise 14). We estimate (6.10) by substituting the appropriate sampling analogs.

We consider an important example of stratified sampling.

Example 6.6 *Stratified Student importance sampling*
Assume that we are able to factor the integrand as mf where $f \geq 0$ is integrable and m is relatively simple with respect to f as described in Section 6.2. As in Example 6.3 we calculate the mode $\hat{x}$ of f and the negative inverse Hessian matrix $\hat{\Sigma}$ at $\hat{x}$. Changing variables from x to $v = \Sigma^{-1/2}(x - \hat{x})$ the integral becomes

$$(\det\hat{\Sigma})^{1/2}\int_{R^k} m(\hat{\Sigma}^{1/2}v+\hat{x})f(\hat{\Sigma}^{1/2}v+\hat{x})\,dv.$$

Having centered and scaled f, we choose the importance sampler w to be the $\text{Student}_k(\lambda, 0, ((\lambda-2)/\lambda)I)$ density for some λ and implement stratified sampling with this density; i.e. we decompose w as described at the start of this section. One way of decomposing w seems quite natural as it reflects the spherical symmetry of w. For this we choose $0 = r_0 < r_1 < \cdots < r_{l-1} < r_l = \infty$ and let

$$A_1 = \{x : r_0 \leq \|x\| < r_1\},$$
$$A_2 = \{x : r_1 \leq \|x\| < r_2\},$$
$$\vdots$$
$$A_l = \{x : r_{l-1} \leq \|x\| < r_l\};$$

i.e. a sequence of annular rings centered at 0 that partition R^k. Then w_i is just the $\text{Student}_k(\lambda, 0, ((\lambda - 2)/\lambda)I)$ conditioned to the annulus A_i. To implement stratified Student importance sampling an algorithm is required to generate from the $\text{Student}_k(0, ((\lambda - 2)/\lambda)I, \lambda)$ distribution conditioned to A_i.

We note that this generation problem can be split into two parts. First (Exercise 8) we have that when $X \sim \text{Student}_k(\lambda, 0, ((\lambda - 2)/\lambda)I)$ then

$$\|X\|^2 \sim k\frac{\lambda - 2}{\lambda}F(k, \lambda).$$

Therefore, because of the rotational symmetry of the $\text{Student}_k(\lambda, 0, ((\lambda - 2)/\lambda)I)$ distribution, we can generate a value from this distribution by generating $r^2 \sim kF(k, \lambda)$, generating a $u \sim U(S^{k-1})$, i.e. u is uniform on the unit sphere in R^k, and then setting $X = ru$ (Exercise 15). It follows that we can generate from the $\text{Student}_k(\lambda, 0, ((\lambda - 2)/\lambda)I)$ conditioned to A_i by generating r^2 from the $kF(k, \lambda)$ distribution conditioned to (r_{i-1}^2, r_i^2), generating $u \sim U(S^{k-1})$ and setting $X = ru$. The challenge here is generating from the $kF(k, \lambda)$ distribution conditioned to (r_{i-1}^2, r_i^2) as simplistic algorithms are too inefficient. An excellent algorithm for this is developed in Evans and Swartz (1998a) using the envelope methods described in Section 3.6. We note that this is an example of a distribution which is not log-concave.

Stratified Student importance sampling was implemented for the integration problem discussed in Example 6.4. For this $l = 25$ strata were chosen with $p_1 = p_2 = \cdots = p_{25}$. We started with proportional allocation with a sample of 100 being taken from each stratum and then switched to Tschuprow–Neyman allocation for the remaining sampling. In this example, the average efficiency increase when estimating the $R(\theta_i)$ over $i = 1, \ldots, 10$ was 2.8 when compared to straight Student importance sampling.

With a general mixture $w = p_1 w_1 + \cdots + p_l w_l$, i.e. $\text{supp}(w_i) \cap \text{supp}(w_j) \neq \emptyset$ for some $i \neq j$, it is sometimes recommended that samples be stratified by taking a proportion np_i of the total sample of size n from w_i yet still use w as the basis for the importance sampling weights; see Hesterberg (1995). Using w rather than w_i to determine the importance sampling weight should prevent the weight from becoming large if the mixture has been chosen appropriately. It is easy to show that such estimates have the correct convergence properties.

6.4.3 ERMAKOV–ZOLOTUKHIN SAMPLING

This is an interesting method that yields a guaranteed variance reducer. The original formulation can be found in Ermakov and Zolotuhkin (1960). Still there are technical obstacles that have prevented its widespread usage.

Suppose then that we have a sequence of functions $\varphi_0, \varphi_1, \ldots$ orthonormal with respect to the density w; i.e.

$$\int_{R^k} \varphi_i(x)\varphi_j(x)w(x)\,dx = \delta_{ij}.$$

Now define the function

$$\Phi\left(x_0, \ldots, x_l; \varphi_0, \ldots, \varphi_l\right) = \det \begin{pmatrix} \varphi_0(x_0) & \cdots & \varphi_l(x_0) \\ \vdots & & \vdots \\ \varphi_0(x_l) & \ldots & \varphi_l(x_l) \end{pmatrix}$$

and let

$$\omega\left(x_0, \ldots, x_l\right) = \frac{\Phi^2\left(x_0, \ldots, x_l; \varphi_0, \ldots, \varphi_l\right)}{(l+1)!}.$$

Then it can be easily shown that if $X_0, \ldots, X_l$ is a sample from w then $E_w\left[\omega(X_0, \ldots, X_l)\right] = 1$ (Exercise 16) which implies that

$$\omega_w\left(x_0, \ldots, x_l\right) = \omega\left(x_0, \ldots, x_l\right) w\left(x_0\right) \cdots w\left(x_l\right)$$

is a density on $R^k \times \cdots \times R^k$. Now suppose that $g = f/w \in L^2(w)$ and define

$$T_g\left(x_0, \ldots, x_l\right) = \frac{\Phi\left(x_0, \ldots, x_l; g, \varphi_1, \ldots, \varphi_l\right)}{\Phi\left(x_0, \ldots, x_l; \varphi_0, \varphi_1, \ldots, \varphi_l\right)}.$$

It can be shown (Exercise 17) that if $(X_0, \ldots, X_l) \sim \varphi_w$ then

$$E_{\omega_w}\left[T_g\left(X_0, \ldots, X_l\right)\right] = \int_{R^k} \varphi_0(x) g(x) w(x)\, dx;$$

i.e. $T_g\left(X_0, \ldots, X_l\right)$ is an unbiased estimator of $\int_{R^k} \varphi_0(x) f(x)\, dx$. Note that by Cramer's rule, $T_g\left(x_0, \ldots, x_l\right) = b_0$ in the system of linear equations

$$\begin{pmatrix} \varphi_0(x_0) & \cdots & \varphi_l(x_0) \\ \vdots & & \vdots \\ \varphi_0(x_l) & \ldots & \varphi_l(x_l) \end{pmatrix} \begin{pmatrix} b_0 \\ \vdots \\ b_l \end{pmatrix} = \begin{pmatrix} g(x_0) \\ \vdots \\ g(x_l) \end{pmatrix}. \tag{6.12}$$

Similarly an unbiased estimator of $\int_{R^k} \varphi_i(x) g(x) w(x)\, dx$ is given by

$$\frac{\Phi\left(x_0, \ldots, x_l; \varphi_0, \ldots, \varphi_{i-1}, g, \varphi_{i+1}, \ldots, \varphi_l\right)}{\Phi\left(x_0, \ldots, x_l; \varphi_0, \ldots, \varphi_l\right)}.$$

Note that these unbiased estimators can be written in the form

$$\sum_{i=0}^{n} w_i\left(x_0, \ldots, x_l\right) f\left(x_i\right) \tag{6.13}$$

for functions w_i for $i = 0, \ldots, n$; i.e. they take the form of a multiple quadrature rule. When the weights and points of a multiple quadrature rule are randomly determined according to some fixed distribution it is known as a randomized quadrature rule. We provide more discussion on such methods in Section 6.4.5.

Now let

$$\alpha_i = \int_{R^k} \varphi_i(x)g(x)w(x)\,dx$$

for $i = 1, 2, \ldots$; i.e. α_i is the i-th Fourier coefficient in the expansion of g with respect to the sequence $\{\varphi_n\}$. Then we have the following result.

Theorem 6.10 Suppose that $\mathrm{Var}_w[f/w] < \infty$. Then we have that

$$\mathrm{Var}_{\varphi_w}\left[T_g\right] = \mathrm{Var}_w\left[\frac{f}{w}\right] - \sum_{i=1}^{l} \alpha_i^2.$$

Proof: Exercise 18.

These results are used as follows. Suppose that our goal is to approximate

$$\int_{R^k} \varphi_0(x) f(x)\,dx;$$

and we have an importance sampler w that mimics f. We have, by Parseval's identity, that

$$\mathrm{Var}_w\left[\frac{f}{w}\right] = \int_{R^k} g^2(x)w(x)\,dx - \alpha_0^2 = \sum_{i=1}^{\infty} \alpha_i^2.$$

This together with Theorem 6.10 says that ω_w and T_g together give an estimation algorithm whose variance can be made as low as we like by an appropriate choice of l.

Of course, there are several difficulties that have to be overcome in order to implement this algorithm. First we have to have an appropriate orthonormal sequence $\{\varphi_n\}$ so that $\sum_{i=1}^{\infty} \alpha_i^2$ converges quickly. Second, and this is typically the major hurdle, we have to be able to generate from the density ω_w. In Bogues, Morrow and Patterson (1981) and Patterson (1987) the issue of sampling from φ_w is addressed via an approximate technique known as Gibbs sampling (see Chapter 7). These papers also contain some applications of the methodology.

6.4.4 SYSTEMATIC SAMPLING

Systematic sampling here refers to the evaluation of an integrand at a set of points as determined by a set of specified transformations applied to a randomly generated point. The integrand is then averaged over these points. The motivation behind this procedure is to make the averaged integrand more like the importance sampler w than the original integrand. So we are changing our point of view here; namely, we have made our best choice of w and we now work on f. A number of variance reduction methods can be considered as instances of systematic sampling. An outline of this approach to systematic sampling applied to importance sampling appeared in Evans and Swartz (1995a, 1995b).

Suppose than that we want to evaluate

$$I(f) = \int_A f(x)\,dx$$

for $A \subset R^k$. Let $T_i : A \to A$ for $i = 1, \ldots, l$ be one-to-one, onto transformations such that $\text{supp}(f) \subset A$ and T_i and T_i^{-1} have continuous Jacobians on A except perhaps on a set of Lebesgue measure 0. Denote the Jacobian of T_i by $J_{T_i} : A \to R$ where $J_{T_i}(x)$ is taken to be 0 at any point x where the Jacobian does not exist. Let $\mathcal{T} = \{T_1, \ldots, T_l\}$ and assume that $f/w \in L^2(w)$ hereafter. An unbiased estimate of the integral $I(f)$, with respect to the probability distribution specified by w, is then given by

$$\frac{f^{\mathcal{T}}(x)}{w(x)} = \frac{1}{l}\sum_{i=1}^{l} \frac{f(T_i(x))J_{T_i}(x)}{w(x)}. \tag{6.14}$$

For a sample $x_1, \ldots, x_n$ from w, (6.14) is averaged over the sample values. This estimate is called a *systematic sampling estimate*.

When $l = 1$ we see that systematic sampling is simply a reparameterization of the integration problem. Therefore we can also think of systematic sampling for $l > 1$ as a generalization of a reparameterization. Typically reparameterizations are carried out so that an integration problem becomes more amenable to standard techniques and this is exactly the motivation here.

The standard examples of systematic sampling satisfy some additional restrictions. We require hereafter that the T_i are *volume preserving symmetries* of the importance sampler. That is, we require that the transformations T_i have Jacobian determinants $J_{T_i} = 1$ and $w \circ T_i = w$ except perhaps on a set of Lebesgue measure 0. We denote the set of all such transformations by G_w and note that this is a group under composition. We call G_w the *symmetry group* of w.

Consider the situation when $\mathcal{T}$ is a finite subgroup of G_w. We have the following result concerning the variance of the systematic sampler in this situation.

Theorem 6.11 If $\mathcal{T}$ is a finite subgroup of G_w then

$$\text{Var}_w\left[\frac{f^{\mathcal{T}}}{w}\right] = \text{Var}_w\left[\frac{f}{w}\right] - E_w\left[\left(\frac{f - f^{\mathcal{T}}}{w}\right)^2\right].$$

Proof: Suppose we are given the information that an unobserved $x \sim w$ is in the set $\mathcal{T}(x) = \{T_1(x), \ldots, T_l(x)\}$. Then we know that $x \in \cup_{i=1}^{l} T_i^{-1}\mathcal{T}(x) = \mathcal{T}\mathcal{T}(x) = \mathcal{T}(x)$ as $\mathcal{T}(x)$ is the orbit of x under $\mathcal{T}$. Therefore we have that

$$E_w\left[\frac{f}{w}\middle|\,\mathcal{T}(x)\right] = \frac{1}{l}\sum_{i=1}^{l}\frac{f(T_i(x))}{w(T_i(x))} = \frac{1}{l}\sum_{i=1}^{l}\frac{f(T_i(x))}{w(x)} = \frac{f^{\mathcal{T}}(x)}{w(x)}$$

and

$$\operatorname{Var}_w\left[\frac{f(x)}{w(x)}\middle|\mathcal{T}(x)\right]=\frac{1}{l}\sum_{i=1}^{l}\left(\frac{f(T_i(x))-f^{\mathcal{T}}(x)}{w(x)}\right)^2.$$

Then

$$\operatorname{Var}_w\left[\frac{f^{\mathcal{T}}}{w}\right]=\operatorname{Var}_w\left[\frac{f}{w}\right]-E_w\left[\frac{1}{l}\sum_{i=1}^{l}\left(\frac{f\circ T_i-f^{\mathcal{T}}}{w}\right)^2\right]$$

follows immediately from the well-known fact that the unconditional variance of a random variable equals the variance of its conditional expectation plus the expectation of its conditional variance.

For $T\in\mathcal{T}$ we have that

$$f^{\mathcal{T}}(T(x))=\frac{1}{l}\sum_{i=1}^{l}f(T_i\circ T(x))=\frac{1}{l}\sum_{i=1}^{l}f\left(T_i(x)\right)=f^{\mathcal{T}}(x)$$

since $\mathcal{T}\mathcal{T}=\mathcal{T}$. Thus $f^{\mathcal{T}}$ is invariant under $\mathcal{T}$. Therefore after a change of variable $x\to T_i(x)$ we have $E_w[((f\circ T_i-f^{\mathcal{T}})/w)^2]=E_w[((f-f^{\mathcal{T}})/w)^2]$ and this gives the result.

We note that the requirement of volume preserving symmetries is essential for Theorem 6.11. Also the restriction to groups is required if we want to condition on $\mathcal{T}(x)$. It seems very difficult to analyze the situation when $\mathcal{T}$ is just an arbitrary class of transformations.

Theorem 6.11 establishes that, when the set $\mathcal{T}$ is a group of volume preserving symmetries of w, then systematic sampling always leads to a variance reduction when compared to importance sampling using w. The worst case, i.e. no variance reduction, occurs if and only if the T_i are not only symmetries of w but are also symmetries of f, i.e. $f\circ T_i=f$ as then $f^{\mathcal{T}}=f$. The best possible case occurs when $\operatorname{Var}_w\left[f^{\mathcal{T}}/w\right]=0$ and this happens if and only if $f^{\mathcal{T}}$ equals $I(f)w$, i.e. if and only if systematic sampling via $\mathcal{T}$ effectively integrates f exactly. Note, however, that the estimator $f^{\mathcal{T}}/w$ requires l times as many function evaluations as the estimator f/w. Thus it is fairer to say that $\mathcal{T}$ provides a *true variance reduction* only when $\operatorname{Var}_w\left[f^{\mathcal{T}}/w\right]<l^{-1}\operatorname{Var}_w[f/w]$. Combining this with the theorem we have the following corollary.

Corollary 6.12 Systematic sampling using $\mathcal{T}$ is a true variance reducer if and only if

$$\operatorname{Var}_w\left[\frac{f^{\mathcal{T}}}{w}\right]<\frac{1}{l-1}E_w\left[\left(\frac{f-f^{\mathcal{T}}}{w}\right)^2\right]. \tag{6.15}$$

Using Theorem 6.11 and (6.15) we also see that $\mathcal{T}$ is a true variance reducer if and only if

$$\frac{l-1}{l}\mathrm{Var}_w\left[\frac{f}{w}\right] < E_w\left[\left(\frac{f-f^{\mathcal{T}}}{w}\right)^2\right] \leq \mathrm{Var}_w\left[\frac{f}{w}\right]. \tag{6.16}$$

We note that (6.16) is very stringent. This has implications for the utility of some commonly advocated variance reduction techniques.

It is interesting to note that $f^{\mathcal{T}}/w$ is the orthogonal projection of f/w onto the subspace of $L^2(w)$ consisting of all functions invariant under the group $\mathcal{T}$. Therefore $(f - f^{\mathcal{T}})/w$ is the residual and we can interpret $E_w[((f - f^{\mathcal{T}})/w)^2]$, the squared length of the residual, as a measure of how noninvariant f is under the group $\mathcal{T}$. This interpretation follows from the fact that the subspace of all functions invariant under $\mathcal{T}$ is a closed subspace of $L^2(w)$ and the map $f/w \to f^{\mathcal{T}}/w$ is idempotent and symmetric; i.e. denoting the map by M we have that $M(f^{\mathcal{T}}/w) = f^{\mathcal{T}}/w$ and

$$\begin{aligned}
\langle M(f), g\rangle &= E_w\left[f^{\mathcal{T}} g\right] \\
&= \frac{1}{l}\sum_{i=1}^{l} E_w\left[(f \circ T_i)\, g\right] \\
&= \frac{1}{l}\sum_{i=1}^{l} E_w\left[f\,(g \circ T_i)\right] \\
&= \langle f, M(g)\rangle
\end{aligned}$$

where for the third equality we have made the transformation $x \to T_i^{-1}(x)$ and noted that $T_i^{-1} = T_j$ for some j. From (6.16) we see that, for the group $\mathcal{T}$ of order l to be a true variance reducer, the noninvariant part of f/w must account for a proportion greater than $(l-1)/l$ of the total variation in f/w when sampling from w. So it is primarily in contexts where w is a bad importance sampler for f, in the sense that it doesn't approximate f well, that the method of systematic sampling can be effective. So systematic sampling can be viewed as a way of fixing up a bad importance sampler or as a way of providing insurance that the importance sampler performs well.

The restriction that $\mathcal{T}$ is a subgroup of G_w has additional significance. Recall that we are trying to choose $\mathcal{T}$ so that $f^{\mathcal{T}}$ and w are more alike than f and w. When $\mathcal{T}$ is a subgroup of G_w then, because $f^{\mathcal{T}}$ is invariant under $\mathcal{T}$, we have $\mathcal{T} \subseteq G_{f^{\mathcal{T}}}$ as well as $\mathcal{T} \subseteq G_w$. Thus $f^{\mathcal{T}}$ and w share the property of both being invariant under $\mathcal{T}$. They are more alike than f and w whenever $\mathcal{T} \nsubseteq G_f$. This indicates that, in a given problem, we should attempt to construct systematic sampling plans $\mathcal{T}$ where f is "far" from being invariant under $\mathcal{T}$. Further we note that choosing an arbitrary $\mathcal{T}$, i.e. not necessarily a group, does not guarantee that $f^{\mathcal{T}}$ and w will be more alike than f and w.

The group property also has another direct computational significance. This is seen by noting that in certain circumstances we have a short-cut for the evaluation of (6.14). For when $\mathcal{T}$ is a subgroup of G_w, we need only average f over the orbit $\mathcal{T}(x)$. The advantage in this is that the cardinality of the orbit is, in particular instances, significantly lower than the cardinality of the group. We note that this points to an alternative way of thinking about the problem of systematic sampling; namely, rather than thinking about appropriate groups for a problem we can think about appropriate sets of points $\mathcal{T}(x)$ to average over. Our discussion indicates, however, that such sets of points will only be useful when the subgroup of G_w leaving each such set invariant is large and reflects some asymmetry of f relative to w. For example, it is only under such circumstances that the stringency of the requirement in (6.16) can be substantially weakened by replacing the order of the group by the order of the orbit, as we no longer need to average over the whole group.

We consider another approach when the order of the group $\mathcal{T}$ is large. For example, consider the group generated by the reflections

$$T_i(x) = (x_1, \ldots, x_{i-1}, -x_i, x_{i+1}, \ldots, x_k)$$

for $i = 1, \ldots, k$. This group has order $l = 2^k$ and thus, for large k, the evaluation of $f^{\mathcal{T}}(x)$ is computationally prohibitive. Further the orbits have the same cardinality as the group. In such a case it may seem intuitively reasonable, for each generated x, to sample $T_1^*, \ldots, T_L^*$ uniformly from the subgroup and consider the estimator

$$\frac{1}{L}\sum_{i=1}^{L}\frac{f\left(T_i^*(x)\right)}{w(x)}. \tag{6.17}$$

We have the following result.

Theorem 6.13 If $\mathcal{T}$ is a finite subgroup of G_w of order l and $T_1^*, \ldots, T_L^*$ is a uniform sample from $\mathcal{T}$ then

$$\mathrm{Var}\left[\frac{1}{L}\sum_{i=1}^{L}\frac{f\circ T_i^*}{w}\right] = \mathrm{Var}_w\left[\frac{f}{w}\right] - \frac{L-1}{L}E_w\left[\left(\frac{f-f^{\mathcal{T}}}{w}\right)^2\right].$$

Proof: First observe that

$$E\left[\frac{1}{L}\sum_{i=1}^{L}\frac{f\left(T_i^*(x)\right)}{w(x)}\,\middle|\, x\right] = \frac{1}{l}\sum_{i=1}^{l}\frac{f(T_i(x))}{w(x)} = \frac{f^{\mathcal{T}}(x)}{w(x)}$$

and

$$\text{Var}\left[\frac{1}{L}\sum_{i=1}^{L}\frac{f\left(T_i^*(x)\right)}{w(x)}\,\middle|\, x\right] = E\left[\left(\frac{1}{L}\sum_{i=1}^{L}\frac{f\left(T_i^*(x)\right)-f^{\mathcal{T}}(x)}{w(x)}\right)^2 \,\middle|\, x\right]$$
$$= \frac{1}{L}\left[\frac{1}{l}\sum_{i=1}^{l}\left(\frac{f\left(T_i(x)\right)-f^{\mathcal{T}}(x)}{w(x)}\right)^2\right].$$

We then apply the conditional variance formula and use Theorem 6.11 to obtain

$$\text{Var}\left[\frac{1}{L}\sum_{i=1}^{L}\frac{f\circ T_i^*}{w}\right] = \text{Var}_w\left[\frac{f^{\mathcal{T}}}{w}\right]+\frac{1}{L}E_w\left[\frac{1}{l}\sum_{i=1}^{l}\left(\frac{f\circ T_i-f^{\mathcal{T}}}{w}\right)^2\right]$$
$$= \text{Var}_w\left[\frac{f}{w}\right]-\frac{L-1}{L}E_w\left[\left(\frac{f-f^{\mathcal{T}}}{w}\right)^2\right].$$

We have then the following result.

Corollary 6.14 The estimator given by (6.17) is never a true variance reducer.

Proof: To be a true variance reducer we must have

$$\text{Var}\left[\frac{1}{L}\sum_{i=1}^{L}\frac{f\circ T_i^*}{w}\right] < \frac{1}{L}\text{Var}_w\left[\frac{f}{w}\right].$$

By Theorem 6.13 this is equivalent to

$$\text{Var}_w\left[\frac{f}{w}\right] < E_w\left[\left(\frac{f-f^{\mathcal{T}}}{w}\right)^2\right]$$

and by Theorem 6.11 we know that this is impossible.

Therefore the estimator (6.17) is never useful. When $f^{\mathcal{T}} \equiv I(f)w$, i.e. the systematic sampling plan has effectively calculated the integral exactly, then the variance of (6.17) is $\text{Var}_w[f/w]/L$ so only in this rare setting there is no loss of efficiency.

There are cases where there is a group $\mathcal{T} \subseteq G_w$ of infinite order; e.g. the orthogonal group $O(k)$ acting on R^k. In such contexts it may be possible to place a uniform measure U on $\mathcal{T}$ and put

$$f^{\mathcal{T}}(x) = \int_{\mathcal{T}} f(T(x))\,U(dT). \tag{6.18}$$

Note that (6.18) is unbiased for $I(f)$ by Fubini's theorem and the proof of Theorem 6.11 is then basically the same. By Corollary 6.14, however, (6.18) will only be useful in the rare cases where we can evaluate it explicitly.

We now consider some specific examples of systematic sampling.

Example 6.7 *Antithetic sampling*
Consider the estimation of the integral $I(f) = \int_0^1 f(x)\,dx$ and let w be the density of the $U(0,1)$ distribution. Hammersley and Morton (1956) introduced the technique of *antithetic sampling* whereby x is generated from the $U(0,1)$ distribution and $I(f)$ is estimated by $f^{\mathcal{T}}(x) = \frac{1}{2}(f(x) + f(1-x))$. To place the above procedure in our framework, define $T_1(x) = x$ and $T_2(x) = 1-x$ and note that $\mathcal{T} = \{T_1, T_2\}$ forms a subgroup of G_w. It follows that

$$\begin{aligned}\mathrm{Var}_w\left[f^{\mathcal{T}}\right] &= \tfrac{1}{2}\mathrm{Var}_w[f] + \tfrac{1}{2}\mathrm{Cov}_w[f(x), f(1-x)] \\ &= \mathrm{Var}_w[f] - \int_0^1 \left(f(x) - f^{\mathcal{T}}(x)\right)^2 dx. \qquad (6.19)\end{aligned}$$

From (6.19) we observe that antithetic sampling provides a true variance reducer if and only if $\mathrm{Cov}_w[f(x), f(1-x)] < 0$. This occurs whenever f is monotone on $(0,1)$ and is strictly monotone on some subinterval of $(0,1)$ (Exercise 19).

The best and worst cases for antithetic sampling can be determined from (6.19). The best case occurs when $f^{\mathcal{T}}$ is constant or equivalently when $f(1-x) = c - f(x)$. The worst case occurs when $f = f^{\mathcal{T}}$ or equivalently f is symmetric about $1/2$. The linear operator $f \to f^{\mathcal{T}}$ projects onto the space of functions symmetric about $1/2$. Thus antithetic sampling is more effective the further f is from being symmetric about 1/2.

Consider now the situation where f has a single maximum or minimum at some known point $q \in (0,1)$. We generalize Hammersley and Morton's method. For this define T_1 and T_2 so that $f \circ T_1$ and $f \circ T_2$ have opposite monotonicities. Let

$$T_1(x) = \begin{cases} q - x & 0 < x < q \\ x & q < x < 1 \end{cases}$$

and

$$T_2(x) = \begin{cases} x & 0 < x < q \\ 1 - x + q & q < x < 1. \end{cases}$$

Then we have that $\frac{1}{2}(f \circ T_1 + f \circ T_2)$ is a true variance reducer.

Note, however, that $\{T_1, T_2\}$ is not a group since it does not contain the identity nor $T_1 \circ T_2$. If we further define $T_3(x) = x$ and

$$T_4(x) = \begin{cases} q - x & 0 < x < q \\ 1 - x + q & q < x < 1 \end{cases}$$

then $\mathcal{T} = \{T_1, T_2, T_3, T_4\}$ is the Klein 4-group. The orbits are given, however, by $\mathcal{T}x = \{T_1 x, T_2 x\}$ and hence $f^{\mathcal{T}} = \frac{1}{2}(f \circ T_1 + f \circ T_2)$. Also notice in this example that the transformations are not smooth. The function $f^{\mathcal{T}}$ is the orthogonal projection of f onto the space of functions symmetric about $q/2$ in $[0, q]$ and symmetric about $(1+q)/2$ in $[q, 1]$.

Consider now a simple example where $f(x) = x(1-x)$ is maximized at $q = 1/2$. Using T_1 and T_2 as defined above it is easy to show that systematic sampling provides a 93.75% reduction in variance compared to a 50% reduction that would be achieved by sampling another value via importance sampling.

This approach can clearly be generalized to give a true variance reducer for functions having a finite number of local maxima and minima in the interval $[0, 1]$. If f has l minima or maxima in (0,1) then $\mathcal{T}$ has 2^{l+1} elements but $\mathcal{T}(x)$ has only two elements. For example, applying this to $f(x) = \cos 2\pi kx$ gives an estimator with infinite efficiency. Of course the practical application of this technique requires that we be able to efficiently compute the extrema of f.

In contexts where we are integrating over $[0, 1]^k$ the transformations $T_1(x) = x$ and $T_2(x) = 1 - x$ yield a guaranteed variance reducer when f is monotone in each coordinate (Exercise 20). If we can efficiently compute the extrema of f as a function of each coordinate x_i with the other variables fixed, then the generalized method can be implemented coordinate-wise to give a true variance reducer. This is only feasible, however, for low dimensions k. For suppose that f has a single extreme point as a function of x_i for each i. Then we need 2^k evaluations of f to implement the method for each generation of X. See Rubinstein (1986) for more discussion of generalizations of antithetic variates.

Example 6.8 *Rotation sampling on the unit cube*
Rotation sampling on the unit interval has been studied by Hammersley and Morton (1956) and Fishman and Huang (1983). Suppose then that w is the $U(0, 1)$ density. Let $T_i(x) = (i-1)/l \oplus x$ where $i = 1, \ldots, l$ and $\oplus$ denotes addition modulo 1. Thus $\mathcal{T} = \{T_1, \ldots, T_l\}$ is a subgroup of G_w, isomorphic to the cyclic group of order l, and

$$f^{\mathcal{T}}(x) = \frac{1}{l}\sum_{i=1}^{l} f\left(\frac{i-1}{l} \oplus x\right).$$

Since $f^{\mathcal{T}}$ is a Riemann sum we have that $f^{\mathcal{T}} \to \int_0^1 f(x)\,dx$ and $\mathrm{Var}_w[f^{\mathcal{T}}] \to 0$ as $l \to \infty$.

We have

$$\begin{aligned}\mathrm{Var}_w\left[\frac{f^{\mathcal{T}}}{w}\right] &= \frac{1}{l}\mathrm{Var}_w\left[\frac{f}{w}\right]\left\{1 + \sum_{i=1}^{l-1}\mathrm{Corr}_w\left[f(x), f\left(\frac{i}{l} \oplus x\right)\right]\right\} \\ &= \mathrm{Var}_w\left[\frac{f}{w}\right] - \int_0^1 \left(f(x) - \frac{1}{l}\sum_{i=1}^{l} f\left(\frac{i-1}{l} \oplus x\right)\right)^2 dx\end{aligned}$$

where the first equality follows from Fishman and Huang (1983) and the second equality is an instance of Theorem 6.11. The worst case for the estimator $f^{\mathcal{T}}$

occurs when f is periodic with period $1/l$; i.e.

$$f(x) = \frac{1}{l}\sum_{i=1}^{l} f\left(\frac{i-1}{l} \oplus x\right).$$

The best case occurs when f is a step function with the length of each step equal to $1/l$. The linear operator $f \to f^{\mathcal{T}}$ projects onto the space of periodic functions with period $1/l$. Therefore rotation sampling is more effective the further f is from being periodic with period $1/l$. Fishman and Huang (1983) establish that every function in $C^1[0, 1]$ has the property

$$l\text{Var}_w\left[f^{\mathcal{T}}/w\right] \to 0$$

as $l \to \infty$ and so $\mathcal{T}$ is a true variance reducer for all l large enough.

We can generalize the above in a number of ways for $[0, 1]^k$. For example, define $T_i(x) = a_i \oplus x$ where $a_i \in [0, 1]^k$ and $\oplus$ now denotes addition modulo 1 in each coordinate. The question arises as to how we should choose $a_1, \ldots, a_l$ so that $\mathcal{T} = \{T_1, \ldots, T_l\}$ forms a group. One possibility is to consider divisions of length $1/p$, and include in $\mathcal{T}$ all vectors a_i whose coordinates take any of the values $0, 1/p, \ldots, (p-1)/p$. In this case $f \to f^{\mathcal{T}}$ is the orthogonal projection onto the subspace of $L^2(w)$ of all functions that are periodic with period $1/p$ in each coordinate. As this group has order $l = p^k$ we see that it suffers from the dimensional effect and will not be practical for large k. Therefore we look for smaller subgroups of this group. Geweke (1991) discusses some practical uses of this group when k is low. To avoid the dimensional effect one possibility is to put $a_i = (i-1)(1, 1, \ldots, 1)^t/p$ for $i = 1, \ldots, p$. In this case $f \to f^{\mathcal{T}}$ is the orthogonal projection onto the set of functions that are periodic with period $1/p$ simultaneously in all coordinates. We note that all of these groups have orbits of the same cardinality as the group.

Example 6.9 *Randomized good lattice point rules*
Randomized good lattice point rules are discussed in Cranley and Patterson (1976) and Sloan and Joe (1994). Recall from Section 5.9 that good lattice point rules are applicable to integrals over $[0, 1]^k$ and the integrand must be periodic with period 1 in each coordinate. Thus whenever the integrand is not periodic we must first make a periodizing transformation such as those discussed in Section 5.9.

For this technique we choose $z \in Z^k$, a k-dimensional vector of integers, we choose a positive integer l and w is the density of the $U([0, 1]^k)$ distribution. We then define $\mathcal{T} = \{T_1, \ldots, T_l\}$ by $T_i(x) = \{((i-1)/l)z\} \oplus x$ where $\{\cdot\}$ denotes the fractional part function applied coordinatewise. Then, as with rotation sampling, $\mathcal{T}$ is a group and $f^{\mathcal{T}}$ is the orthogonal projection of f onto the space of functions periodic with period $1/l$ in the direction given by z modulo 1. Notice that if we project the points $T_i(x)$ onto any axis then we have a set of l equispaced points in [0,1]. When $k = 1$ the randomized good lattice point rule is equivalent to rotation sampling with periodization.

Now suppose that $f \in L^2([0,1]^k)$ has Fourier expansion

$$f(x) = \sum_{h \in Z^k} \hat{f}(h) e^{i2\pi h^t x}$$

where $\hat{f}(h) = \int_{[0,1]^k} e^{-2\pi h^t x} f(x)\, dx$ is the h-th Fourier coefficient and observe that $\hat{f}(0) = \int_{[0,1]^k} f(x)\, dx$. Then it is easily shown (see Section 5.9) that

$$f^{\mathcal{T}}(x) = \sum_{h^t z \equiv 0 \bmod l} \hat{f}(h) e^{i2\pi h^t x}$$

and so the group has integrated out various terms in the Fourier expansion. This is further reflected in the fact that $\mathrm{Var}_w[f/w] = \sum_{h \neq 0} |\hat{f}(h)|^2$ while

$$\mathrm{Var}_w[f^{\mathcal{T}}/w] = \sum_{h \neq 0, h^t z \equiv 0 \bmod l} |\hat{f}(h)|^2.$$

Often $|\hat{f}(h)|$ decreases rapidly with $|h| = |h_1| + \cdots + |h_k|$ and so a good lattice point z is one for which $|h|$ is large for all h satisfying $h^t z \equiv 0 \bmod l$. Conditions can be stated for the rate of decrease of $|\hat{f}(h)|$ with increasing $|h|$ to ensure that we have a true variance reducer for sufficiently large l. Sloan and Joe (1994) give an extensive discussion of methods for determining appropriate z. Notice that if f is a trigonometric polynomial of degree n, and any h satisfying $h^t z = 0 \bmod l$ has $|h| > n$, then $f^{\mathcal{T}}$ integrates f exactly. Also note that if f is additive; i.e. $f = f_1 + \cdots + f_k$ where f_i is a function of the i-th coordinate x_i only, then the use of $\mathcal{T}$ is equivalent to applying one-dimensional rotation sampling, with periodization, to each f_i.

Example 6.10 *Latin hypercube sampling*
A randomized good lattice point rule is closely related to *latin hypercube sampling*. As we noted in Example 6.9, a randomized good lattice point rule projects to a simple Riemann sum on each axis and this is the basic property characterizing latin hypercube sampling. The transformations are of the form $T_i(x) = a_i + x/l$ where the $a_i \in \{0, 1/l, \ldots, (l-1)/l\}^k$ for $i = 1, \ldots, l$ are chosen to satisfy the constraint that each stratum along an axis contains one and only one of the a_i (Exercise 21). Latin hypercube sampling then involves generating l independent $U([0,1]^k)$ values $x_1, \ldots, x_l$ and averaging the integrand at $T_i(x_i)$ for $i = 1, \ldots, l$. With randomized good lattice point rules a single $U([0, 1/l]^k)$ offset is generated for all the selected grid-points and these grid-points are constrained to lie along a single vector modulo 1.

With latin hypercube sampling, and with good lattice point sampling, the additive component of the integrand f, i.e. the orthogonal projection of f onto the space of all $L^2([0,1]^k)$ functions of the form $g(x) = g_1(x_1) + \cdots + g_k(x_k)$, is integrated exactly as $l \to \infty$ and this causes a variance reduction. For we can decompose f as

$$f = \sum_{i=0}^{k} f_i \tag{6.20}$$

where, if L_i is the set of all functions in $L^2([0, 1]^k)$ that can be written as a sum of functions each of which is a function of i variables, then f_i is the orthogonal projection of f onto $L_i \cap L_{i-1}^{\perp}$ for $i = 1, \ldots, k$ and f_0 is the orthogonal projection of f onto $L\{1\}$; i.e. $f_0 = I(f)$ (Exercises 22, 23 and 24). For $i \geq 1$ the f_i have 0 means and are uncorrelated under uniform sampling and $\text{Var}[f] = \sum_{i=1}^{k} \text{Var}[f_i]$ (Exercise 22). For more on this decomposition see Efron and Stein (1981) and Owen (1992). This is referred to as the *ANOVA decomposition* because of its similarity with the analysis of variance technique used in statistics. Stein (1987) proved that the variance of the latin hypercube estimator is $\sum_{i=2}^{k} \text{Var}[f_i] + o(n)$ as $n \to \infty$.

Strictly speaking latin hypercube sampling is not a systematic sampling method but it is clearly closely related to good lattice point sampling. It is not clear, however, that the greater generality of the latin hypercube leads to any additional benefits. References on latin hypercube and orthogonal array sampling, which is concerned with approximately integrating out higher order components of f in (6.20), include McKay, Beckman and Conover (1979), Stein (1987), Owen (1992) and Tang (1993).

Example 6.11 *Orthogonal transformation sampling*
The use of subgroups of the group of orthogonal transformations provides a number of useful estimators when $A = R^k$. In this context the primary family of importance samplers is the multivariate Student family. Note that for every $c > 0$ the $\text{Student}_k(\lambda, 0, cI)$ density w is invariant under orthogonal transformations; i.e. G_w contains the orthogonal group $O(k)$.

Suppose that $|f|$ is unimodal and that we standardize by the transformation $x \to y = \hat{\Sigma}^{-1/2}(x - \hat{x})$ where $\hat{x}$ is the mode of $|f|$ and $\hat{\Sigma}$ is the inverse Hessian matrix of $-\log|f|$ evaluated at $\hat{x}$. Then after making this change of variable we have that

$$I(f) = (\det \hat{\Sigma})^{1/2} \int_{R^k} f(\hat{x} + \hat{\Sigma}^{1/2} y)\, dy = \int_{R^k} f^*(y)\, dy$$

Then an appropriate importance sampler for this problem is a $\text{Student}_k(\lambda, 0, cI)$ density w where c is chosen so that the variance matrix of w is the identity or so that the negative Hessian matrix of $\log w$ at 0 is the identity matrix, whichever is felt to be appropriate. At this point, however, let us just suppose that we have chosen w to be rotationally symmetric, i.e. G_w contains the orthogonal group $O(k)$.

We now make the transformation $y \to (r, v)$ where $r = \|y\|$ and $v = y/\|y\|$. Further we note that any spherically symmetric w can be written as $w(y) = g(r)$

where, denoting the surface area of S^{k-1} by A_{k-1}, $h(r) = A_{k-1}r^{k-1}g(r)$ is a density on R^1 (Exercise 25). Then we can generate y from w by generating $v \sim U(S^{k-1})$ statistically independent of $r \sim h$ and putting $y = rv$ (Exercise 25). When h is the chi-square(k) density then y has the $N_k(0, I)$ distribution. When h is the F(k, λ) density then y has the Student$_k(0, I, \lambda)$ distribution. If H denotes the distribution function associated with h then we also make the transformation $r \to s = H(r)$ and note that $s \sim U(0, 1)$. Then

$$I(f) = \int_0^1 \int_{S^{k-1}} \frac{(H^{-1}(s))^{k-1} f^*(H^{-1}(s)v)}{h(H^{-1}(s))}\, ds\, dv \tag{6.21}$$

where dv denotes surface measure on the unit sphere S^{k-1} in R^k.

So now for (6.21) $w = w_1 w_2$ where w_1 is the $U(0, 1)$ density and w_2 is the $U(S^{k-1})$ density. Then a natural set of symmetries to use is of the form of a direct product group $\mathcal{T} = \mathcal{T}_1(l_1) \times \mathcal{T}_2(l_2)$ where $\mathcal{T}_1 = \{T_1, \ldots, T_{l_1}\}$ is the rotation sampling group acting on $[0, 1]$ and $\mathcal{T}_2$ is a finite subgroup of $O(k)$ acting on S^{k-1}. The search then is for appropriate subgroups $\mathcal{T}_2$.

Accordingly, we now look for finite subgroups $\mathcal{T}$ of $O(k)$ to use for systematic sampling when calculating integrals of the form

$$\int_{S^{k-1}} f(v)\, dv$$

and when the importance sampler w is the density of the $U(S^{k-1})$ distribution. Note that $w(v) = 1/A_{k-1}$ where A_{k-1} is the surface area of S^{k-1}.

Other than the trivial subgroup, the simplest finite subgroup $\mathcal{T}$ of $O(k)$ is generated by the reflection through the origin so that $\mathcal{T}$ has order 2. We refer to this as *central inversion sampling*. This technique was discussed in Geweke (1988). Here $f^{\mathcal{T}}/w$ is the orthogonal projection of f/w onto the space of all even functions in $L^2(S^{k-1})$ and the residual $(f - f^{\mathcal{T}})/w$ is the orthogonal projection of f/w onto the space of all odd functions in $L^2(S^{k-1})$. Thus the worst case performance arises when f/w is even and the best case performance arises when f/w is odd. For example, if f/w is well approximated by a linear function times an even function then we expect the method to work well as $f^{\mathcal{T}}/w$ integrates polynomials of degree 1 exactly on the sphere.

As a simple generalization we consider the subgroup $\mathcal{T}$ of $O(k)$ generated by reflections in the coordinate planes. We refer to this as *reflection sampling* but note that this is only feasible in low dimensions as $\mathcal{T}$ has order 2^k and the orbits have the same cardinality. In this case $f^{\mathcal{T}}/w$ is the orthogonal projection of f/w onto the space of all functions in $L^2(S^{k-1})$ that are even in each coordinate and a similar statement applies to the residual.

Another very natural subgroup to consider is the group of all permutations of the k coordinates. This group has order $k!$ and we refer to this as *permutation sampling*. Again this is only feasible for low dimensions. Alternatively we could take a much smaller subgroup such as the group of order k consisting just of cyclical permutations.

Combining the k cyclical permutations with reflection through the origin gives a group with $2k$ elements and we will refer to this as *cyclic central inversion sampling*. Denoting the surface area of the sphere by A_{k-1}, we have that

$$\frac{f^{\mathcal{T}}(x)}{w(x)} = \frac{A_{k-1}}{2k}\sum_{i=1}^{k}(f(T_i(x)) + f(-T_i(x)) \tag{6.22}$$

and it is simple to show (Exercise 26) that $f^{\mathcal{T}}/w$ exactly integrates all polynomials of degree 3 or less on the sphere.

In general if $f \in L^2(S^{k-1})$ then f has Fourier expansion

$$f(x) = \sum_{i=1}^{\infty} a_i u_i(x)$$

where u_i is a spherical harmonic of degree i and

$$a_i = \int_{S^{k-1}} f(x)u_i(x)\,dx;$$

e.g. see Stein and Weiss (1971). Then $\mathrm{Var}_f[f/w] = \sum_{i=1}^{\infty} a_i^2$ while if $\mathcal{T}$ integrates all polynomials of degree m then $\mathrm{Var}_f[f^{\mathcal{T}}/w] = \sum_{i=m+1}^{\infty} a_i^2$. Whether or not this is a true variance reducer will depend on the number of function evaluations required to implement systematic sampling based on $\mathcal{T}$ and the size of $a_1^2 + \cdots + a_m^2$.

6.4.5 RANDOMIZED QUADRATURE

In Theorem 6.7 it was shown that if we can integrate out one or more dimensions in an integral, and use the corresponding marginal importance sampler, then the estimate has reduced variance. This situation does not occur very often but there may be several difficult dimensions where we feel that using a quadrature rule to get an approximate value for the integral over that dimension would be advantageous. This entails, however, that we give up the convergence properties of importance sampling and its convenient error estimates. Randomized (stochastic) quadrature is a compromise between these approaches whereby we apply randomly generated quadrature rules, and thereby achieve their associated high accuracy, but at the same time retain the benefits of Monte Carlo methods. From a certain point of view these algorithms address what has appeared to be a conflict amongst those who advocate random methods and those who advocate nonrandom methods for integration problems. There are similarities between randomized quadrature methods and the methods of systematic sampling discussed in Section 6.4.4.

We define a *randomized quadrature rule* to be a sum

$$\sum_{i=1}^{l} w_i f(x_i) \tag{6.23}$$

where $R_l^* = \{(x_1, w_1), \ldots, (x_l, w_l)\}$ is a realized value from a specified probability distribution W on the class of such sets. To make any sense, additional properties have to be imposed and it is typical to require that each realization from W be such that (6.23) is exact for some class of functions and that

$$E\left[\sum_{i=1}^{l} w_i f(x_i)\right] = I(f).$$

The idea is that we will obtain multiple independent realizations from W and then average the implementations of these to estimate $I(f)$. We then have available the usual Monte Carlo estimates of error.

We have already seen one example of the construction of a randomized quadrature rule; namely Ermakov–Zolotukhin sampling in Section 6.4.3 when expressed as (6.13). From (6.12) it is easy to see that this rule is exact whenever $f = b_0\varphi_0 + \cdots + b_l\varphi_l$ (Exercise 27). While Ermakov–Zolotukhin sampling may be the first instance of a randomized quadrature rule the definition of these seems to have been formalized in Haber (1966, 1968, 1969), where examples of rules exact for polynomials were also provided. Also see Haber (1970) for more discussion. It is easy to see that some systematic sampling plans satisfy the definition of randomized quadrature rules. Note, however, that systematic sampling plans are devised with the goal of fixing up poor importance sampling schemes; i.e. we are given a particular importance sampler that we are going to use, and this is not the case with randomized quadrature.

In Siegel and O'Brien (1985) randomized quadrature rules of the form $\sum_{i=1}^{l} w_i(x_1, \ldots, x_l) f(x_i)$ are considered where the functions w_i and the distribution $(x_1, \ldots, x_l) \sim W$ are specified to make (6.23) exact for polynomials. The discussion for the most part is restricted to dimension $k = 1$ and so these rules are not really of much practical value. It seems characteristic of randomized quadrature rules that we are quickly faced with difficult sampling problems when we either increase the dimension k or increase the degree of the polynomial for which we want exactness. Still practical use has been made of randomized quadrature rules as demonstrated by the following example.

Example 6.12 *Randomized spherical-radial rules*
Suppose that we wish to evaluate an integral of the form

$$I(mf) = \int_{R^k} m(x) f(x)\, dx$$

where $f \geq 0$ has a single dominant peak $\hat{x}$ and $-\log f$ has inverse Hessian matrix at $\hat{x}$ given by $\hat{\Sigma}$. Then as in Example 6.11 we transform via $x \to y = \hat{\Sigma}^{-1/2}(x - \hat{x})$ and write the integral as

$$\begin{aligned} I(mf) &= (\det \hat{\Sigma})^{1/2} \int_{R^k} m(\hat{x} + \hat{\Sigma}^{1/2} y) f(\hat{x} + \hat{\Sigma}^{1/2} y)\, dy \\ &= \int_{R^k} m^*(x) f^*(x)\, dx \qquad (6.24) \end{aligned}$$

where $f^*(x) = (\det \hat{\Sigma})^{1/2} f(\hat{x} + \hat{\Sigma}^{1/2} y)$. Next we make the transformation $x = rv$ where $r = \|x\|$ and $v = x/\|x\|$. Then (6.24) becomes

$$I(mf) = \int_0^\infty \int_{S^{k-1}} r^{k-1} m^*(rv) f^*(rv)\, dv\, dr \tag{6.25}$$

For the integral over the sphere Genz and Monahan (1998a, 1998b, 1999) and Monahan and Genz (1997) recommend using a randomized quadrature rule. For this let $R_l = \{(x_1, w_1), \ldots, (x_l, w_l)\}$ be a deterministic quadrature rule for the sphere that integrates constants and such that $x_i \in S^{k-1}$ for $i = 1, \ldots, k$. An orthogonal matrix Q is generated from the uniform distribution on $O(k)$ (see Example 3.12) and the randomized quadrature rule is given by $R_l^* = \{(Qx_1, w_1), \ldots, (Qx_l, w_l)\}$. Then, because each $v_i = Qx_i \sim U(S^{k-1})$ (Exercise 28), we have that

$$E\left[\sum_{i=1}^{l} w_i m^*(rQx_i) f^*(rQx_i)\right] = \int_{S^{k-1}} m^*(rv) f^*(rv)\, dv.$$

Further if R_l is exact for all polynomials of degree d or less then R_l^* is exact for all polynomials of degree d or less (Exercise 29). Genz and Monahan (1999) give a listing of rules on the sphere exact for all polynomials of degrees $d = 1, 3, 5$ and 7. The degree 1 and 3 rules are those given in Example 5.9. Note that the degree 1 and 3 rules also arise as systematic sampling plans. The weights are not constant, however, for the degree 5 and 7 rules stated in Genz and Monahan (1999), so these do not correspond to systematic sampling plans.

To complete the specification of the integration procedure we need to deal with the variable of integration r in (6.25). One possibility is to use a standard quadrature rule on this like Simpson's rule or Gauss–Laguerre. A disadvantage of this is that the nice properties of Monte Carlo estimates are lost. Genz and Monahan (1998a) develop randomized quadrature rules for integrals of the form

$$\int_{-\infty}^{\infty} h(r)|r|^{k-1} w(r)\, dr \tag{6.26}$$

where $|r|^{k-1}w$ is the weight function for specified w where w is symmetric about 0. Note that the integration over r in (6.25) can be taken over $(-\infty, \infty)$ if we replace r by $|r|$ and divide the integral by 2. Now suppose that points $x_1, \ldots, x_p$ are selected with each $x_i > 0$. Then a deterministic radial rule is given by

$$\sum_{i=1}^{p} \frac{w_i}{2}(h(x_i) + h(-x_i))$$

where

$$w_i = \int_{-\infty}^{\infty} \prod_{j \neq i} \frac{r^2 - x_j^2}{x_i^2 - x_j^2} |r|^{k-1} w(r)\, dr$$

and w has been chosen so that these integrals are finite. This rule integrates exactly all polynomials h of degree $2p-1$ or less (Exercise 30). To specify a randomized quadrature rule we must specify a joint density g for $(x_1, \ldots, x_p)$ so that

$$E_g\left[\sum_{i=1}^{p} \frac{w_i}{2}(h(x_i) + h(-x_i))\right] = \int_{-\infty}^{\infty} h(r)|r|^{k-1} w(r)\,dr \qquad (6.27)$$

for any h such that (6.27) is finite. When $p = 1$ we take g to be the density proportional to $r^{k-1}w(r)$ for $r > 0$. Note that when $w(r) = \exp(-r)$ then g is a gamma$(k, 1)$ density. When $p = 2$ one point can be taken to be 0 and the remaining point generated from the density proportional to $r^{k+1}w(r)$ (Exercise 31). Genz and Monahan (1998a) also work out the case $p = 3$. In this case one point can be taken to be 0 and the two remaining points generated from a complicated bivariate distribution. It quickly becomes intractable to construct g and moreover the densities become harder to sample from as p increases.

In Yakowitz, Krimmel and Szidarovsky (1978) the method of *weighted Monte Carlo integration* is introduced for integrals on the unit cube. When $k = 1$ and $x_1 < x_2 < \cdots < x_n$ is an ordered sample from the $U(0, 1)$ distribution and $x_0 = 0, x_{n+1} = 1$ then the estimator of $\int_0^1 f(x)\,dx$ takes the form

$$\tfrac{1}{2}\sum_{i=0}^{n}(x_{i+1} - x_i)(f(x_{i+1}) + f(x_i)) \qquad (6.28)$$

which is readily seen to be the trapezoid rule based on these points. Accordingly (6.28) is exact for all polynomials of degree 1 or less. This is close in spirit to a randomized quadrature rule. It is shown in Yakowitz, Krimmel and Szidarovsky (1978) that if $f \in C^2([0, 1])$ then the MSE of (6.28) is bounded by M/n^4 for some constant M for all n. Accordingly the performance of the weighted Monte Carlo approach surpasses the performance of crude Monte Carlo for such functions. Unfortunately, it is also shown there that for natural multidimensional analogues of (6.28) performance deteriorates as dimension rises with crude Monte Carlo becoming inevitably superior.

Clearly error estimation is straightforward with randomized quadrature rules. A weakness with using quasi-random rules is obtaining error estimates. One possibility is to randomize in some fashion the quasi-random rule; e.g. consider Example 6.9 where good lattice point rules are randomized. While these are not strictly speaking examples of randomized quadrature rules, the motivation is the same. See Owen (1997a, 1997b) for a discussion of methods for randomly scrambling (t, m, k)-nets and (t, k)-sequences and some interesting results concerning improvements for this approach over crude Monte Carlo.

6.4.6 CONTROL VARIATES

Suppose that we can exactly evaluate the integral of $g : R^k \to R^1$. Further suppose that $g/w \in L^2(w)$. Then we can write

$$I(f) = \int_{R^k} g(x)\,dx + \int_{R^k} (f(x) - g(x))\,dx$$

and rather than using the importance sampler w to estimate the integral of f, we use w to estimate the integral of $f - g$. In this case g is referred to as an *additive control variate*. We have that

$$\begin{aligned}\mathrm{Var}_w\left[\frac{f-g}{w}\right] &= \mathrm{Var}_w\left[\frac{f}{w}\right] + \mathrm{Var}_w\left[\frac{g}{w}\right]\\ &\quad -2\mathrm{SD}_w\left[\frac{f}{w}\right]\mathrm{SD}_w\left[\frac{g}{w}\right]\mathrm{Corr}_w\left[\frac{f}{w},\frac{g}{w}\right] \qquad (6.29)\end{aligned}$$

so that if $\mathrm{Corr}_w[f/w, g/w]$ is close to 1 and $\mathrm{SD}_w[f/w] > \mathrm{SD}_w[g/w]$ we have

$$\mathrm{Var}_w\left[\frac{f-g}{w}\right] \approx \left(\mathrm{SD}_w\left[\frac{f}{w}\right] - \mathrm{SD}_w\left[\frac{g}{w}\right]\right)^2 < \mathrm{Var}_w\left[\frac{f}{w}\right].$$

When $(f - g)/w$ is less variable than f/w, i.e. $\mathrm{Var}_w[(f - g)/w] < \mathrm{Var}_w[f/w]$, then g is an *effective control variate*. From (6.29) g is an effective control variate if and only if

$$\mathrm{Corr}_w\left[\frac{f}{w},\frac{g}{w}\right]\frac{\mathrm{SD}_w[f/w]}{\mathrm{SD}_w[g/w]} > \frac{1}{2}. \qquad (6.30)$$

Note that the left-hand side of (6.30) is the coefficient of g/w in the orthogonal projection of f/w onto the linear span $L\{1, g/w\} \subset L^2(w)$. Therefore there has to be a strong linear relationship existing between f/w and g/w if the control variate g is to be effective. It is worth remarking that a poorly chosen control variate can be worse than ineffective as it can lead to an increase in the variance of the estimator. The left-hand side of (6.30) can be estimated as part of the simulation to assess whether or not the control variate is indeed effective.

Another possibility with control variates is to consider a number of control variates $g_1, \ldots, g_l$ and ask what linear combination $g = \sum_{i=1}^{l} b_i g_i$ gives the biggest variance reduction. For example, when $l = 1$ the best choice of b_1 is

$$b_1 = \mathrm{Corr}_w\left[\frac{f}{w},\frac{g_1}{w}\right]\frac{\mathrm{SD}_w[f/w]}{\mathrm{SD}_w[g_1/w]}.$$

With this choice we have that the left-hand side of (6.30) equals 1 and

$$\mathrm{Var}_w\left[\frac{f-g_1}{w}\right] = \left(1 - \mathrm{Corr}_w^2\left[\frac{f}{w},\frac{g_1}{w}\right]\right)\mathrm{Var}_w\left[\frac{f}{w}\right].$$

Similarly formulas for the optimal choices of $b_1, \ldots, b_l$ can be developed (Exercise 32) and these values can be estimated as part of the simulation. A downside of this *regression* approach is that the estimates of error become more complicated as we should take into account that the b_i are being estimated as well. Also the regression approach requires the estimation integrals of the form $\int_{R^k} g(x) f(x)\, dx$ which in general may be harder to estimate accurately than the original integral. Still useful applications have been made of this technique; see Rubinstein (1986) for more discussion.

Rather than using g additively we can use it as a *ratio control variate*. For when $\int_{R^k} g(x)\, dx \neq 0$ we can write

$$I(f) = \int_{R^k} f(x)\, dx = \int_{R^k} g(x)\, dx \left\{ \frac{\int_{R^k} f(x)\, dx}{\int_{R^k} g(x)\, dx} \right\} = I(g) \left\{ \frac{I(f)}{I(g)} \right\}$$

and then estimate the ratio $I(f)/I(g)$ using

$$\hat{R}_{w,n} = \frac{\sum_{i=1}^{n} (f(x_i)/w(x_i))}{\sum_{i=1}^{n} (g(x_i)/w(x_i))}$$

based on a sample $x_1, \ldots, x_n$ from w. From Theorem 6.4 we have that the asymptotic variance of this estimator is given by n^{-1} times

$$I^{-2}(g) \left\{ \mathrm{Var}_w \left[\frac{f}{w} \right] + I^2(f) I^{-2}(g) \mathrm{Var}_w \left[\frac{g}{w} \right] - 2I(f) I^{-1}(g) \mathrm{Cov}_w \left[\frac{f}{w}, \frac{g}{w} \right] \right\}.$$

If $I(f) \neq 0$, $I(g) \neq 0$ and have the same sign then g is an effective control variate if and only if

$$\mathrm{Corr}_w \left[\frac{f}{w}, \frac{g}{w} \right] > \frac{1}{2} \frac{\mathrm{CV}_w[g/w]}{\mathrm{CV}_w[f/w]}. \tag{6.31}$$

Note if $\mathrm{CV}_w[f/w]$ is large and $\mathrm{CV}_w[g/w]$ small then the correlation between f/w and g/w does not have to be large for g to be effective. Again we can estimate the components of (6.31) as part of the simulation to assess whether or not g is an effective control variate.

We consider now some techniques for obtaining suitable control variates.

Example 6.13 *Taylor series control variates*
One approach to obtaining an effective additive control variate is to expand f/w about some value x_0 obtaining

$$\frac{f(x)}{w(x)} \approx \frac{f(x_0)}{w(x_0)} + \frac{1}{w^2(x_0)} \left[\frac{\partial f(x_0)}{\partial x} w(x_0) - f(x_0) \frac{\partial w(x_0)}{\partial x} \right] (x - x_0) \tag{6.32}$$

and then define the control variate as

$$g(x) = \left\{ \frac{f(x_0)}{w(x_0)} + \frac{1}{w^2(x_0)} \left[\frac{\partial f(x_0)}{\partial x} w(x_0) - f(x_0) \frac{\partial w(x_0)}{\partial x} \right] (x - x_0) \right\} w(x).$$

This is only useful as a control variate when (6.32) gives a good approximation and of course the first moment of w must exist. Clearly we can extend this approach to using higher order expansions. Note that we are using a Taylor expansion of f/w here as this makes use of the fact that w is in some sense an approximation to w. The hope then is that f/w can be well approximated by a low order polynomial.

Example 6.14 *Asymptotic control variates*
As we noted in Chapter 4 one of the deficiencies of asymptotic approximations is that they do not produce error estimates. Further they are sometimes used in contexts where it is not clear that the asymptotic parameter is sufficiently near its limiting value to provide the required accuracy. One possibility, as proposed in Evans and Swartz (1995a, 1995b), is to use an asymptotic approximation as a control variate. As a natural choice, when it is applicable, we can take $g(x) = I_{\text{Lap}}(f)h(x)$ where $I_{\text{Lap}}(f)$ is the Laplace approximation of $I(f)$ and h is the $N_k(\hat{\mu}, \hat{\Sigma})$ density. This reflects the fact that the multivariate normal distribution is at the heart of the Laplace approximation. We can use g as an additive or ratio control variate.

Now suppose that we want to estimate a ratio

$$R = \frac{I(mf)}{I(f)}$$

where $f \geq 0$ is integrable. Then assuming that we can calculate $I(mg)$ exactly and letting $x_1, \ldots, x_n$ denote a sample from w, we can write the ratio control variate estimator as

$$\frac{I(mg)}{I(g)} \left[\frac{\sum_{i=1}^n m(x_i) f(x_i)/w(x_i)}{\sum_{i=1}^n m(x_i) g(x_i)/w(x_i)} \right] \left[\frac{\sum_{i=1}^n f(x_i)/w(x_i)}{\sum_{i=1}^n g(x_i)/w(x_i)} \right]^{-1}. \tag{6.33}$$

Note that (6.33) is a ratio of ratios. The asymptotic variance of (6.33) is

$$\frac{I^2(mg)}{I^2(g)} \frac{u^t V u}{n}$$

where the vector u is given by

$$u^t = (\mu_2 \mu_3)^{-1} (\mu_4, -\mu_1 \mu_4/\mu_2, -\mu_1 \mu_4/\mu_3, \mu_1),$$

the matrix $V = (\sigma_{ij})$, the indices 1, 2, 3 and 4 refer to the estimators $mf/w, mg/w, f/w$ and g/w respectively and the μ_i, σ_{ij} refer to the means and covariances of the respective estimators (Exercise 33). An expression can also be obtained for the asymptotic variance when we estimate R using additive control variates (Exercise 34).

Introducing these control variates in the linear model example discussed in Example 6.4 (first discussed in Example 4.9) resulted in minor savings. For example, the average efficiency of the additive control variate technique in estimating the vector of posterior means is 1.1 when compared to straight Student importance sampling. For the ratio control variate technique, the average efficiency is 1.2. These small savings are the result of a poor approximation of the unnormalized posterior f by $g = I_{\text{Lap}}(f)h$. Indeed, when we modify the example to have $n = 20$ in each cell instead of $n = 5$, the posterior becomes more normal and the efficiencies increase. For example, the average efficiency in estimating the vector of posterior means is 4.7 using the additive control variate and 5.1 using the ratio control variate. In terms of computing overhead, the control variate methods require only slightly more computing time than Student importance sampling. Thus these control variates are primarily useful when the Laplace approximation is good.

The use of ratio control variates was introduced in Powell and Swann (1966) although g was restricted there to being a density. The method was called *weighted uniform sampling* because the problem was assumed to be such that w could be taken to be the uniform distribution on $[0, 1]^k$. See Spanier (1979), Evans and Swartz (1988a, 1995b) and Hesterberg (1995) for some additional uses of control variates.

6.5 Adaptive importance sampling

As stated at the outset of the chapter, the main problem in importance sampling is finding a "good" importance sampler. For us, a good importance sampler is one which allows efficient generation of variates and one which gives acceptable estimates within practical computing times. In *adaptive importance sampling* we have a family $\mathcal{W} = \{w_\lambda : \lambda \in \Lambda\}$ of potential importance samplers for the integration problem. We begin with an initial importance sampler w_{λ_1}, generate a sample from w_{λ_1} and then use this sample together with estimates of integrals related to the integration problem of interest, to choose a new importance sampler w_{λ_2}. This process is iterated until it stabilizes.

At times, concerns are expressed about whether adaptation really improves the efficiency of the approximation process. In our experience, the answer is an unequivocal "yes" and we provide examples later in this section. If doubts remain, however, we note that the efficiency of a procedure should typically be assessed in a sense broader than the approximation of a single integral. For example, in many statistical applications there are a number of related integrals that we want to approximate where the integrands are of the form mf for functions m where $f \geq 0$ is integrable.

There are many approaches to adaptive importance sampling that have been considered in the literature. For example, see Kloek and van Dijk (1978), Smith, Skene, Shaw and Naylor (1987), Naylor and Smith (1988), Evans (1988, 1991a,

1991b), Oh (1991), Oh and Berger (1992, 1993), West (1992, 1993) and Givens and Raftery (1996). We discuss some of these approaches here.

From a simple point of view it is clear how we should select an element of $\mathcal{W}$; namely choose $\lambda \in \Lambda$ so that

$$E_{w_\lambda}\left[\frac{f^2}{w_\lambda}\right] = \int_{R^k} \frac{f^2(x)}{w_\lambda(x)}\, dx \tag{6.34}$$

is minimized. Of course, this is typically an intractable optimization problem. Alternatively we could try to iteratively minimize (6.34) through the adaptive importance sampling procedure. For example, this could entail estimating

$$\frac{\partial}{\partial\lambda} \int_{R^k} \frac{f^2(x)}{w_\lambda(x)}\, dx$$

and using steepest descent. Alternatively, the minimization could be approached by estimating second derivatives as well and using Newton–Raphson. We return to a discussion of this approach later in this section. For now we focus on a simpler approach which we call *matching characteristics*. For this we suppose that f is nonnegative and integrable and that the density proportional to f would be an excellent importance sampler for the integrands mf that we face. Let us represent this density by p_f. Further let $C : R^k \to R^l$ be such that the vector of characteristics

$$E_{p_f}[C] = \frac{I(Cf)}{I(f)}$$

exists finitely. For example, $E_p[C]$ may be composed of means, variances and probability contents of particular subsets of R^k for p_f. Then we look for $w_\lambda \in \mathcal{W}$ that minimizes

$$\|E_{p_f}[C] - E_{w_\lambda}[C]\|. \tag{6.35}$$

The components of C are chosen so that for each $w_\lambda \in \mathcal{W}$ we can evaluate $E_{w_\lambda}[C]$ in closed form and solve the minimization problem in (6.35) whenever $E_{p_f}[C]$ is known.

Of course, it may also be difficult to minimize (6.35), because we typically do not know the value of $E_{p_f}[C]$, but we can adopt the following approach. First pick $\lambda_1 \in \Lambda$ and generate $x_1, \ldots, x_n$ from w_{λ_1}. Then we estimate $E_{p_f}[C]$ by

$$\hat{R}_{(1)}(C) = \frac{\hat{I}_{w_{\lambda_1},n}(Cf)}{\hat{I}_{w_{\lambda_1},n}(f)}$$

and find $\lambda_2 \in \Lambda$ that minimizes

$$\left\| \hat{R}_{(1)}(C) - E_{w_\lambda}[C] \right\|.$$

Next we generate a sample $x_{n+1}, \ldots, x_{2n}$ from w_{λ_2} and estimate $E_{p_f}[C]$ by

$$\hat{R}_{(2)}(C) = \frac{\hat{I}_{w_{\lambda_1,n}}(Cf) + \hat{I}_{w_{\lambda_2,n}}(Cf)}{\hat{I}_{w_{\lambda_1,n}}(f) + \hat{I}_{w_{\lambda_2,n}}(f)}$$

and find $\lambda_2 \in \Lambda$ that minimizes

$$\left\| \hat{R}_{(2)}(C) - E_{w_\lambda}[C] \right\|.$$

This process is iterated until the estimates $\hat{R}_{(i)}(C)$ stabilize.

The analysis of convergence issues for adaptive importance sampling is more difficult than importance sampling because of the lack of independence between iterations. Almost sure convergence of $\hat{R}_{(i)}(C)$ to $E_{p_f}[C]$, an asymptotic normality result for $\hat{R}_{(n)}(C)$ and a consistent estimator for the asymptotic variance were established in Oh and Berger (1992). These results required several restrictive assumptions, however, that do not conform to the usage of the algorithm. For example, uniform boundedness of a weight function was assumed; namely there must exist B such that $|m(x)|f(x)/w_\lambda(x) \leq B$ for every λ and x, and every function m of interest. This almost never holds in an application. Here we prove these convergence results but with weaker conditions. As with importance sampling, where we want estimates with finite variances, these are nearly impossible to check in practice but at least we know that convergence holds under fairly general conditions. The proofs use well-known results from the theory of martingales, see Hall and Heyde (1980), and follow the general lines of the proofs in Oh and Berger (1992).

We establish a strong law for adaptive importance sampling. Without loss of generality we assume that $l = 1$; i.e. $C(x) \in R^1$. For an integrand mf define

$$p_{|m|f}(x) = \frac{|m(x)|f(x)}{\int_{R^k} |m(x)|f(x)\,dx}$$

and note that this is the optimal importance sampler for the integral $I(mf)$.

Theorem 6.15 If for the matching characteristics adaptive importance sampling procedure

(i) $$E\left\{\sum_{j=1}^{\infty} \frac{1}{j^2} E_{w_{\lambda_j}}\left[\left(\frac{p_{|m|}(X) - w_{\lambda_j}(X)}{w_{\lambda_j}(X)}\right)^2\right]\right\} < \infty$$

for $m = f$ and $m = Cf$ then $\hat{R}_{(i)}(C) \to E_{p_f}[C]$ almost surely as $i \to \infty$.

Proof: Denote the sequence of generated random variables by $X_1, X_2, \ldots$. We show that the numerator

$$\sum_{j=1}^{i} \hat{I}_{w_{\lambda_j},n}(Cf)$$

of $\hat{R}_{(i)}(C)$ converges almost surely to $\int_{R^k} C(x)f(x)\,dx$ as $i \to \infty$ and the result for the denominator follows similarly. Let

$$U_i = \sum_{j=1}^{i} \frac{1}{j}\left(\hat{I}_{w_{\lambda_j},n}(Cf) - \int_{R^k} C(x)f(x)\,dx\right)$$

and note that

$$E[U_i|X_1, \ldots, X_{(i-1)n}] = U_{i-1}.$$

Therefore $\{(U_i, \sigma(X_1, \ldots, X_{in}))\colon i = 1, 2, \ldots\}$ (where $\sigma(X_1, \ldots, X_{in})$ denotes the sigma algebra generated by $X_1, \ldots, X_{in}$) is a martingale with $E[U_i] = 0$.

Now using the result from Exercise 35,

$$\begin{aligned}
E[U_i^2] &= \sum_{j=1}^{i} \frac{1}{j^2} E\left[\left(\hat{I}_{w_{\lambda_j},n}(Cf) - \int_{R^k} C(x)f(x)\,dx\right)^2\right] \\
&= \frac{1}{n}\sum_{j=1}^{i} \frac{1}{j^2} E\left[\int_{R^k} \frac{C^2(x)f^2(x)}{w_{\lambda_j}(x)}\,dx - \left(\int_{R^k} C(x)f(x)\,dx\right)^2\right] \\
&= \frac{\left(\int_{R^k} |C(x)|f(x)\,dx\right)^2}{n} E\left\{\sum_{j=1}^{i} \frac{1}{j^2} E_{w_{\lambda_j}}\left[\left(\frac{p_{|C|f}(X) - w_{\lambda_j}(X)}{w_{\lambda_j}(X)}\right)^2\right]\right\} \\
&\quad + \frac{1}{n}\sum_{j=1}^{i} \frac{1}{j^2}\left[\left(\int_{R^k} |C(x)|f(x)\,dx\right)^2 - \left(\int_{R^k} C(x)f(x)\,dx\right)^2\right]
\end{aligned}$$

where the last equality follows as in the proof of Theorem 6.5. Therefore using assumption (i), $E[U_i^2]$ is bounded above and

$$\sup_i E[U_i^+] \leq \sup_i E[|U_i|] \leq \sup_i \{E[U_i^2]\}^{1/2}.$$

By the martingale convergence theorem (see Durrett 1991) U_i converges almost surely to a random variable U with $E[|U|] < \infty$. Then by *Kronecker's lemma* (if $a_j \uparrow \infty$ and $\sum_{j=1}^{\infty} y_j/a_j$ converges then $a_i^{-1}\sum_{j=1}^{i} y_j \to 0$, see Durrett 1991) applied to $a_j = j$ and

$$y_j = \hat{I}_{w_{\lambda_j},n}(Cf) - \int_{R^k} C(x)f(x)\,dx$$

we have that

$$\frac{1}{i}\left(\sum_{j=1}^{i} \hat{I}_{w_{\lambda_j},n}(Cf)\right) \to \int_{R^k} C(x)f(x)\,dx$$

almost surely as $i \to \infty$.

Note that (i) in Theorem 6.15 states that the chi-square distances between the optimal importance samplers and the adaptively selected importance samplers cannot grow too large as we iterate if we want convergence. If

$$|C(x)|f(x) \leq Bw_\lambda(x)$$

for all x and λ then

$$\left|\frac{p_{|C|f}(x) - w_\lambda(x)}{w_\lambda(x)}\right| \leq \left(\int_{R^k} |C(z)|f(z)\,dz\right)^{-1} B + 1$$

for all x and λ. Therefore uniform boundedness is a much stronger condition than (i). The chi-square distance can grow slowly with j and convergence still takes place. Obviously whether or not condition (i) holds depends on the family $\mathcal{W}$, n and w_{λ_1}. If we choose a poor family, n too small or a poor starting value we cannot expect the condition to apply.

Now suppose that Λ is an open set and there is an open set $\mathcal{R}$ which contains all possible values for the $\hat{R}_{(i-1)}(C)$. The following result ensures that the sequence of importance samplers converge to the importance sampler in $\mathcal{W}$ that minimizes (6.35).

Theorem 6.16 Consider the matching characteristics adaptive importance sampling procedure. If in addition to (i), we have that

(ii) for each $r \in \mathcal{R}$ there is a unique $\alpha(r) \in \Lambda$ such that

$$\left\| E_{w_{\alpha(r)}}[C] - r \right\| = \inf_\lambda \left\| E_{w_\lambda}[C] - r \right\|$$

(iii) $\alpha : \mathcal{R} \to \Lambda$ is continuous

then $\lambda_i = \alpha(\hat{R}_{(i-1)}(C)) \to \alpha(E_{p_f}[C]) = \lambda_*$ almost surely as $i \to \infty$.

Proof: Exercise 36.

We now prove a central limit theorem for adaptive importance sampling.

Theorem 6.17 For the matching characteristics adaptive importance sampling procedure suppose that (i)–(iii) hold and assume

(iv) $\int g(x)/w_\lambda(x)\,dx$ is continuous in λ for $g = f^2, Cf^2$ and $C^2 f^2$,

(v) $\int g(x)/w_\lambda^2(x)\,dx$ is continuous in λ for $g = f^3, |C|f^3, C^2 f^3$ and $|C|^3 f^3$.

Then $\hat{R}_{(i)}(C) \sim AN(E_{p_f}[C], \sigma^2/in)$ as $i \to \infty$ where σ^2 equals

$$I^{-2}(f)\left\{\mathrm{Var}_{w_{\lambda*}}\left[\frac{Cf}{w_{\lambda*}}\right] + I^2(Cf)I^{-2}(f)\mathrm{Var}_{w_{\lambda*}}\left[\frac{f}{w_{\lambda*}}\right]\right.$$
$$\left. -2I(Cf)I^{-1}(f)\mathrm{Cov}_{w_{\lambda*}}\left[\frac{Cf}{w_{\lambda*}}, \frac{f}{w_{\lambda*}}\right]\right\}. \tag{6.36}$$

Proof: The proof establishes the conditions necessary to apply Corollary 3.1 and Theorem 3.2 in Hall and Heyde (1980). We show that an arbitrary linear combination of the numerator and denominator of $\hat{R}_{(i)}(C)$ is asymptotically normal.

For constants a_1, a_2 let

$$Y_{ij} = \frac{1}{\sqrt{i}}\left(a_1 \hat{I}_{w_{\lambda_j},n}(f) + a_2 \hat{I}_{w_{\lambda_j},n}(Cf) - a_1 I(f) - a_2 I(Cf)\right).$$

Then $E[Y_{ij}] = E[E[Y_{ij}|X_1, \ldots, X_{(j-1)n}]] = 0$ and

$$\begin{aligned}
&E\left[Y_{ij}^2 | X_1, \ldots, X_{(j-1)n}\right] \\
&= \frac{1}{in}\left\{a_1^2 \mathrm{Var}_{w_{\lambda_j}}\left[\frac{f}{w_{\lambda_j}}\right] + a_2^2 \mathrm{Var}_{w_{\lambda_j}}\left[\frac{Cf}{w_{\lambda_j}}\right] + 2a_1 a_2 \mathrm{Cov}_{w_{\lambda_j}}\left[\frac{f}{w_{\lambda_j}}, \frac{Cf}{w_{\lambda_j}}\right]\right\}.
\end{aligned}$$

Therefore

$$\begin{aligned}
&\sum_{j=1}^{i} E\left[Y_{ij}^2 | X_1, \ldots, X_{(j-1)n}\right] \\
&= \frac{1}{n}\left\{\frac{a_1^2}{i}\sum_{j=1}^{i} \mathrm{Var}_{w_{\lambda_j}}\left[\frac{f}{w_{\lambda_j}}\right] + \frac{a_2^2}{i}\sum_{j=1}^{i} \mathrm{Var}_{w_{\lambda_j}}\left[\frac{Cf}{w_{\lambda_j}}\right]\right. \\
&\qquad \left. + \frac{2a_1 a_2}{i}\sum_{j=1}^{i} \mathrm{Cov}_{w_{\lambda_j}}\left[\frac{f}{w_{\lambda_j}}, \frac{Cf}{w_{\lambda_j}}\right]\right\} \\
&\le \frac{1}{n}\left\{\frac{a_1^2}{i}\sum_{j=1}^{i} \mathrm{Var}_{w_{\lambda_j}}\left[\frac{f}{w_{\lambda_j}}\right] + \frac{a_2^2}{i}\sum_{j=1}^{i} \mathrm{Var}_{w_{\lambda_j}}\left[\frac{Cf}{w_{\lambda_j}}\right]\right. \\
&\qquad \left. + \frac{2|a_1||a_2|}{i}\sum_{j=1}^{i} \mathrm{Var}_{w_{\lambda_j}}^{1/2}\left[\frac{f}{w_{\lambda_j}}\right] \mathrm{Var}_{w_{\lambda_j}}^{1/2}\left[\frac{Cf}{w_{\lambda_j}}\right]\right\} \\
&= \frac{1}{n}\left\{\frac{1}{i}\sum_{j=1}^{i}\left(|a_1|\mathrm{Var}_{w_{\lambda_j}}^{1/2}\left[\frac{f}{w_{\lambda_j}}\right] + |a_2|\mathrm{Var}_{w_{\lambda_j}}^{1/2}\left[\frac{Cf}{w_{\lambda_j}}\right]\right)^2\right\} \\
&\le \frac{4}{n}\left\{\frac{a_1^2}{i}\sum_{j=1}^{i} \mathrm{Var}_{w_{\lambda_j}}\left[\frac{f}{w_{\lambda_j}}\right] + \frac{a_2^2}{i}\sum_{j=1}^{i} \mathrm{Var}_{w_{\lambda_j}}\left[\frac{Cf}{w_{\lambda_j}}\right]\right\}
\end{aligned}$$

where the first inequality is the Cauchy–Schwarz inequality and the second follows from $(|x| + |y|)^2 \le 4\max\{|x|^2, |y|^2\} \le 4(|x|^2 + |y|^2)$.

Now by Theorem 6.16 and (iv) we have that $\mathrm{Var}_{w_{\lambda_j}}[f/w_{\lambda_j}]$ converges almost surely to $\mathrm{Var}_{w_{\lambda *}}[f/w_{\lambda *}]$ and $\mathrm{Var}_{w_{\lambda_j}}[Cf/w_{\lambda_j}]$ converges almost surely to $\mathrm{Var}_{w_{\lambda *}}[Cf/w_{\lambda *}]$ as $j \to \infty$. Therefore by the Cesàro summation theorem

$$\frac{a_1^2}{i}\sum_{j=1}^{i}\mathrm{Var}_{w_{\lambda_j}}\left[\frac{f}{w_{\lambda_j}}\right]+\frac{a_2^2}{i}\sum_{j=1}^{i}\mathrm{Var}_{w_{\lambda_j}}\left[\frac{Cf}{w_{\lambda_j}}\right]$$

converges almost surely to $a_1^2\mathrm{Var}_{w_{\lambda *}}[f/w_{\lambda *}] + a_2^2\mathrm{Var}_{w_{\lambda *}}[Cf/w_{\lambda *}]$ as $i \to \infty$. From this we get the almost sure convergence of $\sum_{j=1}^{i} E[Y_{ij}^2|X_1,\ldots,X_{(j-1)n}]$. Again applying the Cesàro summation theorem we have that

$$\begin{aligned}
&\lim_{i\to\infty}\sum_{j=1}^{i} E\left[Y_{ij}^2|X_1,\ldots,X_{(j-1)n}\right]\\
&=\frac{1}{n}\left\{a_1^2\mathrm{Var}_{w_{\lambda *}}\left[\frac{f}{w_{\lambda *}}\right]+a_2^2\mathrm{Var}_{w_{\lambda *}}\left[\frac{Cf}{w_{\lambda *}}\right]+2a_1a_2\mathrm{Cov}_{w_{\lambda *}}\left[\frac{f}{w_{\lambda *}},\frac{Cf}{w_{\lambda *}}\right]\right\}\\
&=\frac{1}{n}\begin{pmatrix}a_1\\a_2\end{pmatrix}^t\mathrm{Var}_{w_*}\left[\begin{pmatrix}\dfrac{f}{w_{\lambda *}}\\[2mm]\dfrac{Cf}{w_{\lambda *}}\end{pmatrix}\right]\begin{pmatrix}a_1\\a_2\end{pmatrix}\\
&=\frac{\eta^2}{n}
\end{aligned}$$

which is one of the conditions required in order to apply the results in Hall and Heyde (1980).

Now let

$$S_{i,j}=\sum_{l=1}^{j} Y_{il}$$

for $j = 1,\ldots,i$ and note that $E[S_{i,j}|X_1,\ldots,X_{(j-1)n}] = S_{i,j-1}$ and $E[S_{i,j}] = 0$. Using (Exercise 37)

$$E\left[S_{i,j}^2\right]=\sum_{l=1}^{j}E\left[Y_{il}^2\right]$$

this is finite because

$$\begin{aligned}
\sum_{l=1}^{j}E\left[Y_{il}^2\right]&=E\left[\sum_{l=1}^{j}E\left[Y_{il}^2|X_1,\ldots,X_{(j-1)n}\right]\right]\\
&\le E\left[\frac{4}{n}\left\{\frac{a_1^2}{i}\sum_{l=1}^{j}\mathrm{Var}_{w_{\lambda_l}}\left[\frac{f}{w_{\lambda_l}}\right]+\frac{a_2^2}{i}\sum_{l=1}^{j}\mathrm{Var}_{w_{\lambda_l}}\left[\frac{Cf}{w_{\lambda_l}}\right]\right\}\right]
\end{aligned}$$

and

$$E\left[\mathrm{Var}_{w_{\lambda_l}}\left[\frac{f}{w_{\lambda_l}}\right]\right] = \left(\int_{R^k} f(x)\,dx\right)^2 E\left[E_{w_{\lambda_l}}\left[\left(\frac{p_f(X) - w_{\lambda_l}(X)}{w_{\lambda_l}(X)}\right)^2\right]\right] < \infty$$

by (i) and similarly $E[\mathrm{Var}_{w_{\lambda_l}}[Cf/w_{\lambda_l}]] < \infty$. Therefore

$$\{(S_{i,j}, \sigma(X_1, \dots, X_{nj})) \colon j = 1, \dots i \text{ and } i = 1, 2, \dots\}$$

is a zero mean, square integrable martingale array which is another condition necessary to apply the results in Hall and Heyde (1980).

Finally we have to show that for every $\epsilon > 0$

$$\sum_{j=1}^{i} E\left[Y_{ij}^2 I_{\{|Y_{ij}|>\epsilon\}} \middle| \sigma\left(X_1, \dots, X_{n(j-1)}\right)\right] \to 0 \tag{6.37}$$

almost surely as $i \to \infty$. We have that

$$E\left[Y_{ij}^2 I_{\{|Y_{ij}|>\epsilon\}} \middle| \sigma\left(X_1, \dots, X_{n(j-1)}\right)\right] \le \epsilon^{-1} E\left[|Y_{ij}|^3 \middle| \sigma\left(X_1, \dots, X_{n(j-1)}\right)\right].$$

and

$$\begin{aligned} |Y_{ij}|^3 \le \frac{1}{i^{3/2}} \Big\{ & |a_1|^3 \left|\hat{I}_{w_{\lambda_j},n}(f) - I(f)\right|^3 \\ & + 3|a_1|^2|a_2| \left|\hat{I}_{w_{\lambda_j},n}(f) - I(f)\right|^2 \left|\hat{I}_{w_{\lambda_j},n}(Cf) - I(Cf)\right| \\ & + 3|a_1||a_2|^2 \left|\hat{I}_{w_{\lambda_j},n}(f) - I(f)\right| \left|\hat{I}_{w_{\lambda_j},n}(Cf) - I(Cf)\right|^2 \\ & + |a_2|^3 \left|\hat{I}_{w_{\lambda_j},n}(Cf) - I(Cf)\right|^3 \Big\}. \end{aligned} \tag{6.38}$$

Considering the first term on the left-hand-side of (6.38) we then have that

$$\begin{aligned} & \left|\hat{I}_{w_{\lambda_j},n}(f) - I(f)\right|^3 \\ & \le \left(\left|\hat{I}_{w_{\lambda_j},n}(f)\right| + |I(f)|\right)^3 \\ & = \left|\hat{I}_{w_{\lambda_j},n}(f)\right|^3 + 3\left|\hat{I}_{w_{\lambda_j},n}(f)\right|^2 |I(f)|^2 + 3\left|\hat{I}_{w_{\lambda_j},n}(f)\right| |I(f)|^2 + |I(f)|^3 \end{aligned} \tag{6.39}$$

and

$$
\begin{aligned}
&\left|\hat{I}_{w_{\lambda_j},n}(f)\right|^3 \\
&\quad \le \frac{1}{n^3}\left\{\sum_{l=1}^{n}\left|\frac{f(X_{n(j-1)+l})}{w_{\lambda_j}(X_{n(j-1)+l})}\right|^3\right. \\
&\qquad + \sum_{l_1\neq l_2}\left|\frac{f(X_{n(j-1)+l_1})}{w_{\lambda_j}(X_{n(j-1)+l_1})}\right|^2\left|\frac{f(X_{n(j-1)+l_2})}{w_{\lambda_j}(X_{n(j-1)+l_2})}\right| \\
&\qquad \left. + \sum_{l_1\neq l_2\neq l_3}\left|\frac{f(X_{n(j-1)+l_1})}{w_{\lambda_j}(X_{n(j-1)+l_1})}\right|\left|\frac{f\left(X_{n(j-1)+l_2}\right)}{w_{\lambda_j}(X_{n(j-1)+l_2})}\right|\left|\frac{f(X_{n(j-1)+l_3})}{w_{\lambda_j}(X_{n(j-1)+l_3})}\right|\right\}.
\end{aligned}
$$

Therefore

$$
\begin{aligned}
&\frac{1}{i^{3/2}}\sum_{j=1}^{i} E\left[\left|\hat{I}_{w_{\lambda_j},n}(f)\right|^3 |\sigma(X_1,\ldots,X_{n(j-1)})\right] \\
&\quad \le \frac{1}{i^{3/2}}\sum_{j=1}^{i}\left\{\frac{1}{n^2}E_{w_{\lambda_j}}\left[\left|\frac{f(X)}{w_{\lambda_j}(X)}\right|^3\right]\right. \\
&\qquad + \frac{n-1}{2n^2}E_{w_{\lambda_j}}\left[\left|\frac{f(X)}{w_{\lambda_j}(X)}\right|^2\right]E_{w_{\lambda_j}}\left[\left|\frac{f(X)}{w_{\lambda_j}(X)}\right|\right] \\
&\qquad \left. + \frac{(n-1)(n-2)}{6n^2}\left(E_{w_{\lambda_j}}\left[\left|\frac{f(X)}{w_{\lambda_j}(X)}\right|\right]\right)^3\right\} \\
&\quad \to 0
\end{aligned}
$$

almost surely as $i \to \infty$ by Theorems 6.15 and 6.16 and hypotheses (iv) and (v). Arguing in a similar fashion for the other terms in (6.39) we conclude that

$$
\frac{1}{i^{3/2}}\sum_{j=1}^{i} E\left[\left|\hat{I}_{w_{\lambda_j},n}(f) - I(f)\right|^3\middle|\sigma(X_1,\ldots,X_{n(j-1)})\right] \to 0
$$

almost surely as $i \to \infty$. The other terms on the left-hand side of (6.38) can be dealt with in exactly the same fashion and we therefore conclude that

$$
\sum_{j=1}^{i} E\left[\left|Y_{ij}\right|^3 |\sigma(X_1,\ldots,X_{n(j-1)})\right] \to 0
$$

almost surely as $i \to \infty$ which implies (6.37).

Therefore by the results cited in Hall and Heyde (1980) we have that

$$S_{i,i} \sim AN\left(0, \frac{\eta^2}{n}\right)$$

and since a_1 and a_2 are arbitrary this implies that

$$\begin{pmatrix} \frac{1}{i}\sum_{j=1}^{i} \hat{I}_{w_{\lambda_j},n}(f) \\ \frac{1}{i}\sum_{j=1}^{i} \hat{I}_{w_{\lambda_j},n}(Cf) \end{pmatrix} \sim AN_2\left(\begin{pmatrix} I(f) \\ I(Cf) \end{pmatrix}, \frac{1}{in}\mathrm{Var}_{w_*}\left[\begin{pmatrix} \frac{f}{w_{\lambda *}} \\ \frac{Cf}{w_{\lambda *}} \end{pmatrix}\right]\right)$$

as $i \to \infty$ and Theorem 6.17 follows from the delta theorem.

It is interesting to observe that Theorem 6.17 appears to allow us to allocate the in generations in any way we choose; e.g. take $n = 1$. But recall condition (i). It seems unlikely that this condition will hold if we allow for too much variability in the choice of the w_{λ_j}. For this reason we advocate taking n large enough so that the estimates $\hat{R}_{(i)}(C)$ have small standard errors.

Finally we need an almost sure consistent estimate of σ^2. For this we let

$$\hat{E}_{w_{\lambda*}}\left[\frac{f}{w_{\lambda*}}\right] = \frac{1}{i}\sum_{j=1}^{i} \hat{I}_{w_{\lambda_j,n}}(f),$$

$$\hat{E}_{w_{\lambda*}}\left[\frac{Cf}{w_{\lambda*}}\right] = \frac{1}{i}\sum_{j=1}^{i} \hat{I}_{w_{\lambda_j,n}}(Cf),$$

$$\widehat{\mathrm{Var}}_{w_{\lambda*}}\left[\frac{f}{w_{\lambda*}}\right] = \frac{1}{i}\sum_{j=1}^{i} \hat{I}_{w_{\lambda_j,n}}\left(\frac{f^2}{w_{w_{\lambda_j}}}\right) - \hat{E}^2_{w_{\lambda*}}\left[\frac{f}{w_{\lambda*}}\right],$$

$$\widehat{\mathrm{Var}}_{w_{\lambda*}}\left[\frac{Cf}{w_{\lambda*}}\right] = \frac{1}{i}\sum_{j=1}^{i} \hat{I}_{w_{\lambda_j,n}}\left(\frac{C^2f^2}{w_{w_{\lambda_j}}}\right) - \hat{E}^2_{w_{\lambda*}}\left[\frac{Cf}{w_{\lambda*}}\right],$$

$$\widehat{\mathrm{Cov}}_{w_{\lambda*}}\left[\frac{Cf}{w_{\lambda*}}, \frac{f}{w_{\lambda*}}\right] = \frac{1}{i}\sum_{j=1}^{i} \hat{I}_{w_{\lambda_j,n}}\left(\frac{Cf^2}{w_{w_{\lambda_j}}}\right) - \hat{E}_{w_{\lambda*}}\left[\frac{f}{w_{\lambda*}}\right]\hat{E}_{w_{\lambda*}}\left[\frac{Cf}{w_{\lambda*}}\right].$$

We then compute $\hat{\sigma}^2_{(i)}$, the estimate of the asymptotic variance, by substituting these expressions into (6.36). The following result gives a condition that guarantees the almost sure consistency of this estimate.

Theorem 6.18 If for the matching characteristics adaptive importance sampling procedure (i)–(v) hold and in addition

(vi) $$E\left\{\sum_{j=1}^{\infty} \frac{1}{j^2} E_{w_{\lambda_j}}\left[\left(\frac{p_{|m|} - w_{\lambda_j}}{w_{\lambda_j}}\right)^4\right]\right\} < \infty$$

holds for $m = f$ and $m = Cf$ then $\hat{\sigma}^2_{(i)} \to \sigma^2$ almost surely as $i \to \infty$.

Proof: Exercise 38.

We consider now an application of adaptive importance sampling.

Example 6.15 *Multivariate Student adaptive importance sampling*
Consider the family $\mathcal{W}$ specified in Example 6.3; i.e. the Student$_k(\eta, \mu, \Sigma)$ family with the restriction that $\eta > 2$. This restriction allows us to use coordinate means, variances and covariances for the matching characteristics. We assume that f is such that p_f has a mean vector and variance matrix and then take $E_{p_f}[C]$ to be the vector of dimension $k + k(k+1)/2$ composed of means, variances and covariances of the coordinates. We also specify a value of $\eta = \eta_0$ and fix this through the adaptation so that we are in effect adapting only to the mean and variance. Of course we want to choose η_0 low enough so that all estimates have finite variance. Note that it is very easy to solve $\|\hat{R}_{(i)}(C) - E_{w_\lambda}[C]\|$ for w_λ with these choices.

To start multivariate Student adaptive importance sampling we use the Student$_k(\eta_0, \hat{\mu}, \hat{\Sigma})$ distribution where $\hat{\mu}$ is the mode of f and $\hat{\Sigma}$ is the inverse Hessian matrix of $-\log f$ at $\hat{\mu}$. The suitability of this family for importance sampling and adaptive importance sampling then presupposes that f has a dominant peak and at least approximately elliptical contours. If f diverges strongly from this behavior then we cannot expect importance sampling, or adaptive importance sampling based on the multivariate Student family, to perform very well.

For a specific case consider the linear model considered first in Example 4.9 and for which importance sampling was used based on the Student$_{10}(5, \hat{\mu}, \hat{\Sigma})$ distribution in Example 6.4. Here we use a sample of $n = 1000$ in each adaptive step. Table 6.5 contains the results at the end of the 100-th iteration. These computations took about the same amount of CPU time as straight importance sampling and therefore the adaptation added no real computational burden. There is a clear benefit to adaptation as the estimated absolute coefficients of variation are lower than in Table 6.3. This is particularly noteworthy for $i = 4$ where the reductions exceed 50%. By adapting to the mean and variance we are allowing the skewness in f to have an influence on the choice of importance sampler.

Table 6.5. Adaptive importance sampling estimates and estimated absolute coefficients of variation for Example 6.15 where R denotes the exact value and $\hat{R}_{(100)}$ denotes the adaptive importance sampling estimate.

i	$R(\theta_i)$	$\hat{R}_{(100)}(\theta_i)$	$R(\theta_i^2)$	$\hat{R}_{(100)}(\theta_i^2)$
1	2.043	2.042 (0.001)	4.263	4.262 (0.001)
2	0.095	0.093 (0.013)	0.081	0.080 (0.007)
4	0.018	0.020 (0.057)	0.069	0.070 (0.007)
10	−0.073	−0.073 (0.011)	0.033	0.033 (0.008)

One difficulty with the adaptive importance sampling approach is the necessity of obtaining a good starting value α_1. With a poor choice, convergence may not take place in a practical number of iterations. The technique of *chaining* as described in Evans (1988, 1991a, 1991b) can be useful in dealing with this problem. Suppose that we have an additional parameter $\eta \in \mathcal{N}$ such that $f = f_{\eta*}$ and a good starting importance sampler w_{λ_1} for the problem $I(f_{\eta_1})$. We run the adaptive importance sampling algorithm for this problem and obtain the importance sampler w_{λ_2}. We then make a small change from η_1 to η_2 and run adaptive importance sampling for $I(f_{\eta_2})$, starting with w_{λ_2} and obtain the importance sampler w_{λ_3}. We continue this process and eventually construct a chain from η_1 to η_* where η_* specifies the problem of interest. Often there are parameters in a problem that can be used for chaining but in a general problem we can chain on $\eta = (\eta_1, \eta_2) \in R^2$ where $f_\eta = f^{1/\eta_1} w_{\lambda_1}^{1/\eta_2}$. For large η_1 and $\eta_2 = 1$, f_η is like w_{λ_1}. For $\eta_1 = 1$ and large η_2, f_η is like f. Chaining is an example of a *homotopy method* for integration. There are other applications of homotopy techniques in integration that we will mention in Chapter 7.

As already mentioned, one of the limitations of importance sampling is the dearth of useful families and this applies equally well to adaptive importance sampling. We would like families that reflect a wide variety of shapes and yet still be suitable for the adaptive algorithm. One natural family to consider is the family of mixtures of multivariate Student distributions as discussed in Example 6.5; i.e. $w = p_1 w_1 + \cdots + p_l w_l$ where w_i is the Student$_k(\eta_i, \hat{x}_i, ((\eta_i - 2)/\eta_i)\Sigma_i)$ density. In addition to using mixtures of Students, Oh and Berger (1993) also introduced adaptively selecting the p_i, μ_i, Σ_i to minimize (6.34). The parameters η_i are all set equal to a fixed value η_0 and the starting choices of the parameters are those specified in Example 6.5. Convergence to the minimum of (6.34) is established under the fairly stringent conditions that the set Λ, containing all possible values for the parameters of the mixture, be compact.

West (1992, 1993) has proposed a *kernel density estimation* approach for constructing an importance sampler based on a mixture of Student distributions. For this, let c be a density on R^k, called the *kernel*, and let $x_1, \ldots, x_n$ be a sample from an unknown density g; e.g. we might take c to be the Student$_k(\eta, \hat{x}, ((\eta-2)/\eta)\Sigma_i)$ density for a particular η. Then based on the sample and the kernel we estimate the unknown g by the kernel density estimate

$$\hat{g}_n(x) = \frac{1}{nh^k(n)} \sum_{i=1}^{n} c\left(\frac{x - x_i}{h(n)}\right)$$

where the quantity $h(n)$ is a user specified quantity known as the *bandwidth*. It can be shown that if $h(n) \to 0$ and $nh^k(n) \to \infty$ as $n \to \infty$ (e.g. $h(n) \propto n^{-1/(1+4k)}$) then

$$\int_{R^k} |\hat{g}_n(x) - g(x)|\, dx$$

converges almost surely to 0 as $n \to \infty$; see Devroye (1987).

For importance sampling we choose an initial importance sampler w_1 and generate $x_1, \ldots, x_n$ from this and then approximate the optimal importance sampler for the integrand f by the weighted kernel density estimate

$$w_{1n}(x) = \frac{1}{nh^k(n)} \frac{\sum_{i=1}^n \frac{|f(x_i)|}{w_1(x_i)} c\left(\frac{x - x_i}{h(n)}\right)}{\sum_{i=1}^n \frac{|f(x_i)|}{w_1(x_i)}} \tag{6.40}$$

and, under similar conditions, we will have the same kind of convergence of w_{1n} to the optimal importance sampler as $n \to \infty$. We then iterate this procedure by generating a sample $x_{n+1}, \ldots, x_{2n}$ from $w_2 = w_{1n}$ and construct a new weighted kernel density estimate w_{2n} as the importance sampler as in (6.40). We continue this process until the importance sampler stabilizes. We note that to carry this out we need to be able to generate from densities of the form (6.40) but this is easy to do by the method of composition provided that we can easily generate from the kernel. Once the process has stabilized we can estimate $I(f)$. One drawback of this method is that n typically needs to be large to achieve accuracy particularly as the dimension rises. Another drawback is the computational cost of computing the values of w_{in}. This method is excellent, however, at picking up features of an irregularly shaped f. This approach has been extended by Givens and Raftery (1996). Zhang (1996) establishes some results concerning this approach and in particular argues that expending all the computational resources in obtaining w_{1n} rather than iterating the procedure is preferable.

6.6 Exercises

(1) Prove that if $w_1, \ldots, w_n \geq 0$ with $\sum_{i=1}^n w_i = 1$ then $1/n \leq \sum_{i=1}^n w_i^2 \leq 1$ with $\sum_{i=1}^n w_i^2 = 1/n$ if and only if all the $w_i = 1/n$ and $\sum_{i=1}^n w_i^2 = 1$ if and only if one of the w_i equals 1 and the rest are 0. Verify the expression for the asymptotic variance in Theorem 6.3.

(2) Verify the expression for $\sigma_{R,w}^2$ in Theorem 6.4.

(3) Prove that $\sigma_w^2 = 0$ in (6.3) if and only if $P_w(f(X)/w(X) = c) = 1$ for the constant $c = I$. Note we are implicitly assuming that $\text{supp}(f) \subset \text{supp}(w)$.

(4) Consider approximating $\int_{-\infty}^{\infty} (1 + x^2)^{-1}\, dx$ using importance sampling with w the standard normal density. Monitor the output of the importance sampling estimate and its standard error. Explain what is happening.

(5) Prove that the mean of the $\text{Student}_k(\lambda, \mu, \Sigma)$ density exists if and only if $\lambda > 1$ and when it exists it is equal to μ.

(6) Prove that the variance matrix of the $\text{Student}_k(\lambda, \mu, \Sigma)$ density exists if and only if $\lambda > 2$ and when it exists is equal to $(\lambda/(\lambda - 2))\Sigma$.

(7) Prove that if $(X_1, \ldots, X_k) \sim \text{Student}_k(\lambda, 0, I)$ then $(X_1, \ldots, X_{k-1}) \sim \text{Student}_{k-1}(\lambda, 0, I)$. Establish the analogous result when $(X_1, \ldots, X_k) \sim \text{Student}_k(\lambda, \mu, \Sigma)$.

(8) Prove that if $(X_1, \ldots, X_k) \sim \text{Student}_k(\lambda, 0, I)$ then $Y = X_1^2 + \cdots + X_k^2 \sim F(k, \lambda)$.

(9) If $(X_1, \ldots, X_k) \sim \text{Student}_k(\lambda, 0, I)$ then prove that $\text{Corr}[X_i, X_j] = \delta_{ij}$ but X_i and X_j are not statistically independent. Compare this to the situation when $(X_1, \ldots, X_k) \sim N_k(0, I)$.

(10) Consider the integral $I(m) = \int_0^\infty \cdots \int_0^\infty m(x) f(x)\, dx$ where $f(x) = \prod_{j=1}^n \prod_{i=1}^k x_i \exp\{-x_i y_{ij}\}$. Make the transformation $v_i = \ln(x_i), i = 1, \ldots, k$ and obtain the Student distribution that should be used in Student importance sampling.

(11) Suppose that we want to estimate

$$\int_{R^k} f(x)\, dx \int_{R^k} g(x)\, dx$$

and we use the same importance sampler w for each integral. Use the delta theorem to obtain the asymptotic variance of the product estimator.

(12) Suppose that we are able to evaluate $\int_D f(x)\, dx$ and $p = \int_D w(x)\, dx > 0$ for some $D \subset R^k$ and also generate from w conditioned to D^c (denoted w_{D^c}). Establish that

$$\text{Var}_{w_{D^c}}\left[\int_D f(x)\, dx + \frac{f}{w_{D^c}}\right] \leq (1-p)\text{Var}_w\left[\frac{f}{w}\right].$$

(13) Provided that $\sigma_{w_i}^2 < \infty$ for $i = 1, \ldots, l$ in stratified sampling then prove that

$$\frac{\hat{I}_{\text{strat}} - I(f)}{s_{\text{strat}}} \to^D Z$$

where $Z \sim N(0, 1)$ as $\min(n_1, \ldots, n_l) \to \infty$.

(14) Establish formulas (6.10) and (6.11) under suitable assumptions.

(15) Establish that the algorithm described in Example 6.6 for generating from a $\text{Student}_k(\lambda, 0, (\lambda - 2)/\lambda)I)$ distribution is correct.

(16) Show that if $X_0, \ldots, X_l$ is a sample from w then $E_w[\omega(X_0, \ldots, X_l)] = 1$ where ω is as defined in Section 6.4.3.

(17) Show that if $(X_0, \ldots, X_l) \sim \varphi_w$ then

$$E_{\varphi_w}[T_g(X_0, \ldots, X_l)] = \int_{R^k} \varphi_0(x) g(x) w(x)\, dx$$

where φ_w and T_g are as defined in Section 6.4.3.

(18) Prove Theorem 6.10.

(19) Suppose that $X \sim U(0, 1)$ and $f, g : [0, 1] \to R^1$. Prove (i) if f, g are both increasing then $\text{Cov}[f(X), g(X)] \geq 0$, (ii) if f is increasing and g is

decreasing then $\mathrm{Cov}[f(X), g(X)] \le 0$, (iii) if both f and g are monotone on $[0, 1]$ and strictly monotone in some subinterval then the inequalities in (i) and (ii) are strict.

(20) Suppose that $X \sim U(0, 1)^k$ and $f : (0, 1)^k \to R^1$. Prove that the transformations $T_1(x) = x$ and $T_2(x) = 1 - x$ yield a guaranteed variance reducer via systematic sampling when f is monotone in each coordinate.

(21) For $l = 4$ construct a latin hypercube systematic sampling plan that does not correspond to a good lattice point rule.

(22) Establish (6.22) (show each L_i is closed) and that when $X \sim U([0, 1]^k)$ then $E[f_0(X)] = I(f)$, $E[f_i(X)] = 0$ when $i \ge 1$ and $\mathrm{Var}[f(X)] = \sum_{i=1}^k \mathrm{Var}[f_i(X)]$.

(23) Suppose that $f \in L^2([0, 1]^k)$. Prove that the orthogonal projection of f onto the set of functions in $L^2([0, 1]^k)$ that depend only on $x_1, \ldots, x_i$ is given by $E[f(X_1, \ldots, X_k)|x_1, \ldots, x_i]$.

(24) In Example 6.10 prove that the orthogonal projection of f onto L_1 is given by $I(f) + \sum(f_{1i}(x_i) - I(f))$ where

$$f_{1i}(x_i) = \int_0^1 \cdots \int_0^1 \int_0^1 \cdots \int_0^1 f(x_1, \ldots, x_k)\, dx_1 \cdots dx_{i-1} dx_{i+1} \cdots dx_k.$$

Establish the form of the orthogonal projection of f onto L_2.

(25) Show that if w is a spherically symmetric density on R^k then $w(y) = g(r)$ where, denoting the surface area of S^{k-1} by A_{k-1}, $h(r) = A_{k-1} r^{k-1} g(r)$ is a density on R^1. Further show that we can generate X from w by generating $v \sim U(S^{k-1})$ statistically independent of $r \sim h$ and putting $X = rv$.

(26) Prove that (6.22) exactly integrates all polynomials of degree 3 or less on S^{k-1}.

(27) Prove that Ermakov–Zolotukhin sampling gives a randomized quadrature rule according to the definition of Section 6.4.5.

(28) Prove that if $x \in S^{k-1}$ and Q is a uniformly distributed orthogonal matrix then $u = Qx \sim U(S^{k-1})$.

(29) Prove that if the quadrature rule $R_l = \{(x_1, w_1), \ldots, (x_l, w_l)\}$ is exact on the sphere for all polynomials of degree d or less then for any orthogonal matrix Q, the quadrature rule $R_l^* = \{(Qx_1, w_1), \ldots, (Qx_l, w_l)\}$ is exact on the sphere for all polynomials of degree d or less.

(30) Show that the deterministic radial rule described in Example 6.12 exactly integrates all polynomials of degree $2p - 1$ or less.

(31) In Example 6.12 establish the form of the randomized quadrature rule when $p = 2$.

(32) Determine the optimal choice of $b_1, \ldots, b_l$ in the additive control variate $g = \sum_{i=1}^l b_i g_i$ for the integral $\int_{R^k} f(x)\, dx$.

(33) Establish the asymptotic variance of (6.33) under suitable assumptions.

(34) Write down the expression for the additive control variate for estimating R in Example 6.14 and determine its asymptotic variance.

(35) In the proof of Theorem 6.15 establish that

$$E[U_i^2] = \sum_{j=1}^{i} \frac{1}{j^2} E\left[\left(\hat{I}_{w_{\lambda_j},n}(Cf) - \int_{R^k} C(x) f(x)\, dx\right)^2\right].$$

(36) Prove Theorem 6.16.

(37) In the proof of Theorem 6.17 establish that

$$E\left[S_{i,j}^2\right] = \sum_{l=1}^{j} E\left[Y_{il}^2\right].$$

(38) Prove Theorem 6.18.

(39) (*Conditional Monte Carlo*) Suppose that $X \sim f$ on R^k and we want to estimate $E[g(X)|Y = y_0]$ for some $Y : R^k \to \mathcal{Y}$ with marginal density f_Y with respect to support measure $\mu_{\mathcal{Y}}$. It can happen that it is extremely difficult to generate from the conditional distribution of X given $Y = y_0$ or to implement importance sampling because the set on which the conditional distribution is concentrated is difficult to work with and has probability 0. Suppose, however, that we have a set $\mathcal{W}$ and a 1–1, onto smooth map $T : R^k \to \mathcal{W} \times \mathcal{Y}$ and a density h on $\mathcal{W} \times \mathcal{Y}$ with respect to a support measure $\mu_{\mathcal{W}} \times \mu_{\mathcal{Y}}$ such that we can generate $(W, Y) \sim h$. Denote the Jacobian of the transformation T^{-1} by J_T. Then prove that

$$g(T^{-1}(w, y_0)) \frac{f(T^{-1}(w, y_0))}{f_Y(y_0)} \frac{J_T(w, y_0)}{h_W(w)}$$

is an unbiased estimator of $E[g(X)|Y = y_0]$ when (w, y) is generated from h and h_W is the marginal density of W. See Granovsky (1981) for more details on conditional Monte Carlo and for examples.

7

Markov chain methods

7.1 Introduction

Markov chain methodology, or *Markov chain Monte Carlo* (MCMC), applies to contexts where there is a μ-integrable function $f \geq 0$ and we want to approximate

$$R(m) = \frac{\int m(x) f(x)\, \mu(dx)}{\int f(x)\, \mu(dx)} = \int m(x) p(x)\, \mu(dx) \tag{7.1}$$

typically for a variety of functions m. We suppose that the functions of interest m are simple in the sense that if we could generate from the distribution with density p (with respect to μ) proportional to f then this would lead to good approximations within reasonable computation times. Here μ is a general support measure which may be Lebesgue measure defined on some Euclidean space or, for example, μ may be counting measure on a set of combinatorial objects. Since we are concerned with approximating integrals we focus on applications where μ is Lebesgue measure. On the other hand, there are many applications of this methodology where this is not the case and the reader should be aware of these; see, for example, Hammersley and Handscomb (1964), Sinclair (1993) and Madras and Slade (1993).

Under these circumstances, if we can generate samples from p then we can consider the problem solved. The difficulty, of course, is that for an arbitrary f this is hard to do and it becomes harder as the dimension k rises. Markov chain methods relax the requirement that we be able to generate exactly from p. For this we construct a stochastic process $\{X_i : i = 0, 1, \ldots\}$ such that

$$\lim_{n\to\infty} \frac{1}{n+1} \sum_{i=0}^{n} m(X_i) = \int_{R^k} m(x) p(x)\, dx \tag{7.2}$$

for all m in some class of functions and $X_n \to^D p$ as $n \to \infty$. In particular, the process is an ergodic Markov chain with stationary distribution given by p.

In Section 7.2 we present some of the relevant Markov chain theory. In Section 7.3 we consider the main devices for constructing a Markov chain with the requisite properties. In Section 7.4 we discuss some methods related to MCMC and some extensions. In Section 7.5 we discuss issues associated with the assessment of convergence and error estimation. In Section 7.6 we discuss some related issues and problems. Note that a key difference between the methods of Chapter 6 and this chapter is that the random variables $X_0, X_1, \ldots$ are no longer required to be independent. This of course broadens the class of methods that we can employ to approximate integrals via Monte Carlo but it should also be noted that there is a price paid in terms of the complexity of the analysis of the behavior of the algorithms. Accordingly it makes sense to treat the methods separately.

There are numerous issues associated with MCMC that go well beyond what is discussed here. For more information on the theory and applications of MCMC the interested reader is directed to Neal (1993), Besag and Green (1993), Smith and Roberts (1993), Tierney (1994), Gilks, Richardson and Spiegelhalter (1996), Gamerman (1997), Robert (1998), Roberts and Rosenthal (1998), Geweke (1996, 1997, 1999) and Robert and Casella (1999).

7.2 Markov chains

We consider some basic concepts and results from Markov chain theory that are needed for the discussion of MCMC. For the most part we ignore measurability considerations and treat functions that are equal almost surely, with respect to an underlying probability measure P, as being equal. For more details on the theory of Markov chains see Nummelin (1984), Meyn and Tweedie (1993) and Tierney (1994).

Suppose then that $(\Omega, \mathcal{A}, P)$ is a probability model and $X_i : (\Omega, \mathcal{A}) \to (R^k, \mathcal{B}^k)$ for $i = 0, 1, \ldots$ where $\mathcal{B}^k$ is the Borel class on R^k. Then the stochastic process $\{X_i : i = 0, 1, \ldots\}$ is a *Markov chain* if

$$P(X_n \in B | X_0, \ldots, X_{n-1}) = P(X_n \in B | X_{n-1})$$

for every n and every $B \in \mathcal{B}^k$; i.e. the conditional distribution of the state of the process at the future time n, given the past and present states, depends only on the present state. We call a set $E \in \mathcal{B}^k$ an *effective state space* for the chain if $P(X_i \in E) = 1$ for $i \in \{0, 1, \ldots\}$.

The Markov chain is *time homogeneous* if for every $B \in \mathcal{B}^k$ the quantity $P(X_n \in B | X_{n-1})$ does not depend on n. We restrict our attention to time homogeneous Markov chains. We refer to the function $K : R^k \times \mathcal{B}^k \to R^1$ given by

$$K(x, B) = P(X_1 \in B | X_0 = x)$$

as the *transition probability function* or *kernel* of the chain and we refer to the distribution P_0 of X_0 as the *initial distribution* of the chain. The kernel and initial distribution determine the Markov chain (Exercise 1) since

$$\begin{aligned} &P(X_0 \in B_0, \ldots, X_n \in B_n) \\ &= \int_{B_0} \int_{B_1} \cdots \int_{B_n} K(x_{n-1}, dx_n) \cdots K(x_0, dx_1)\, P_0(dx_0) \end{aligned} \tag{7.3}$$

for every $B_0, \ldots, B_n \in \mathcal{B}^k$. We call

$$K^n(x, B) = P(X_n \in B | X_0 = x)$$

the *n-step probability transition function.*

The following is a central example in the theory of Markov chains.

Example 7.1 *Random walks*
Suppose that $X_1, X_2, \ldots$ are iid random variables with density g with respect to Lebesgue measure on R^k statistically independent of $X_0 \sim k_0$. Let

$$W_n = \sum_{i=0}^{n} X_i.$$

Then $\{W_n : n = 0, 1, 2, \ldots\}$ is a Markov chain (Exercise 2) with

$$K(x, B) = \int_B g(y - x)\, dy.$$

This Markov chain is known as a *random walk.*

Of considerable importance in applications of Markov chains is the long-run behavior of the system being modeled. In particular we would like to establish the long-run behavior of sample averages, as in (7.2) and the distribution of X_n when n is large. These are characterized in terms of the stationary distribution of the chain should a unique stationary distribution exist. A probability measure π on $\mathcal{B}^k$ is called a *stationary distribution* for the Markov chain if

$$\pi(B) = \int_{R^k} K(x, B)\, \pi(dx) \tag{7.4}$$

for every $B \in \mathcal{B}^k$. Note that if the initial distribution P_0 is stationary then $P(X_n \in B) = P_0(B)$ (Exercise 3); i.e. the distribution of the state at each time n is the same as the distribution of the initial state. An arbitrary Markov chain may not have a stationary distribution or may have more than one stationary distribution. We want conditions such that a stationary distribution π exists for a chain and is unique. Moreover, we want conditions such that the distribution of X_n converges to π and such that sample averages converge to expectations with respect to π where these sample averages have finite variance.

A Markov chain is *ϕ-irreducible* for non-zero, σ-finite measure ϕ on $\mathcal{B}^k$ if for every $B \in \mathcal{B}^k$ with $\phi(B) > 0$ we have that $P(\tau_B < \infty | X_0 = x) > 0$ for every $x \in E$ where

$$\tau_B = \inf\{n \geq 1 : X_n \in B\}$$

is the *hitting time* of B. Note that by the Lebesgue decomposition, see Royden (1968), the measure $K^n(x, \cdot)$ can be uniquely decomposed into the sum of two components where one component is absolutely continuous with respect to ϕ and the other component is singular with respect to ϕ. If there exists n such that the absolutely continuous component of $K^n(x, \cdot)$ has a positive density with respect to ϕ on E for every $x \in E$ then the chain is ϕ-irreducible. The Markov chain is *aperiodic* if, for any integer $d > 1$, we cannot find a partition $E = \cup_{i=1}^d S_i$ such that $K(x, S_{i+1}) = 1$ when $x \in S_i$ for $i = 1, \ldots, d-1$ and $K(x, S_1) = 1$ when $x \in S_d$. For probability measures P and Q on $\mathcal{B}^k$ the *total variation distance* between them is defined to be

$$\|P - Q\| = \sup\{|P(B) - Q(B)| : B \in \mathcal{B}^k\}.$$

We then have the following result.

Theorem 7.1 If $\{X_i : i = 0, 1, \ldots\}$ is a ϕ-irreducible, aperiodic Markov chain having stationary distribution π, then for π-almost every $x \in E$ we have

$$\lim_{n\to\infty} \|K^n(x, \cdot) - \pi\| = 0. \tag{7.5}$$

Proof: See, for example, Rosenthal (1999).

Then by the dominated convergence theorem

$$\lim_{n\to\infty} P(X_n \in B) = \lim_{n\to\infty} \int_{R^k} K^n(x, B)\, P_0(dx) = \pi(B)$$

and so we have that $X_n \to^D \pi$. Therefore if it is difficult to generate from π but it is easy to generate from P_0 and each $K(x, \cdot)$, then we can approximately generate from π by generating the chain and returning X_n for large n. Of course, we do not know in general how large n has to be for the approximation to be reasonable. Also note that Theorem 7.1 establishes that if a stationary distribution exists for a ϕ-irreducible, aperiodic Markov chain then it is unique.

A ϕ-irreducible Markov chain is *recurrent* if for each $B \in \mathcal{B}^k$ with $\phi(B) > 0$

$$P\left[\sum_{n=1}^{\infty} I_B(X_n) = \infty \,\middle|\, X_0 = x\right] > 0$$

for all $x \in E$ and

$$P\left[\sum_{n=1}^{\infty} I_B(X_n) = \infty \,\middle|\, X_0 = x\right] = 1$$

for ϕ-almost every x. If a chain is recurrent then there is a measure π, not necessarily a finite measure, unique up to constant multiples that satisfies (7.4); see, for example, Meyn and Tweedie (1993), Theorem 10.4.4. A ϕ-irreducible Markov chain is *Harris recurrent* if for each $B \in \mathcal{B}^k$ with $\phi(B) > 0$ then $P\left[\sum_{n=1}^{\infty} I_B(X_n) = \infty | X_0 = x\right] = 1$ for every $x \in E$. If a chain is Harris recurrent and aperiodic then it has a unique stationary distribution π and furthermore (7.5) holds for every $x \in E$; see, for example, Meyn and Tweedie (1993), Theorem 13.0.1. Note that if a ϕ-irreducible chain has a stationary distribution π and there exists n, such that a component of $K^n(x, \cdot)$ has a positive density with respect to π on E for every $x \in E$ then the chain is Harris recurrent.

The following result gives a strong law of large numbers for Markov chains.

Theorem 7.2 If $\{X_i : i = 0, 1, \ldots\}$ is a Harris recurrent Markov chain having stationary distribution π and $E_\pi[|m|] < \infty$ then

$$\lim_{n\to\infty} \frac{1}{n+1} \sum_{i=0}^{n} m(X_i) = \int_{R^k} m(x)\pi(x)\, dx$$

almost surely.

Proof: See, for example, Meyn and Tweedie (1993), Theorem 17.1.7.

Theorem 7.2 says that if our goal is to estimate $\int_{R^k} m(x)\pi(x)\, dx$ we can use

$$\bar{m}_n = \frac{1}{n+1} \sum_{i=0}^{n} m(X_i) \tag{7.6}$$

for large n. This says nothing about how large n has to be for this to be an accurate approximation. For this we need a central limit theorem for Markov chains. Therefore we want the existence of $\sigma_m^2 > 0$ such that

$$\frac{\bar{m}_n - E_\pi[m]}{\sigma_m/\sqrt{n}} \to^D Z$$

where $Z \sim N(0, 1)$ as $n \to \infty$. For such a result to be usable for error assessment we also need a consistent estimator σ_m^2.

A chain which satisfies (7.5) for all x is called *ergodic*. In Theorem 17.3.6 of Meyn and Tweedie (1993) a central limit theorem is proven for a general ergodic chain under quite complicated conditions. With additional conditions a simpler result is obtained. For a nonnegative function $f : E \to R^1$ and any signed measure ν define the f-norm by

$$\|\nu\|_f = \sup\left\{\left|\int_E g(x)\, \nu(dx)\right| : |g| \le f\right\}.$$

For nonnegative $V : R^k \to R^1$ satisfying $V \ge 1$ and transition kernel K define

$$|||K^n - \pi|||_V = \sup_{x\in E} \frac{\|K^n(x, \cdot) - \pi\|_V}{V(x)}.$$

A Markov chain is called V-*uniformly ergodic* if $|||K^n - \pi|||_V \to 0$ as $n \to \infty$. If $V \equiv 1$ then the chain is said to be *uniformly ergodic*. Then the following central limit theorem holds.

Theorem 7.3 If $\{X_i : i = 0, 1, \dots\}$ is a Harris recurrent Markov chain having stationary distribution π which is also V-uniformly ergodic and $m^2 \leq V$ then

$$\sigma_m^2 = \mathrm{Var}_{P^*}[m(X_0)] + 2\sum_{i=1}^{\infty} \mathrm{Cov}_{P^*}[m(X_0), m(X_i)] \tag{7.7}$$

is nonnegative and finite and

$$\lim_{n\to\infty} n\mathrm{Var}_{P^*}[\bar{m}_n] = \sigma_m^2$$

where P^* refers to the Markov chain with initial distribution π and kernel K.

(i) If $\sigma_m^2 = 0$ then

$$\lim_{n\to\infty} \sqrt{n}(\bar{m}_n - E_\pi[m]) = 0$$

almost surely P^*.

(ii) If $\sigma_m^2 > 0$ then

$$\sqrt{n}(\bar{m}_n - E_\pi[m]) \to^D \sigma_m Z$$

as $n \to \infty$ where $Z \sim N(0, 1)$.

Proof: See Meyn and Tweedie (1993), Theorem 17.0.1.

We see immediately that if a chain is uniformly ergodic and m is bounded then Theorem 7.3 can be applied. In Section 7.4 we discuss methods for estimating the quantity σ_m^2.

Note that if $\{X_i : i = 0, 1, \dots\}$ is a ϕ-irreducible, aperiodic Markov chain having stationary distribution π then the chain is V-uniformly ergodic if and only if there exist positive constants R and $r > 1$ such that $|||K^n - \pi|||_V \leq Rr^{-n}$; see for example Meyn and Tweedie (1993), Theorem 16.0.2. When $V \equiv 1$ then this says that $K^n(x, \cdot)$ converges to π at a geometric rate and this is uniform in x. A weaker form of ergodicity is given by *geometric ergodicity* which requires the existence of a function $R : E \to R^1$ with $R > 1$ such that $\|K^n(x, \cdot) - \pi\| \leq R(x)r^{-n}$. See, for example, Roberts and Rosenthal (1998) for a discussion of the utility of geometric ergodicity.

The Markov chain $\{X_i : i = 0, 1 \dots\}$ is said to be *reversible* with respect to probability measure Q on $\mathcal{B}^k$ if for every $A, B \in \mathcal{B}^k$ we have that

$$\int_A K(x, B)\, Q(dx) = \int_B K(x, A)\, Q(dx); \tag{7.8}$$

i.e. the probability of being in B after one transition, if we start in A according to Q, is equal to the probability of being in A after one transition, if we start in B.

Equation (7.8) is sometimes referred to as *detailed balance*. We see immediately that

$$\int_{R^k} K(x, B)\, Q(dx) = \int_B K(x, R^k)\, Q(dx) = Q(B)$$

and so reversibility with respect to Q establishes that Q is stationary for the chain. In Roberts and Rosenthal (1997) (Corollary 2.1) it is established that $\sqrt{n}(\bar{m}_n - E_\pi[m]) \to^D \sigma_m Z$ as $n \to \infty$ where $Z \sim N(0, 1)$ for a Harris recurrent, aperiodic, geometrically ergodic, reversible Markov chain when $\mathrm{Var}_\pi[m] < \infty$. This is somewhat more useful than Theorem 7.3. Kipnis and Varadhan (1986), Tierney (1994) and Chan and Geyer (1994) provide more discussion of central limit theorems for Markov chains.

There are many nice features associated with reversibility not the least of which being the ease with which it can be established that a particular π is a stationary distribution for the chain. Also of some importance is the fact that for such a chain the transformation $T_K : L^2(\pi) \to L^2(\pi)$ given by

$$T_K(g)(x) = \int_{R^k} g(z) K(x, dz)$$

satisfies

$$\langle T_K(g), h\rangle_\pi = \int_{R^k} \left[\int_{R^k} g(z) K(x, dz)\right] h(x)\pi(dx) = \langle g, T_K(h)\rangle_\pi$$

for every $g, h \in L^2(\pi)$; i.e. T_K is self-adjoint. This permits spectral theory to be used in the analysis of the chain so that the theory of reversible chains tends to be somewhat simpler than that for nonreversible chains. Still nonreversible chains, e.g. the Gibbs sampling chains discussed in Section 7.3.2, are important in practical applications.

There are many more results concerning the convergence of Markov chains. In particular there are a number of results concerned with obtaining bounds for $\|K^n(x, \cdot) - \pi\|$ as a function of n; see, for example, Tierney (1994) for a discussion of some of these. These bounds allow a determination of the number of iterations required for the Markov chain to effectively converge to its stationary distribution. While this is useful information it is not so important for the integration problem we are considering because it is really the quantity $\sigma_m/\sqrt{n}$ that we wish to bound. Also, for the most part, such bounds are not available for the typical integration problems we are faced with in practice.

There are in general many chains that have p as stationary distribution and in Section 7.3 we discuss some methods for constructing some of these. So in a particular application we need to select one of these. Natural criteria for selecting amongst the various possibilities are the asymptotic variances σ_m^2 for various functions m of interest. As mentioned in Chapter 6 the time necessary to generate a value should also be part of this comparison as well. Peskun (1973), see also

Tierney (1995), introduces the partial ordering of chains with stationary distribution π as $K_1 \preceq_{\text{Peskun}} K_2$ if

$$K_1(x, B\backslash\{x\}) \le K_2(x, B\backslash\{x\}) \tag{7.9}$$

for every x and $B \in \mathcal{B}^k$ and shows that (7.9) implies that

$$\sigma_m^2(K_1) \ge \sigma_m^2(K_2) \tag{7.10}$$

for every $m \in L^2(\pi)$ where $\sigma_m^2(K)$ now denotes the dependence of the asymptotic variance on the kernel K. Tierney (1995) shows that $K_1 \preceq_{\text{Peskun}} K_2$ implies

$$0 \le \langle (T_{K_1} - T_{K_2})(g), g \rangle_\pi \tag{7.11}$$

for any $g \in L^2(\pi)$ and so (7.10) implies a more general partial ordering since (7.11) can hold even when (7.9) does not. In Mira (1998) it is shown that (7.11) is equivalent to (7.10) for reversible chains. Now

$$\langle (T_{K_1} - T_{K_2})(g), g \rangle_\pi = \text{Cov}_{P_1}[g(X_0), g(X_1)] - \text{Cov}_{P_2}[g(X_0), g(X_1)]$$

where P_i is the chain with initial distribution π and kernel K_i and so the partial ordering for reversible chains is based on the sizes of the lag 1 autocovariances in the stationary chain. Accordingly the partial ordering based on (7.11) is referred to as the *covariance ordering* in Mira (1998). The nonreversible case is also discussed in Mira (1998).

We note that there are many conditions that a Markov chain must satisfy if we want it to be useful for the integration problem. As we will see it is relatively easy to ensure that p is a stationary distribution for the chain but then we also want the chain to be Harris recurrent, aperiodic and appropriate conditions to hold so that we can apply a central limit theorem. Strictly speaking it is not always feasible to check these conditions and algorithms are often used without verifiying the conditions which, of course, must affect the degree of confidence we can have in the accuracy of any approximations we quote.

7.3 Markov chain sampling algorithms

We now consider a number of MCMC algorithms. We restrict our attention to those which are most generally useful in integration problems and note that there are many other algorithms discussed in the literature which are useful in specific problems. Rather than give detailed conditions under which the convergence results discussed in Section 7.2 hold for these chains we provide references to the relevant literature when these are available. Hereafter we suppose that f and p are as specified in Section 7.1.

7.3.1 METROPOLIS CHAINS

Metropolis *et al.* (1953) developed a basic version of what is now known as the *Metropolis algorithm.* We present a more general version due to Hastings (1970) which is sometimes referred to as the *Metropolis–Hastings algorithm.* The ingredients consist of a transition kernel Q for a Markov chain and an initial density k_0 and from these we construct a Markov chain with stationary distribution p. Let $q(x, \cdot)$ be the density of $Q(x, \cdot)$ with respect to Lebesgue measure. If Q is reversible with respect to p then it will have p as a stationary distribution. Typically, however, Q will not be reversible and the Metropolis algorithm is a technique for modifying the initial chain so that it is reversible with respect to p.

We start with $X_0 \sim k_0$. Suppose now that we have generated $X_0, X_1, \ldots, X_i$. The iterative step in the algorithm is then:

(i) generate $Y_i \sim Q(X_i, \cdot)$ and $W_i \sim \text{Bernoulli}(\alpha(X_i, Y_i))$ where

$$\alpha(X_i, Y_i) = \min\left\{1, \frac{f(Y_i)}{f(X_i)} \frac{q(Y_i, X_i)}{q(X_i, Y_i)}\right\}; \tag{7.12}$$

(ii) if $W_i = 1$ return $X_{i+1} = Y_i$; otherwise return $X_{i+1} = X_i$.

We will refer to the density $q(x, \cdot)$ as the *proposal density* at x and $\alpha(x, \cdot)$ as the *acceptance probability function* at x. At this point we only require that we be able to generate from k_0 and $q(x, \cdot)$ for each x so that we are otherwise free to choose these ingredients as we please. Since convergence, as specified in (7.2), is dependent on the chain visiting all the regions where f is nonzero, intuitively it makes sense to require that if we are at the state x and $f(y) > 0$ then we want a positive probability of making this transition; i.e. we require $q(x, y) > 0$ for every x and every y in $\text{supp}(f)$.

It is clear that $\{X_i : i = 0, 1, 2, \ldots\}$ is a time homogeneous Markov chain and

$$\begin{aligned}
&P(X_{i+1} \in B | X_i = x) \\
&\quad = P(Y_i \in B, W_i = 1 | X_i = x) + P(X_i \in B, W_i = 0 | X_i = x) \\
&\quad = \int_B \alpha(x, y) q(x, y)\, dy \\
&\qquad + I_B(x) \int_{\{y: f(y)q(y,x) < f(x)q(x,y)\}} \{1 - \alpha(x, y)\} q(x, y)\, dy \\
&\quad = \int_B \alpha(x, y) q(x, y)\, dy + r(x) I_B(x) \\
&\quad = K(x, B)
\end{aligned}$$

is the transition kernel of this chain where

$$r(x) = \frac{1}{f(x)} \int_{\{y: f(y)q(y,x) < f(x)q(x,y)\}} \{f(x)q(x, y) - f(y)q(y, x)\}\, dy$$

satisfies $0 \le r(x) \le 1$. Note that $K(x, \cdot)$ is a mixture of a discrete distribution, which puts mass $r(x)$ at x, and an absolutely continuous distribution which distributes $1 - r(x)$ of the mass over R^k with density $s(\cdot|x) = \alpha(x, \cdot)q(x, \cdot)$. Further since (Exercise 5)

$$\alpha(x, y)q(x, y)p(x) = \alpha(y, x)q(y, x)p(y) \tag{7.13}$$

we have that

$$\begin{aligned}
&\int_A K(x, B)p(x)\,dx \\
&\quad = \int_A \left\{ \int_B \alpha(x, y)q(x, y)\,dy + r(x)I_B(x) \right\} p(x)\,dx \\
&\quad = \int_A \int_B \alpha(x, y)q(x, y)p(x)\,dy\,dx + \int_A r(x)I_B(x)p(x)\,dx \\
&\quad = \int_B \int_A \alpha(y, x)q(y, x)p(y)\,dx\,dy + \int_B r(x)I_A(x)p(x)\,dx \\
&\quad = \int_B K(y, A)p(y)\,dy
\end{aligned}$$

and the chain $\{X_i : i = 0, 1, \ldots\}$ is reversible with respect to p. This immediately implies that p is a stationary distribution of the chain as required.

It doesn't seem possible to give general conditions for irreducibility but if we take ϕ to be the measure given by p, then it is clear that ϕ-irreducibility of Q, and $\alpha(x, y)$ never 0, implies the ϕ-irreducibility of K. Tierney (1994) shows that, with this choice of ϕ, a ϕ-irreducible Metropolis chain is Harris recurrent. Also Tierney (1994) gives conditions for the Metropolis chain to be aperiodic. See Roberts and Tweedie (1996) for more discussion of convergence issues for Metropolis chains. Peskun (1973) shows that (7.12) leads to an optimal chain in the Peskun ordering in a class determined by possible choices for the acceptance probability.

There are many possible choices for Q. We now discuss some of the possibilities.

Example 7.2 *Symmetric chains*
In the original paper by Metropolis *et al.* (1953), the proposal density was required to be symmetric; i.e. $q(x, y) = q(y, x)$ which implies that

$$\alpha(x, y) = \min\left\{1, \frac{f(y)}{f(x)}\right\}$$

so that a move is made for sure whenever $f(y) > f(x)$. A popular choice for a symmetric proposal density $q(x, \cdot)$ is the density corresponding to the $N_k(x, \Sigma)$ distribution although a Student$_k(\lambda, x, \Sigma)$ density could also be used. The specification of the matrix Σ is guided by considering the following extreme cases. On the one hand, overdispersed values of Σ relative to f lead to many proposals y

that are rejected since they are distant from the bulk of the probability of f. This slows the convergence of the algorithm. On the other hand, underdispersed values of Σ relative to f lead to proposals that are only a short distance from the current value x. This induces strong and unwanted correlations in the chain and the chain simply doesn't move enough around the region where f is significant, i.e. it doesn't mix enough for the sample averages to give reasonable approximations to integrals. One suggestion is to match Σ to the variance matrix Σ_f corresponding to f. The idea is that if x is near the mean value of the distribution corresponding to f, then proposals y should in some way resemble realizations from f; neither systematically too far nor too close to x. Since Σ_f is typically unavailable, we may set it equal to the inverse Hessian matrix of $-\log f$ evaluated at the mode of f or alternatively estimate it in a preliminary run as in Müller (1991).

Example 7.3 *Independence chains*
A proposal density of the form $q(x, y) = q(y)$ produces what is known as an *independence chain* as the new variate Y is generated independently of the current variate x. In this case the acceptance probability is

$$\alpha(x, y) = \min\left\{1, \frac{f(y)\,q(x)}{f(x)\,q(y)}\right\}.$$

A useful consideration is to choose $q(y)$ so as to mimic f. For if $q(y) = f(y)$, then $f(y)q(x)/(f(x)q(y)) = 1$, and we are then generating and always accepting independent proposals from the target distribution with density f. It is also generally thought that it is better to have q somewhat overdispersed relative to f. One possibility is to choose q equal to the density of a $N_k(\mu, \Sigma)$ or a $\text{Student}_k(\lambda, x, ((\lambda - 2)/\lambda)\Sigma)$ distribution where μ is the mode of f and Σ is the inverse Hessian matrix of $-\log f$ evaluated at μ multiplied by some constant greater than 1. Tierney (1994) shows that independence chains are aperiodic and ϕ-irreducible whenever $q > 0$ on the support of f. For stronger results, see Liu (1996). As an alternative to independence chain sampling we note that we could use importance sampling with the importance sampler $w = q$. Liu (1996) contains some discussion concerning the relative merits of these algorithms but there doesn't appear to be any definitive conclusions at this time or much guidance in determining which is better to use in a given context. While the independence chain has the property that we are inevitably sampling, at least approximately, from the stationary distribution importance sampling has the benefit of a relatively simple error analysis because of the iid nature of the samples.

Example 7.4 *Rejection chains*
We recall the rejection algorithm for generating from a distribution as discussed in Section 3.4. For this we require a density g that we can generate from and a constant c such that $f \leq cg$. If these conditions hold then we can generate iid samples from p.

Typically, however, we will not have such a g and c but perhaps we feel we can make reasonable choices so that we can generate from g and the inequality

$f \le cg$ holds over large parts of the sample space; i.e. $C = \{x : f(x) \le cg(x)\}$ is such that $\int_C p(x)\,dx \approx 1$. Now for fixed x let

$$q \propto \min\{f, cg\}.$$

Then we can generate from q as follows (Exercise 6): generate pairs (Y, U) where $Y \sim g$ statistically independent of $U \sim U(0, 1)$ until $f(Y) \ge cUg(Y)$ and then return Y. Note that if $f(Y) \ge cg(Y)$ then there are no rejections. The *rejection chain* uses the density q in an independence chain. Under these circumstances we have that (Exercise 7)

$$\alpha(x, y) = \min\left\{1, \frac{f(y)\min\{f(x), cg(x)\}}{f(x)\min\{f(y), cg(y)\}}\right\}$$

$$= \begin{cases} 1 & \text{if } x, y \in C \text{ or } x \in C, y \in C^c \\ \dfrac{cg(x)}{f(x)} & \text{if } x \in C^c, y \in C \\ \min\left(1, \dfrac{f(y)\,g(x)}{f(x)\,g(y)}\right) & \text{if } x, y \in C^c. \end{cases}$$

Liu (1996) has shown that the rejection chain is asymptotically less efficient than the independence chain using $q = g$ in the case when $f \le cg$ holds. So interestingly there does not seem to be a good reason to use the rejection chain even in the case when we are generating iid samples from the stationary distribution. Note that this analysis takes into account the total number of values generated from g in generating values from f via rejection.

Example 7.5 *Random walk chains*
A proposal density of the form $q(x, y) = q(y - x)$ produces what is known as a *random walk chain* as $Y_i = X_i + X$ where Z is generated according to q. Popular choices for q include the normal and Student distributions having mean zero. We note that these two particular chains can also be constructed through the consideration of symmetric proposal densities.

The essential problem faced by the user of the Metropolis algorithm is the selection of the proposal density q. A good proposal density is one that leads to a chain that visits the entire support of f and does so relatively quickly. As the dimension rises it typically becomes increasingly difficult to characterize a good q. Again the dimensional effect is apparent.

7.3.2 GIBBS SAMPLING

Gibbs sampling is perhaps the simplest of the Markov chain algorithms. Although it was named by Geman and Geman (1984) in the context of image processing using Gibbs distributions, its applicability extends to more general settings and is not restricted to Gibbs distributions. Extensive use of Gibbs sampling in statistics, where it often arises in a natural way, began with the paper of Gelfand and Smith (1990). Typically a special structure is required in a problem if Gibbs sampling is to be applicable, at least in a straight-forward way.

The attraction of Gibbs sampling is based upon the fact that even though it may be difficult to generate from the density given by p often it is easy to generate from the conditional distributions of one coordinate given the others. If $X = (X_1, \ldots, X_k) \sim p$ we write $X_j|x_1, \ldots, x_{j-1}, x_{j+1}, \ldots, x_k$ for the conditional distribution of X_j given the remaining coordinates of X. Let us denote such a conditional density by

$$p_j(x_j|x_1, \ldots, x_{j-1}, x_{j+1}, \ldots, x_k).$$

Note that as a function of x_j we have that

$$p_j(x_j|x_1, \ldots, x_{j-1}, x_{j+1}, \ldots, x_k) \propto f(x_1, \ldots, x_{j-1}, x_j, \ldots, x_k).$$

Recall from our discussion of random variate generation in Chapter 3 that it is often possible to develop an algorithm to generate from p_j even when we don't know the normalizing constant. We assume hereafter that we have such algorithms although it must be remembered that this is not always the case.

For the Gibbs sampling algorithm we start by generating a starting value $x^{(0)} = (x_1^{(0)}, \ldots, x_k^{(0)})$ from k_0. Suppose then that we have generated

$$x^{(0)}, \ldots, x^{(i-1)}.$$

The iterative step for obtaining the i th state of the chain is then to generate

$$x_j^{(i)} \sim p_j\left(\cdot \,|x_1^{(i)}, \ldots, x_{j-1}^{(i)}, x_{j+1}^{(i-1)}, \ldots, x_k^{(i-1)}\right)$$

for $j = 1, \ldots, k$. This clearly gives a time homogeneous Markov chain with

$$\begin{aligned} K(x, B) &= \int_B \prod_{j=1}^{k} p_j(y_j|y_1, \ldots, y_{j-1}, x_{j+1}, \ldots, x_k)\, dy_1 \cdots dy_k \\ &= \int_B k(y_1, \ldots, y_k|x_1, \ldots, x_k)\, dy_1 \cdots dy_k. \end{aligned}$$

It is straightforward to verify that $k(\cdot|x_1, \ldots, x_k)$ is a density (Exercise 9) and that p is a stationary distribution for this chain (Exercise 10). If $k(y|x) > 0$ for all $x, y \in \text{supp}(p)$ then the chain is ϕ-irreducible where ϕ is Lebesgue measure concentrated on $\text{supp}(p)$. This also implies that the chain is aperiodic and Harris recurrent. These conditions are commonly met in applications. Schervish and Carlin (1992) discuss convergence issues associated with the Gibbs sampler. Also see Chan (1993), Smith and Roberts (1993), Roberts and Polson (1994), Roberts and Smith (1994) and Liu, Wong and Kong (1995).

As a simple example where Gibbs sampling solves a hard generating problem consider the following.

Example 7.6 *$N_k(\mu, \Sigma)$ conditioned to a rectangle*
Suppose that we want to generate x from a $N_k(\mu, \Sigma)$ distribution conditioned to a rectangle $\prod_{i=1}^k (a_i, b_i)$. Then is easy to show (Exercise 11) that

$$x_i | x_1, \ldots, x_{i-1}, x_{i+1}, \ldots, x_k$$

is distributed as a univariate normal conditioned to an interval. Since the univariate normal distribution is log-concave the adaptive rejection methods of Section 3.6 can be used to generate from this conditional distribution.

The Gibbs sampler can be generalized in a number of ways. One simple generalization is to allow the x_i in $x = (x_1, \ldots, x_l)$ to be multivariate rather than univariate. The developments are then identical to the above.

One major stumbling block in the use of the Gibbs sampling algorithm is the necessity of generating from the required conditional distributions. The attractiveness of the Gibbs sampler in many statistical problems is that it is often straightforward to do this using software packages such as NAG or IMSL which have a good coverage of random variable generators for the common distributions. Alternatively we can develop generators using rejection or adaptive rejection. For example, many classical distributions are log-concave and therefore so are truncations of these distributions. Accordingly the algorithm of Gilks and Wild (1992) has found many applications in Gibbs samplers.

One other technique can often be brought into play to address the generation issue. This is the technique known as *auxiliary variables, latent variables* or *data augmentation.* This approach has its origins in some related Monte Carlo algorithms developed in Tanner and Wong (1987) and Tanner (1993). The basic idea behind this technique is that new variables are introduced into the problem which can be exactly integrated out but when left in make the problem of generating from the conditionals relatively straightforward. Note that this seems to run contrary to the result on dimension reduction developed in Section 6.4.1 as we are effectively increasing the dimension of the integrals we face. There is no question, however, that the technique has proven invaluable in leading to effective Gibbs sampling algorithms for problems which are otherwise intractable; also see Polson (1996) concerning the effectiveness of this technique.

Suppose then that we want to evaluate the integral

$$\int_{R^k} m(x)p(x)\,dx \tag{7.14}$$

where p is a density function. In addition suppose that m is simple in the sense that if we could generate from p then sample averages of m provide adequate approximations. Suppose that we can neither generate directly for p nor carry out Gibbs sampling directly because there are no readily available generating algorithms. In some of these contexts there are densities $p(\cdot|\eta)$ on R^k, with $\eta \in \Xi$ a subset of R^l, and q a density on Ξ with respect to a support measure ν such that

$$p(x) = \int_{\Xi} p(x|\eta)q(\eta)\,\nu(d\eta)$$

and such that we can implement Gibbs sampling on $p(x|\eta)q(\eta)$. Then it is clear that (7.14) equals

$$\int_{\Xi}\int_{R^k} m(x)p(x|\eta)q(\eta)\,dx\,\nu(d\eta)$$

and therefore Gibbs sampling with respect to (x, η) produces estimates of (7.14). Note that this is not mixture sampling; i.e. we are not supposing that we can generate η from q and x from $p(\cdot|\eta)$ which directly gives a method for generating from p. Note that we can also write the joint density of (x, η) as $p(x)q(\eta|x)$ and we have that

$$p(x) = \int_{\Xi} p(x)q(\eta|x)\,\nu(d\eta)$$

which presents another approach to adding variables to a problem. See the discussion on slice samplers in Section 7.3.5 for an application of this.

Sometimes we can achieve exact sampling by the addition of latent variables as the following example demonstrates.

Example 7.7 *Exact sampling using latent variables*
As a simple specific example suppose that p is proportional to

$$\prod_{i=1}^{n}\left\{1 + \left(\sum_{j=1}^{k} a_{ij}x_j\right)^2\right\}^{-1} \tag{7.15}$$

where $A = (a_{ij}) \in R^{n\times k}$ has rank k. Calculating integrals with respect to p is a very difficult problem and generating directly from p is extremely difficult. Further, a direct Gibbs sampling algorithm for (7.15) seems unavailable. But note that (7.15) is proportional to

$$\begin{aligned}
&\prod_{i=1}^{n}\int_0^{\infty} \exp\left\{-\eta_i\left(\sum_{j=1}^{k} a_{ij}x_j\right)^2\right\}\exp(-\eta_i)\,d\eta_i \\
&= \int_{[0,\infty)^k} \exp\left\{-\tfrac{1}{2}x'(2A'\operatorname{diag}(\eta_1,\ldots,\eta_n)A)x\right\} \\
&\quad\times \exp(-\eta_1)\cdots\exp(-\eta_n)\,d\eta_1\cdots d\eta_n.
\end{aligned} \tag{7.16}$$

From (7.16) we see that

$$x|\eta_1,\ldots,\eta_n \sim N_k\big(0, (2A'\operatorname{diag}(\eta_1,\ldots,\eta_n)A)^{-1}\big)$$

and

$$\eta_1, \ldots, \eta_n \sim \text{exponential}\left(\left\{1 + \left(\sum_{j=1}^{k} a_{ij}x_j\right)^2\right\}^{-1}\right)$$

and therefore we have a straightforward sampling algorithm for integrals with respect to p. In fact we do not even need Gibbs sampling as the additional variables have made possible iid sampling from the distribution.

Example 7.8 *Gibbs sampling using latent variables*
As a simple example where the addition of latent variables permits approximate generation from a distribution via Gibbs sampling consider the distribution with density in $(p, \theta, \phi) \in [0, 1]^3$ proportional to

$$\prod_{i=1}^{n}[(1-\theta)^{x_i}\theta^{m_i-x_i}p + \phi^{x_i}(1-\phi)^{m_i-x_i}(1-p)] \tag{7.17}$$

where n is typically large and fixed and x_i, m_i are fixed nonnegative integers with $0 \leq x_i \leq m_i$; see Evans, Guttman, Haitovsky and Swartz (1996) for discussion of an application where this density arises. If we introduce the latent binary variables $\eta_1, \ldots, \eta_n$, where these are independent and $\eta_i \sim \text{Bernoulli}(p)$, then we have that (7.17) equals

$$\prod_{i=1}^{n}\sum_{\eta_i=0}^{1}[(1-\theta)^{x_i}\theta^{m_i-x_i}]^{\eta_i}[\phi^{x_i}(1-\phi)^{m_i-x_i}]^{1-\eta_i}p^{\eta_i}(1-p)^{1-\eta_i}.$$

This implies that the joint density of $(p, \theta, \phi, \eta_1, \ldots, \eta_n)$ is proportional to

$$\prod_{i=1}^{n}[(1-\theta)^{x_i}\theta^{m_i-x_i}]^{\eta_i}[\phi^{x_i}(1-\phi)^{m_i-x_i}]^{1-\eta_i}p^{t}(1-p)^{n-t} \tag{7.18}$$

where $t = \sum_{i=1}^{n}\eta_i$. Therefore the relevant conditional distributions obtained from (7.18) are (Exercise 13) given by

$$p|\theta, \phi, \eta_1, \ldots, \eta_n \sim \text{beta}(1+t, 1+n-t)$$

$$\theta|p, \phi, \eta_1, \ldots, \eta_n \sim \text{beta}\left(1+\sum_{i=1}^{n}\eta_i(m_i-x_i), 1+\sum_{i=1}^{n}\eta_i x_i\right)$$

$$\phi|p, \theta, \eta_1, \ldots, \eta_n \sim \text{beta}\left(1+\sum_{i=1}^{n}(1-\eta_i)x_i, 1+\sum_{i=1}^{n}(1-\eta_i)(m_i-x_i)\right)$$

$$\eta_i|p, \theta, \phi, \eta_1, \ldots, \eta_{i-1}, \eta_{i+1}, \ldots, \eta_n$$
$$\sim \text{Bernoulli}\left(\frac{p(1-\theta)^{x_i}\theta^{m_i-x_i}}{p(1-\theta)^{x_i}\theta^{m_i-x_i} + (1-p)\phi^{x_i}(1-\phi)^{m_i-x_i}}\right)$$

and Gibbs sampling can be easily implemented.

Quite often Gibbs sampling with or without the addition of auxiliary variables is available in Bayesian problems. Sometimes you have to be quite inventive to find appropriate auxiliary variables. See, for example, Gelfand and Smith (1990) and Escobar (1994). One of the advantages of Gibbs sampling via auxiliary variables is that it can make seemingly intractable integration problems straightforward. Of course, the existence of auxiliary variables is no guarantee that the Markov chain will converge in a reasonable amount of computation time. There is a general impression, however, that the addition of auxiliary variables helps the chain mix faster and this is at least intuitively reasonable. However, one should not be blinded by the ease with which a Gibbs samplers can be constructed. For example, Bélisle (1998) has shown that Gibbs samplers can be constructed that have total-variation distance between the n-th step kernel and the stationary distribution diminishing at a slower rate than any specified decreasing sequence. See also Matthews (1991) for a description of the *witch's hat distribution* which is a distribution with a sharp peak and flat tails and for which the Gibbs sampler performs poorly.

Sometimes we are interested in calculating marginal densities of coordinates of p; i.e. integrals of the form

$$p_i(x_i) = \int_{R^{k-1}} p(x_1, \ldots, x_k)\, dx_1 \cdots dx_{i-1} dx_{i+1} \cdots dx_k.$$

There are a number of possibilities for this but the best seems to be what Gelfand and Smith (1990) call *Rao–Blackwellization*. For example, to estimate the marginal density p_1 we calculate

$$\hat{p}_1(x_1) = \frac{1}{n} \sum_{j=0}^{n} p_1\left(x_1 | x_2^{(j)}, \ldots, x_k^{(j)}\right)$$

based on n steps of the chain. Of course this requires a closed form for the conditional density but this is often available. See also Chen (1994) where a method for approximating marginal densities is presented that avoids this necessity.

7.3.3 HYBRID CHAINS

There are various techniques for utilizing several Markov chains in a problem. For example, suppose that $K_1, K_2, \ldots$ are transition kernels of Markov chains, all having the same stationary distribution. We consider several ways of combining these together to form a new chain.

Suppose $p_1, p_2, \ldots$ are nonnegative numbers satisfying $\sum p_i = 1$. Then in the j-th step of the *mixture chain*, we choose an index i with probability p_i, and use K_i to choose the next step. The transition kernel of this new chain is clearly

$$K(x, B) = \sum p_i K_i(x, B).$$

It is immediate that the mixture chain has the same stationary distribution as the component chains.

Suppose that the transition kernels $K_1, K_2, \ldots, K_m$ are repeated in fixed order. The transition kernel of this *cyclic chain* over the entire cycle is then

$$K(x, B) = \int_{R^k} \cdots \int_{R^k} K_m(x_{m-1}, B) K_{m-1}(x_{m-2}, dx_{m-1}) \cdots K_1(x, dx_1).$$

If π is stationary for each chain we have that π is stationary for the new chain (Exercise 15).

Tierney (1994) shows that as long as one of the component chains satisfies a certain type of desirable convergence property (e.g. uniform ergodicity), then the mixture chain also shares this property. Cyclic chains are also good if at least one of the component chains is good; see Tierney (1994) for the technical details.

In fact the Gibbs sampler can be seen as an example of Metropolis algorithms applied cyclically. For this consider the move from

$$\left(x_1^{(i)}, \ldots, x_{j-1}^{(i)}, x_j^{(i-1)}, x_{j+1}^{(i-1)}, \ldots, x_k^{(i-1)}\right)$$

to

$$\left(x_1^{(i)}, \ldots, x_{j-1}^{(i)}, x_j^{(i)}, x_{j+1}^{(i-1)}, \ldots, x_k^{(i-1)}\right).$$

This chain has kernel given by

$$K_j(x, B) = \int_{B(x_1, \ldots, x_{j-1}, x_{j+1}, \ldots, x_k)} p_j(w | x_1, \ldots, x_{j-1}, x_{j+1}, \ldots, x_k)\, dw$$

where $B(x_1, \ldots, x_{j-1}, x_{j+1}, \ldots, x_k) = \{w | (x_1, \ldots, x_{j-1}, w, x_{j+1}, \ldots, x_k) \in B\}$ is the *slice* through B at $(x_1, \ldots, x_{j-1}, x_{j+1}, \ldots, x_k)$. Then p is a stationary distribution for this chain in the sense that

$$\int_{R^k} K_j(x, B) p(x)\, dx = \int_B p(x)\, dx$$

although it is not a unique stationary distribution as any other distribution with the same conditional of x_j given the remaining coordinates will also satisfy this equation. The cyclic chain based on the kernels $K_1, \ldots, K_k$ is then the Gibbs chain. But now observe that if we let the proposal density be

$$q_j(x, \cdot) = p_j(\cdot | x_1, \ldots, x_{j-1}, x_{j+1}, \ldots, x_k)$$

then

$$\min\left\{1, \frac{p(x_1, \ldots, x_{j-1}, w, x_{j+1}, \ldots, x_k)}{p(x_1, \ldots, x_{j-1}, x_j, x_{j+1}, \ldots, x_k)} \frac{p_j(x_j | x_1, \ldots, x_{j-1}, x_{j+1}, \ldots, x_k)}{p_j(w | x_1, \ldots, x_{j-1}, x_{j+1}, \ldots, x_k)}\right\}$$
$$= 1$$

so that each step in the cycle can be thought of as an application of the Metropolis algorithm. Note that an alternative Gibbs sampling chain can be obtained by choosing K_i with probability $1/k$. This mixture chain is called the *random scan Gibbs sampler*. See Amit and Grenander (1991) for a comparison of different scanning or sweep strategies. Also see Roberts and Sahu (1997) and MacEachern and Perrugia (1999).

Obviously this can generalized in a number of ways. For example, if we break $X \in R^k$ into l blocks $X = (X_1, \ldots, X_l)$ where $X_i \in R^{j_i}$ and $j_1 + \cdots + j_l = k$ then we can construct chains on each block with stationary distribution p and cycle through them *one block at a time* or perform a random scan. Also these chains need not be all Gibbs chains or all Metropolis chains but some of each. One of the disadvantages of Gibbs chains is that there is often a high level of correlation between consecutive states and this manifests itself via slow convergence to stationarity. Breaking the chain into blocks is one technique used to avoid this problem. Also, if in applying Gibbs sampling we face some difficult generation problems, then these can be avoided by choosing a Metropolis chain on the difficult parts.

We consider a typical example of this.

Example 7.9

Suppose that we want to compute expectations with respect to the density in $(\theta, w) \in R^{l+1}$

$$p(\theta, w) \propto \prod_{i-1}^{N} \left[\frac{e^{y_i v_i' \theta}}{(1 + e^{v_i' \theta})^{n_i}} \right] w^{l/2 + a_0 - 1} e^{-w(1/b_0 + \theta'\theta/2)}$$

where $y_i \in \{0, 1, \ldots, n_i\}$, $v_i \in R^l$, $a_0 > 0$ and $b_0 > 0$ are all known for $i = 1, \ldots, N$. This function arises as a posterior density in a Bayesian hierarchical model in statistics. To implement a version of Gibbs sampling, we need to be able to generate variates from the conditional distributions $p(w|\theta)$ and $p(\theta|w)$. Whereas $w|\theta \sim \text{gamma}(a_0 + l/2, (\theta'\theta/2 + 1/b_0)^{-1})$ leads to straightforward variate generation, the density of $\theta|w$ has a nonstandard form which is proportional to

$$\prod_{i=1}^{N} \left[\frac{e^{y_i v_i' \theta}}{(1 + e^{v_i' \theta})^{n_i}} \right] e^{-w\theta'\theta/2}.$$

A successful procedure here uses Gibbs for w and implements a Metropolis algorithm for θ where the $N_l(0, w^{-1} I)$ proposal distribution is used in the Metropolis step via an independence chain.

7.3.4 HIT-AND-RUN CHAINS

Target distributions with highly correlated variables often cause convergence difficulties for MCMC algorithms. This is especially true of the Gibbs sampler since univariate coordinates are generated one at a time along line segments that are parallel to their respective axes. When variables are highly correlated, the range of movement along these line segments can be very narrow.

The *hit-and-run algorithm*, Smith (1984), Chen and Schmeiser (1993, 1996), alleviates this difficulty to a certain extent by generating variates along random directions. We start the chain by selecting $X_0 \sim k_0$. The iterative step is then:

(i) the hit: generate a unit-length direction vector $d_i \sim U(S^{k-1})$;

(ii) the run: generate $\lambda_i \in R^1$ according to the density proportional $p(X_{i-1} + \lambda_i d_i)$ and let $X_i = X_{i-1} + \lambda_i d_i$.

Note that the generation of λ_i, while only a one-dimensional problem, can be difficult. This clearly gives a time homogeneous Markov chain with transition kernel (Exercise 17)

$$K(x, B) = A_{k-1}^{-1} \int_B \frac{2p(y)}{\|y - x\|^{k-1}} \left\{ \int_{R^1} p\left(x + s\frac{y - x}{\|y - x\|}\right) ds \right\}^{-1} dy$$

where A_{k-1} is the surface area of S^{k-1}. It can be shown directly from this expression that p is stationary for the hit-and-run chain (Exercise 18).

When the correlations amongst variables are very high, the hit-and-run algorithm can still have difficulties traversing the sample space since at any point x very few of the generated direction vectors d lie in the important directions. Consequently, generalizations of the hit-and-run sampler have been developed. Particular generalizations allow for nonuniform direction distributions and for adaptive selection of the direction vectors d. These latter methods are known as *adaptive direction sampling* and are discussed in Gilks, Roberts and George (1994). Bélisle, Romeijn and Smith (1993) discuss conditions for the convergence of hit-and-run chains. Bélisle (1999) has shown that hit-and-run samplers can be constructed that have total-variation distance between the n-step kernel and the stationary distribution diminishing at a slower rate than any specified decreasing sequence.

7.3.5 SLICE SAMPLING

Slice sampling is an instance of adding auxiliary variables in a particular way to a problem to hopefully simplify the generation. The simplest slice sampling algorithm arises by introducing a variable η such that

$$\eta | x \sim U(0, f(x));$$

i.e. the conditional density $q(\cdot|x)$ of η is the uniform density on $(0, f(x))$. We then have that (η, x) has joint density $p(x)q(\eta|x)$ and this is the uniform distribution on the set

$$D = \{(\eta, x) \colon 0 \leq \eta \leq f(x)\}.$$

To implement Gibbs sampling on the pair (η, x) then we need to be able to generate from the conditional distribution of x given η which is uniform on $\{x \colon f(x) \geq \eta\} = f^{-1}[\eta, \infty)$.

In general it is difficult to generate a uniform value in $f^{-1}[\eta, \infty)$ and so a different approach is required. Suppose that we can write

$$f(x) = \prod_{i=0}^{l} f_i(x)$$

where the f_i are positive functions. Let $\eta = (\eta_1, \ldots, \eta_l)$ where these variables are conditionally independent given x and $\eta_i | x \sim U(0, f_i(x))$. Then we have that $x|\eta$ (Exercise 19) has density proportional to

$$f_0(x) I_{\{z: f_i(z) \geq \eta_i, i=1,\ldots,l\}}(x) = f_0(x) \prod_{i=1}^{l} I_{f_i^{-1}[\eta_i, \infty)}(x).$$

Hopefully we can choose $f_0, \ldots, f_k$ so that this it is easy to generate from $x|\eta$.

We consider a particular generating problem where slice sampling is an easy approach to implement.

Example 7.10
Suppose that we want to generate from the density proportional to

$$(x + y) \exp\left\{-\tfrac{1}{2}(x^2 + y^2)\right\}$$

for $(x, y) \in [0, 1]^2$. Then take $f_0(x, y) = I_{[0,1]^2}(x, y)$, $f_1(x, y) = x + y$, $f_2(x, y) = \exp\left\{-\frac{1}{2}x^2\right\}$ and $f_3(x, y) = \exp\left\{-\frac{1}{2}y^2\right\}$. Therefore

$$f_0(x, y) \prod_{i=1}^{3} I_{f_i^{-1}[\eta_i, \infty)}(x, y)$$

is the indicator function for the convex polytope

$$\left\{(x, y) \in [0, 1]^2 : x + y \geq \eta_1, x \leq \sqrt{-2 \log \eta_2}, y \leq \sqrt{-2 \log \eta_3}\right\}$$

and it is straightforward to generate a uniform value from this set.

Slice sampling was introduced into simulation by Swendsen and Wang (1987) and extended in Edwards and Sokal (1988). Also see Besag and Green (1993), Higdon (1996), Neal (1997), Damien, Wakefield and Walker (1999), Mira and Tierney (1999), Mira (1998), Fishman (1999) and Roberts and Rosenthal (1999) where convergence and other issues are addressed.

7.3.6 PERFECT SAMPLING

Perfect or *exact sampling* refers to a technique whereby a value is generated from a Markov chain that is exactly distributed according to the stationary distribution of the chain. Perfect sampling therefore alleviates the need to assess convergence which is one of the major practical difficulties of MCMC. Perfect sampling was first presented in Propp and Wilson (1996).

The main idea of perfect sampling can be most easily seen in the context of a discrete target distribution p with a finite number m of possible outcomes, say $S = \{s_1, \ldots, s_m\}$; i.e. we want to generate $X \sim p$ where p is a probability function on S. To do this we use a backwards simulation technique based on an ergodic (irreducible, aperiodic) Markov chain that has p as its stationary distribution. Let K denote the kernel of the chain. We use the following technique to construct random maps $g_{-i} : S \to S$ for $i = -1, -2, \ldots$. For g_{-i} we generate m values $g_{-i}(s_1), \ldots, g_{-i}(s_m)$ with $g_{-i}(s_j) \sim K(s_j, \cdot)$ for all $j = 1, \ldots, m$. The perfect sampling procedure then proceeds as follows. To start we generate the random map g_{-1}. If $g_{-1}(s_1) = \cdots = g_{-1}(s_m)$ then we output this common value as X and stop. If this condition is not satisfied we independently generate g_{-2}. If $g_{-1} \circ g_{-2}$ is constant then we output the common value as X and stop. In general we stop at the first i such that $g_{-1} \circ g_{-2} \circ \cdots \circ g_{-i}$ is constant. Note that at each iteration we need only store the values of the composite map rather than each g_{-i}.

The following result is due to Propp and Wilson (1996).

Theorem 7.4 For the above backward simulation process we have that the process stops after finitely many steps with probability 1 and $X \sim p$.

Proof: We show first that this process stops after a finite number of steps with probability 1; i.e. with probability 1 a value X is returned. First we note that, because the chain is irreducible, there is a j such that

$$\epsilon = P(g_{-1} \circ g_{-2} \circ \cdots \circ g_{-j} \text{ is constant}) > 0.$$

Also we have that $g_{-1} \circ g_{-2} \circ \cdots \circ g_{-j}, g_{-(j+1)} \circ g_{-(j+2)} \circ \cdots \circ g_{-2j}, \ldots$ are independent and have the same distribution. Therefore

$$\begin{aligned}
&P(g_{-1} \circ g_{-2} \circ \cdots \circ g_{-i} \text{ is constant for some } i) \\
&\quad \geq P(g_{-(kj+1)} \circ g_{-(kj+2)} \circ \cdots \circ g_{-(k+1)j} \text{ is constant for some } k) \\
&\quad = \sum_{k=0}^{\infty} P\begin{pmatrix} g_{-(kj+1)} \circ g_{-(kj+2)} \circ \cdots \circ g_{-(k+1)j} \text{ is constant and} \\ g_{-(lj+1)} \circ g_{-(lj+2)} \circ \cdots \circ g_{-(l+1)j} \text{ is not constant for } l < k \end{pmatrix} \\
&\quad = \sum_{k=0}^{\infty} \epsilon(1-\epsilon)^k = 1.
\end{aligned}$$

This implies that $\lim_{i \to \infty} g_{-1} \circ g_{-2} \circ \cdots \circ g_{-i}$ exists almost surely as a constant function on S. It is immediate that

$$P(X = s) = P\Big(\lim_{i \to \infty} g_{-1} \circ g_{-2} \circ \cdots \circ g_{-i} = s\Big).$$

Then we have that

$$
\begin{aligned}
&\sum_{i=1}^{m} P(X = s_i)K(s_i, \{s_*\}) \\
&\quad = \sum_{i=1}^{m} P\Big(\lim_{l\to\infty} g_{-1} \circ g_{-2} \circ \cdots \circ g_{-l} = s_i\Big) K(s_i, \{s_*\}) \\
&\quad = \sum_{i=1}^{m} P\Big(\lim_{l\to\infty} g_{-2} \circ \cdots \circ g_{-l} = s_i\Big) K(s_i, \{s_*\}) \\
&\quad = \sum_{i=1}^{m} P\Big(\lim_{l\to\infty} g_{-1} \circ \cdots \circ g_{-l} = s_*\Big) \\
&\quad = P(X = s_*)
\end{aligned}
$$

which establishes that X is indeed distributed according to the stationary distribution.

This approach is sometimes also called *coupling from the past* because we can think of starting m chains all with transition kernel K, one starting at each state $s_i \in S$, at time $t = -\infty$ based on the same underlying $\omega \in \Omega$ from P. When these chains coalesce or couple we know that they remain coupled thereafter and this happens in a finite number of transitions with probability 1. Since the state we observe at time 0 is the result of an infinite number of transitions its distribution must correspond to the stationary distribution of the chain. In practice one can think of starting m copies of the chain at time $-T$ and running the chains and seeing if they have coalesced at time 0. If they have coalesced then this state is a value from the stationary distribution. If they have not coalesced then we go back further in time, say time $-2T$, while retaining the paths already generated from time $-T$ to 0, and run m copies of the chain from time $-2T$ to $-T$ and see if coalescence has been achieved at time 0. We continue this until we have coalescence. Carrying out the simulations in this fashion leads to some efficiencies as we need only generate new states from states that have appeared.

The structure that we have described for backwards simulation leads in many examples to a prohibitive amount of computation if S is large. Suppose, however, that there is a *structure function* ϕ and random quantity U such that $P(\phi(s_i, U) = s_j) = K(s_i, \{s_j\})$. The backward simulation is then carried out by using independent versions $U_{-1}, U_{-2}, \ldots$ and putting $g_{-t}(s_i) = \phi(s_i, U_{-t})$. Notice that

$$g_{-1} \circ g_{-2}(s) = \phi(\phi(s, U_{-2}), U_{-1}),$$

etc. Now if the elements of the state space are ordered with least element s_L and maximum element s_M then some simplifications arise when ϕ satisfies the monotone property

$$\phi(x, U) \le \phi(y, U)$$

whenever $x \le y$. Then we need only look at the sequences

$$(\phi(s_L, U_{-1}), \phi(\phi(s_L, U_{-1}), U_{-2}), \dots)$$

and

$$(\phi(s_M, U_{-1}), \phi(\phi(s_M, U_{-1}), U_{-2}), \dots)$$

to find the time t when they coalesce as this implies that $g_{-1} \circ g_{-2} \circ \cdots \circ g_{-t}$ is constant (Exercise 20). This is called *monotone Monte Carlo*. Propp and Wilson (1996) present a number of significant examples where this structure exists.

For more complicated state spaces such as Euclidean state spaces we generate independent values $U_{-T}, \dots, U_{-1}$ and check whether

$$\phi(\cdots \phi(\phi(E, U_{-T}), U_{-(T-1)}) \cdots, U_{-1})$$

results in coalescence, where E is the effective state space. If not, we go further back in time. There are some technical details involved in this and the choice of ϕ is crucial for effective implementation. At this point it is fair to say that perfect simulation is not a technique that can be used in general integration problems but the methodology holds promise for useful results. See Green and Murdoch (1999), Murdoch and Green (1998), Corcoran and Tweedie (1998a, 1998b) and Moller (1999) for results on perfect sampling.

7.4 Related techniques and issues

7.4.1 REPARAMETERIZATION

As with all integration methods we may consider transformations in MCMC to try and make the integral in question more amenable to a particular technique. For example, we may be able to transform a problem so that a Gibbs sampler is easy to implement or some particular Metropolis chain is rapidly convergent. It is a well-known fact that MCMC methods are very sensitive to parameterization.

In particular if there are some high correlations amongst coordinates in $X \sim p$ then we can expect a Gibbs chain to be slowly convergent. Similarly without a careful choice of the proposal chain, which reflects the correlations in p, a Metropolis chain will also be slowly convergent. To avoid this it is typically recommended that a reparameterization of p be undertaken to help reduce correlations. For example, if p is unimodal with approximate ellipsoidal contours then transforming to $y = \hat{\Sigma}^{-1/2}(x - \hat{x})$, where $\hat{x}$ is the mode of p and $\hat{\Sigma}$ is the Hessian matrix of $-\log p$ at $\hat{x}$, is usually recommended and can be effective. Alternatively, a preliminary run of the chain can be used to estimate the mean of p and its variance matrix. A location-scale transformation based on these quantities can then be used. To be effective this depends on the preliminary run having mixed at least reasonably well. For another approach to avoiding this problem with the random walk Metropolis chain, see Neal (1998a) where the technique of *overrelaxation* due to

Adler (1981) is discussed. This entails generating each coordinate in sequence, e.g. using Gibbs sampling or using a Metropolis chain on each coordinate, and ensuring that any generated coordinate is on the opposite side of the mode of the conditional density for that coordinate when compared to the previously generated value of the coordinate. See also Green and Han (1992).

A much more serious problem for MCMC algorithms arises when p is multimodal with widely separated modes and we do not know the location of the modes. In this case, or even if p is unimodal, it is almost always a good idea to run a maximizer with many different starting values to try and get some idea of where p is high. The price for not doing this can be severe as it is common for a chain to get stuck in a region about a single mode and thus invalidate the subsequent approximations. Even worse is that the chain will have appeared to have converged. The best way to prevent this is to have knowledge of where the modes are and then design the sampler so that all regions of high probability are visited. The technique of simulated tempering, which we subsequently discuss, can be helpful in this regard. We note that this phenomenon can occur even with Gibbs sampling; see, for example, Evans, Guttman and Olkin (1993) and the following simple example.

Example 7.11 *Mixture of normals*
Suppose that $X \sim 0.5N(0, 1) + 0.5N(c, 1)$ where we vary the constant c. Obviously we can simulate directly from this distribution but for illustrative purposes we introduce the latent variable $U \sim \text{Bernoulli}(0.5)$ such that $X|U = 0 \sim N(0, 1)$ and $X|U = 1 \sim N(c, 1)$. Then for Gibbs sampling we have that

$$U|X = x \sim \text{Bernoulli}\left(\frac{\exp\left\{-\frac{1}{2}(x-c)^2\right\}}{\exp\left\{-\frac{1}{2}x^2\right\} + \exp\left\{-\frac{1}{2}(x-c)^2\right\}}\right). \tag{7.19}$$

Suppose that we start the chain at $U = 1$, $X = c$. It is then clear from (7.19) that if c is far from 0 the chain will only visit the region near $X = 0$, where 1/2 of the probability resides, with a very small probability. For example, in Table 7.1 we give the probability of the chain remaining in the region near $X = c$ when we are in the region near $X = c$ and a value of X has occurred that is three standard deviations towards 0 from c. This clearly shows that even in one dimension a stationary distribution with widely separated modes can be disastrous unless these modes are detected and a sampler is implemented that permits switching amongst the modes. As the dimension rises this becomes more difficult to ensure.

The issue of reparameterization of a stationary distribution to ensure reasonably rapid convergence and avoid failure is obviously an important one and yet there is not much guidance on this beyond what we have mentioned here. With a multimodal p, or a p with very irregular contours, we can expect problems to arise unless we have an appropriate sampler. More discussion on these issues can be found in Achcar and Smith (1990), Hills and Smith (1992), Wakefield (1992) and Gilks and Roberts (1996).

Table 7.1. Probability of remaining near the value $X = c$ in Example 7.11.

c	$P(U = 1 \mid X = c - 3)$
5	0.0758582
7	0.9706878
9	0.9999986
10	1.0000000

7.4.2 IMPORTANCE SAMPLING

When considering MCMC methods it is not necessary to construct a chain with stationary distribution p. For example, we could construct a chain with stationary distribution w. Then, when the necessary conditions apply for the convergence of the chain and $\operatorname{supp}(f) \subset \operatorname{supp}(w)$ we have that

$$\lim_{n\to\infty} \frac{\sum_{i=1}^{n} m(X_i)\{f(X_i)/w(X_i)\}}{\sum_{i=1}^{n} f(X_i)/w(X_i)} = \int_{R^k} m(x)p(x)\,dx.$$

This technique is known as importance sampling and has obvious similarities to the approach of Chapter 6. As with the methods of Chapter 6 the importance sampler w should be a good approximation to p if we want the estimate to be reasonable within acceptable computation times.

The utility of this approach arises in several contexts. For example, we may have a chain which rapidly converges to the stationary distribution w and it may be difficult to find a chain that rapidly converges to p. Also importance sampling gives us the freedom of choosing w to try and minimize the asymptotic variance of the estimate for a particular m. Finally, and perhaps most importantly, we are sometimes confronted with the situation where we want to compute

$$\int_{R^k} m(x)p_\theta(x)\,dx \tag{7.20}$$

for a family of probability densities $\{p_\theta : \theta \in \Omega\}$. The idea then is to find, if possible, a chain with stationary distribution w such that

$$\frac{\sum_{i=1}^{n} m(X_i)\{f_\theta(X_i)/w(X_i)\}}{\sum_{i=1}^{n} f_\theta(X_i)/w(X_i)}$$

gives an appropriate estimate of (7.20) for every value of $\theta \in \Omega$. If we can find such a w then there is a tremendous saving in computation over running a chain for each θ. Of course, in general it will be difficult to find such a w. One possibility is to take w to be a mixture of the p_θ. See Madras and Piccioni (1998) for a discussion of importance sampling in this context. Of course all of this supposes that we can't simulate directly from w otherwise we can use the methods of Chapter 6.

7.4.3 VARIANCE REDUCTION TECHNIQUES

It has at times been recommended that only every N-th iteration of a Markov chain sampler be used when computing estimates of integrals. The idea behind this is that we choose N so that after N iterations there is minimal correlation between the states; i.e. assuming we have achieved stationarity, then $X_{jN}, X_{(j+1)N}, \ldots$ can be thought of as approximately iid from the stationary distribution. Irrespective of the fact that we do not typically know an appropriate value of N it has been shown in MacEachern and Berliner (1994) that the variance of the sample mean is smaller if every state is used (Exercise 14). This result applies to any second order stationary process. Still it is often recommended in applications of MCMC that the first few iterations of the chain be discarded to allow the chain to *burn-in*. In essence the motivation for this is to allow the chain to achieve stationarity before estimation takes place.

Some of the variance reduction methods discussed in Chapter 6 can be used in conjunction with MCMC. For example, suppose that we can calculate $E_p[g]$ for some prescribed function g. Then

$$E_p[g] + \frac{1}{n+1}\sum_{i=0}^{n}(m(X_i) - g(X_i))$$

is an estimator of $E_p[m]$ using g as an additive control variate. If $\mathrm{Corr}_p[m, g]$ is high then we expect this to give a more accurate estimate than just averaging m. If $E_p[g] \neq 0$ then

$$E_p[g]\frac{\sum_{i=0}^{n} m(X_i)}{\sum_{i=0}^{n} g(X_i)}$$

is an estimator of $E_p[m]$ using g as a ratio control variate. Much of the discussion in Section 6.4.6 is applicable here but with the additional complication that we no longer have iid sampling.

We can also make use of systematic sampling in this context. Again the idea is to average m with respect to a finite group of orthogonal transformations that leave $w = p$ invariant. If the group is $\mathcal{T} = \{T_1, \ldots, T_l\}$ then we use the estimator

$$\frac{1}{n+1}\sum_{i=0}^{n} m^{\mathcal{T}}(X_i)$$

where

$$m^{\mathcal{T}}(x) = \frac{1}{l}\sum_{i=1}^{l} m(T_i x).$$

The discussion of Section 6.4.4 is applicable here but with the additional complication that we no longer have iid sampling. For a discussion of the use of some systematic sampling techniques in the context of the Gibbs sampler see Greenwood, McKeague and Wefelmeyer (1996).

7.4.4 SIMULATED TEMPERING

Simulated tempering is another example of a homotopy method for integration problems. This technique was introduced in Marinari and Parisi (1992). See also Geyer and Thompson (1995) for an extensive discussion. Simulated tempering is similar in spirit to the technique of chaining introduced for adaptive importance sampling in Evans (1991a, 1991b) and discussed in Section 6.5. Also see the method of path sampling presented in Section 7.4.5. All of these techniques derive their motivation from the method of *simulated annealing* which is a Markov chain Monte Carlo method for the global optimization of a function; see Kirkpatrick, Gelatt and Vecchi (1983). The basic idea behind homotopy methods is that, when confronted with a very hard problem, e.g. evaluate an integral, maximize a function, etc., we look for an easier instance of the problem for which we can develop an effective algorithm. In addition we suppose that these problems can be embedded in a class of problems indexed by some parameter, say $\lambda \in \Lambda$, such that small changes in λ correspond to small changes in the problem. The idea then is to choose $\lambda_0, \ldots, \lambda_l \in \Lambda$ so that the change from λ_i to λ_{i-1} is small enough that a solution to the problem given by λ_i can be used towards constructing a solution for the problem given by λ_{i-1}. Further we suppose that λ_l is the initial problem that we can solve and λ_0 is the problem of interest.

With simulated tempering we suppose that we have $p \in \{p_\lambda : \lambda \in \Lambda\}$ and let $p_{\lambda_0}, \ldots, p_{\lambda_l}$ be chosen such that p_{λ_l} has a rapidly convergent MCMC algorithm, $p_{\lambda_0} = p$ and p_{λ_i} is close to $p_{\lambda_{i-1}}$. Let f_{λ_i} denote a function proportional to p_{λ_l}. Now let $r_0, \ldots, r_l$ be a probability distribution on $L = \{0, 1, \ldots, l\}$. We then run a chain $\{(X_n, I_n) : n = 0, 1, \ldots\}$ with stationary distribution proportional to $r_i f_{\lambda_i}$ and so the conditional stationary distribution of X for all those states where $I = 0$ is p. The idea is to run the joint chain and then keep only the states for which $I = 0$. For this to be useful we construct the joint chain in such a way that I only makes one-step transitions and so, if the p_{λ_i} are close together, the chain should mix well as we chain through the effective supports of these densities.

For an implementation of simulated tempering let $K^{(i)}$ be the kernel for a chain with stationary distribution given by p_{λ_i} and let $Q = (q_{ij}) \in R^{(l+1)\times(l+1)}$ be a transition matrix on L, with rows and columns indexed from $0, \ldots, l$. We set

$$q_{ij} = \begin{cases} 1 & (i, j) = (0, 1) \text{ or } (i, j) = (l, l-1) \\ \frac{1}{2} & j = i \pm 1 \\ 0 & \text{otherwise;} \end{cases}$$

i.e. we move to a (nearest) neighboring state with probability 1. Now suppose that we are at (x, i) at time $n - 1$. Then at time n we generate the state I_n according to the Metropolis chain on L given by generating $j \sim q_i$ and putting $I_n = j$ with acceptance probability

$$\min\left\{1, \frac{f_{\lambda_j}(x)}{f_{\lambda_i}(x)} \frac{r_j}{r_i} \frac{q_{ji}}{q_{ij}}\right\}$$

or else putting $I_n = i$. Then X_n is generated from $K^{(I_n)}(x, \cdot)$. Note that if two of the p_{λ_i} and $p_{\lambda_{i+1}}$ are too far apart then we will observe too few transitions from $i \to i+1$ or $i+1 \to i$ which will tell us we need to introduce an intermediate distribution. Note that this chain has the correct stationary distribution (Exercise 21).

The marginal stationary distribution of I is given by

$$P(I = i) \propto r_i \int_{R^k} f_{\lambda_i}(x)\, dx. \tag{7.21}$$

To be effective we want the chain to spend about the same proportion of its time generating from each of the component chains as this will help ensure that in the end we have thoroughly traversed the support of p_{λ_0}. To guarantee this means that we have to have reasonable approximations to each of the normalizing constants

$$\int_{R^k} f_{\lambda_i}(x)\, dx$$

so that we can choose r_i so that (7.21) is approximately constant. The methods of Section 7.4.5 are available for this, and other methods as discussed in Geyer and Thompson (1995), but this necessity can still be viewed as a weakness of the technique. In Madras and Piccioni (1998) it is shown that in certain circumstances simulated tempering is inferior to importance sampling, as described in Section 7.4.2, with respect to the density $\sum_{i=1}^{l} r_i p_{\lambda_i}$.

Neal (1996) has proposed the method of *tempered transitions* as an improvement on simulated tempering as it avoids the need to determine reasonable choices of the r_i; i.e. we don't need to approximate the normalizing constants. For this let $p_0 = p$ and suppose that we have density p_{λ_l} for which we have a rapidly convergent sampler with kernel $\hat{K}_{\lambda_l}$, interpolating densities $p_{\lambda_1}, \ldots, p_{\lambda_{l-1}}$ and kernels $\hat{K}_{\lambda_1}, \ldots, \hat{K}_{\lambda_l}, \check{K}_{\lambda_1}, \ldots, \check{K}_{\lambda_l}$ which satisfy p_{λ_i} is a stationary distribution for both $\hat{K}_{\lambda_i}$ and $\check{K}_{\lambda_i}$ and

$$p_{\lambda_i}(x)\hat{K}_{\lambda_i}(x, dy) = p_{\lambda_i}(y)\check{K}_{\lambda_i}(y, dx) \tag{7.22}$$

for $i = 1, \ldots, l$ for all x, y. We can take $\hat{K}_{\lambda_i} = \check{K}_{\lambda_i}$ and in this case (7.22) implies that this kernel is reversible with respect to p_{λ_i}. The algorithm then proceeds to generate a new state y from state x as follows:

$$\begin{aligned}
&\text{generate } \hat{x}_1 \sim \hat{K}_{\lambda_1}(x, \cdot),\\
&\text{generate } \hat{x}_2 \sim \hat{K}_{\lambda_2}(\hat{x}_1, \cdot),\\
&\vdots\\
&\text{generate } \hat{x}_{l-1} \sim \hat{K}_{\lambda_{l-1}}(\hat{x}_{l-2}, \cdot),\\
&\text{generate } \bar{x}_l \sim \hat{K}_{\lambda_l}(\hat{x}_{l-1}, \cdot),\\
&\text{generate } \check{x}_{l-1} \sim \check{K}_{\lambda_l}(\bar{x}_l, \cdot),\\
&\text{generate } \check{x}_{l-2} \sim \check{K}_{\lambda_{l-1}}(\check{x}_{l-1}, \cdot),\\
&\vdots\\
&\text{generate } \check{x}_1 \sim \check{K}_{\lambda_2}(\check{x}_2, \cdot),\\
&\text{generate } y \sim \check{K}_{\lambda_1}(\check{x}_1, \cdot).
\end{aligned}$$

Then accept y with probability

$$\min\left\{1, \frac{p_{\lambda_1}(x)}{p(x)}\frac{p_{\lambda_2}(\hat{x}_1)}{p_{\lambda_1}(\hat{x}_1)}\cdots\frac{p_{\lambda_l}(\hat{x}_{l-1})}{p_{\lambda_{l-1}}(\hat{x}_{l-1})}\frac{p_{\lambda_{l-1}}(\check{x}_{l-1})}{p_{\lambda_l}(\check{x}_{l-1})}\cdots\frac{p_{\lambda_1}(\check{x}_2)}{p_{\lambda_2}(\check{x}_1)}\frac{p(y)}{p_{\lambda_1}(y)}\right\} \tag{7.23}$$

else return the value x. Note that because each p_{λ_i} appears in a ratio in (7.23) we do not need the normalizing constant. It can be shown that this prescription leads to a reversible chain with stationary distribution given by p (Exercise 22). See Neal (1996) for examples of the use of tempered transitions.

We note that another method, similar to simulated tempering, involves running the chain p_{λ_l} until it has converged. We then use the final state of this chain to start the chain for $p_{\lambda_{l-1}}$ and run this chain until it has converged, etc. This is in fact the Markov chain technique that underlies simulated annealing. In simulated annealing, however, the final distribution p_0 is in effect degenerate at the global maximum of the function we are trying to maximize. In the integration contexts we are discussing, the p_{λ_i} are not necessarily becoming degenerate but rather may have regions of positive mass that are widely separated. Accordingly, this technique will not work effectively in such a context; i.e. at some stage we will fail to obtain convergence from a single starting point. Neal (1998b) develops the technique of *annealed importance sampling* to address this problem.

7.4.5 ESTIMATING NORMALIZING CONSTANTS AND RATIOS OF NORMALIZING CONSTANTS

Quite often we only know f and do not know the inverse normalizing constant

$$\int_{R^k} f(x)\,dx \tag{7.24}$$

used to obtain p. This is not a problem for MCMC methods as only f is needed to construct a Markov chain with stationary distribution given by p. Still there are a number of applications where it is important to be able to estimate (7.24).

The importance sampling methods of Chapter 6 or of Section 7.4.1 are available for the estimation of (7.24). For example, this was the technique used in Geyer and Thompson (1991, 1992) for the estimation of a set of normalizing constants.

It is also of interest to estimate this quantity based on a Markov chain $\{X_i : i = 0, 1, \dots\}$ with stationary distribution given by p. For this let q be any normed density function on R^k; i.e. q is nonnegative and integrates to 1, such that $\mathrm{supp}(q) \subset \mathrm{supp}(f)$. Then, assuming the necessary conditions for convergence, we have that

$$\lim_{n\to\infty} \sum_{i=1}^{n} \frac{q(X_i)}{f(X_i)} = \int_{R^k} \frac{q(x)}{f(x)} p(x)\, dx = \left(\int_{R^k} f(x)\, dx \right)^{-1}.$$

Estimators based on various choices of q were suggested in Gelfand and Dey (1994) and Newton and Raftery (1994). This density q is a kind of *inverse importance sampler* and we see that it is important that q have lighter tails than p. It is natural then to choose q to have compact support located in a region where p is significant.

In Chib (1995) it is noted that

$$\int_{R^k} f(x)\, dx = \frac{f(x_*)}{p(x_*)}$$

at any value x_*. Therefore, if we have a good estimate $\hat{p}(x_*)$ of $p(x_*)$ then $f(x_*)/\hat{p}(x_*)$ gives an estimate of the inverse normalizing constant. When we are using the Gibbs chain a simple estimate of $p(x_*)$ is given by

$$\hat{p}(x_*) = \hat{p}_1(x_{1*})\hat{p}_2(x_{2*}|x_{1*}) \cdots \hat{p}_k(x_{k*}|x_{1*}, \dots, x_{k-1*})$$

where

$$\hat{p}_1(x_{1*}) = \frac{1}{n} \sum_{j=1}^{n} p_1(x_{1*}|x_2^{(j)}, \dots, x_k^{(j)})$$

and the remaining estimates are specified as follows. To calculate $\hat{p}_2(x_{2*}|x_{1*})$ run a conditional Gibbs chain, where the first coordinate is fixed at x_{1*}, using the density

$$p(x_2, \dots, x_k|x_{1*}); \tag{7.25}$$

e.g. the full conditional for x_2 from (7.25) is just $p_2(x_2|x_{1*}, x_3, \dots, x_k)$. This is continued for each variable except that we note that $\hat{p}_k(x_{k*}|x_{1*}, \dots, x_{k-1*})$ is taken to be $p_k(x_{k*}|x_{1*}, \dots, x_{k-1*})$. Further it is suggested that x_* be chosen to be a mode or at least a point where p is relatively large. There are several limitations of this method, the first being that it requires more than just the output of a single chain. Further it is reliant on the full conditional densities being available to us.

A method for estimating ratios of normalizing constants, called *path sampling*, is discussed in Gelman and Meng (1998) with origins in the work of Ogata (1989, 1990). This is another example of a homotopy method; see the discussion in Sections 6.5 and 7.4.4. For this we consider the integrand f as a member of a family of integrands $\{f_{\lambda(t)} : t \in [0, 1]\}$ where $\lambda : [0, 1] \to R^l$ is a path in R^l such that $f = f_{\lambda(0)}$ and $f_{\lambda(1)}$ is some other unnormalized density of interest. Then, under assumptions that make the dominated derivative theorem applicable, we have that

$$\frac{d}{dt} \log \left[\int_{R^k} f_{\lambda(t)}(x)\, dx \right] = \frac{\int_{R^k} \frac{\partial f_{\lambda(t)}(x)}{\partial t}\, dx}{\int_{R^k} f_{\lambda(t)}(x)\, dx}$$

which, letting $p_{\lambda(t)}$ denote the density proportional to $f_{\lambda(t)}$, implies that

$$\begin{aligned} \log \left[\frac{\int_{R^k} f_{\lambda(1)}(x)\, dx}{\int_{R^k} f_{\lambda(0)}(x)\, dx} \right] &= \int_0^1 \left\{ \frac{\int_{R^k} \frac{\partial f_{\lambda(t)}(x)}{\partial t}\, dx}{\int_{R^k} f_{\lambda(t)}(x)\, dx} \right\} dt \\ &= \int_0^1 \left\{ \int_{R^k} \frac{\partial \log f_{\lambda(t)}(x)}{\partial t}\, p_{\lambda(t)}(x)\, dx \right\} dt \\ &= E \left[\frac{\partial \log f_{\lambda(T)}(X)}{\partial t} \right] \end{aligned} \tag{7.26}$$

where E refers to expectation where $X|T \sim p_{\lambda(T)}$ and $T \sim U(0, 1)$. Then we can estimate (7.26) by running a chain on (X, T) with this stationary distribution. Note that if $f_{\lambda(1)}$ is a density chosen such that we know its normalizing constant then this leads to a method for estimating the normalizing constant for f. There are various considerations with path sampling. For example, what path or homotopy should be used? For example, if we let f_0 denote the other density whose normalizing constant is desired then we could take $f_{\lambda(t)} = f^t f_0^{1-t}$. Ogata (1989) uses the homotopy $f_{\lambda(t)}(x) = f((1-t)x)$ in the context of integrating over a cube. Gelman and Meng (1998) discuss methods for obtaining the optimal path.

Related to path sampling is the method of *bridge sampling* introduced in Meng and Wong (1996) and also discussed in Gelman and Meng (1998). For this we note that we can write, for any unnormalized densities f_0 and f_1,

$$\frac{\int_{R^k} f_1(x)\, dx}{\int_{R^k} f_0(x)\, dx} = \frac{E_0[f_1(X)\alpha(X)]}{E_1[f_0(X)\alpha(X)]} \tag{7.27}$$

for any function α such that

$$\int_{\text{supp}(f_0)\cap\text{supp}(f_1)} f_0(x) f_1(x)\alpha(x)\, dx < \infty$$

and where E_i refers to expectation with respect to the density $p_i \propto f_i$. Then if we run two chains $\{X_i^{(j)}: i = 0, 1, 2, \ldots\}$ with stationary distribution p_j for $j = 0, 1$ we can estimate (7.27) by

$$\frac{\sum_{i=1}^{n} f_1(X_i^{(0)})\alpha(X_i^{(0)})}{\sum_{i=1}^{n} f_0(X_i^{(1)})\alpha(X_i^{(1)})}.$$

Some discussion is given in Gelman and Meng (1998) concerning the optimal choice of α and of methods related to bridge sampling; e.g. $\alpha = 1/f_0$ gives importance sampling with respect to the density p_0.

A variety of different techniques for estimating normalizing constants is discussed in DiCiccio, Kass, Raftery and Wasserman (1997). This paper also contains a simulation study comparing many of the available methods. Also Chen and Shao (1997) also make some comparisons amongst different methods for approximating ratios of normalizing constants.

As mentioned at the beginning of this chapter MCMC has many uses besides the approximation of integrals as we have presented the problem in this text. For example, there are problems in statistical applications where there is no simple underlying support measure (such as Lebesgue or counting measure) for a posterior distribution. Green (1995) has developed methods known as *reversible jump Markov chain Monte Carlo* that are appropriate for many such contexts. The reversible jump MCMC approach has also been considered in Dellaportas and Forster (1999), Richardson and Green (1997) and Giudici and Green (1998). Special considerations and difficulties arise with the use of reversible jump algorithms. For example, Brooks and Giudici (1999) consider convergence diagnostics for reversible jump algorithms, and Giudici and Roberts (1999) discuss the choice of the reversible jump proposal distribution.

7.5 Assessing convergence and error

The issue of error estimation is somewhat more difficult with MCMC than with the iid methods discussed in Chapter 6. Still there are many problems where a MCMC algorithm easily comes to mind and an appropriate iid importance sampler does not.

Our primary interest, at least in this book, is whether or not

$$\frac{1}{n+1}\sum_{i=0}^{n} m(X_i) \tag{7.28}$$

reasonably approximates $\int_{R^k} m(x)p(x)\,dx$ for a specific n. To assess this we need to estimate the variance of (7.28). Under stationarity this variance is given by

$$\frac{1}{n+1}\left\{\text{Var}_{P^*}\left[m(X_0)\right] + 2\sum_{i=1}^{n}\frac{(n-i+1)}{n+1}\text{Cov}_{P^*}\left[m(X_0), m(X_i)\right]\right\} \tag{7.29}$$

where P^* refers to the Markov chain with initial distribution p and kernel K. For large n (7.29) is approximated by $\sigma_m^2/(n+1)$ where

$$\sigma_m^2 = \mathrm{Var}_{P^*}[m(X_0)] + 2\sum_{i=1}^{\infty} \mathrm{Cov}_{P^*}[m(X_0), m(X_i)]. \tag{7.30}$$

Note that (7.29) and (7.30) refer to the stationary distribution of the chain so it makes sense that any estimator we use be based on values from the chain when the chain has achieved stationarity. Our first task then is to determine whether the Markov chain has achieved its steady state or stationarity. If the chain has achieved stationarity then we estimate σ_m^2 to assess the accuracy of the estimate.

7.5.1 DIAGNOSTICS FOR STATIONARITY

The approach to assessing whether or not stationarity has been achieved is to examine the output from the chain and see if it exhibits appropriate characteristics. Often this means computing various *convergence diagnostics* based on this output. A common recommendation is to discard the initial part of the output, called the burn-in, and base the diagnostics on the remaining output.

There are two points that need to be made when discussing convergence diagnostics. First, no convergence diagnostic is fail-safe. That is, for any diagnostic, there are situations where a recommendation of convergence will be made even when the Markov chain has not converged. The second point is that since no convergence diagnostic is fail-safe, it is wise to have several at hand that assess different kinds of stationary behavior. This decreases the chances of being fooled. Of course, the converse is also possible, a chain may have achieved stationarity and yet some diagnostic may indicate otherwise. One possibility for convergence diagnostics are statistical tests of the null hypothesis that the output from the chain is from a process that has achieved the stationary state. The power of such a test, i.e. the probability of rejecting the null hypothesis when it is false, then gives us some idea of how sensitive and useful the test is when measured against a specific alternative. Selection of a low size for the test also protects us against rejecting a true null. Not all convergence diagnostics correspond to statistical tests as some are data analytic in character; e.g. they involve the examination of various plots of the output for appropriate appearance.

One of the simplest convergence diagnostics is the *trace plot*. A trace plot is a graph of some function of the state of the chain plotted against the iteration number. We expect the trace plot to settle within a band of plausible values upon reaching stationarity. Trace plots should be displayed for all coordinates of X although this is not always feasible when k is large. It is also recommended to construct trace plots for the log of the target density. Naturally, a recommendation of convergence would require that all trace plots reach a convergent state. Figure 7.1 is an example of a trace plot of a coordinate based on the output of the Gibbs sampling algorithm in Heard and Swartz (1999). In this example, it appears that the chain reaches stationarity very quickly.

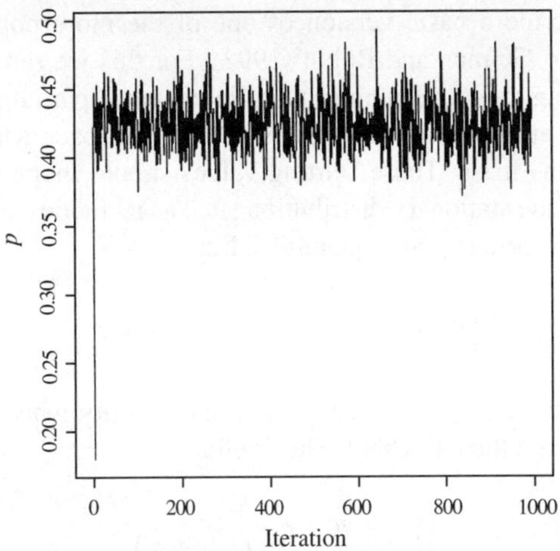

Fig. 7.1. Trace plot of realizations of a coordinate in a Gibbs sampling chain.

As the chain is running, one may also periodically calculate various estimates and their standard errors as discussed in Section 7.5.2. When deviations from the expected theory present themselves, this is an indication of lack of convergence to stationarity. For example, when quadrupling the number of iterations, we expect the standard error of averages of nearly independent quantities to decrease by a factor of 2.

Another simple technique, related to the investigation of convergence to stationarity, is the calculation of correlation coefficients. When autocorrelations are large, i.e. correlations between states a fixed number of time intervals apart, it is generally the case that the chain will take longer to mix or move entirely about the state space as these indicate small steps in the chain will be common. In such a context one may consider the modification of estimators. For example, if lag-1, ..., lag-k autocorrelations are present, but the lag-$(k+1)$ autocorrelation is nearly zero, one may skip over and exclude blocks of k realizations in the estimator although we recall the discussion in Section 7.4.3 concerning the accuracy of such an estimator.

One of the best general strategies that can be used with all of the convergence diagnostics is to simply replicate chains using disparate starting values. If the diagnostics for each of the chains agree, this provides evidence in favor of convergence. When the diagnostics disagree, it is not usually clear which chains, if any, have converged.

Cowles and Carlin (1996) give a comparative review of 13 of the more sophisticated convergence diagnostics. Although the diagnostics are not ranked in order of efficacy, valuable commentary is provided on issues such as implementation details, general applicability, ease of use, weaknesses, etc. Brooks and Roberts (1999) review a slightly larger set of convergence diagnostics and give stronger guidelines as to when the various diagnostics are appropriate.

We now describe a basic version of one of the more popular convergence diagnostics due to Gelman and Rubin (1992). For this we run l different chains at different starting values. Running multiple chains is partial protection against the possibility of entirely missing regions of the state space where the stationary distribution assigns mass. These starting values should themselves be based on characteristics of the stationary distribution such as its modes, etc. The diagnostic $\hat{R}$ is computed on various scalar quantities. Let

$$\left\{v_j^{(i)}: i = 1, \ldots, l, j = 1, \ldots, n\right\}$$

denote Markov chain output on one such scalar quantity where $v_j^{(i)}$ refers to the j-th realization from the i-th chain. Then define

$$B = \frac{n}{l-1} \sum_{i=1}^{l} \left(\bar{v}^{(i)} - \bar{\bar{v}}\right)^2$$

as the between-sequence variance of the $v_j^{(i)}$ where $\bar{v}^{(i)} = \sum_{j=1}^{n} v_j^{(i)}/n$ and $\bar{\bar{v}} = \sum_{i=1}^{l} \bar{v}^{(i)}/l$. Also, define the within-sequence variance of the $v_j^{(i)}$ by

$$W = \frac{1}{l} \sum_{i=1}^{l} s_i^2$$

where $s_i^2 = \sum_{j=1}^{N} \left(v_j^{(i)} - \bar{v}^{(i)}\right)^2/(n-1)$. The diagnostic is then given by

$$\hat{R} = \frac{(1/n)B + ((n-1)/n)W}{W}$$

which should be approximately 1 when the chain has converged to stationarity. The motivation behind $\hat{R}$ is that the numerator initially overestimates variability since starting values are typically overdispersed. Conversely, the denominator W should initially underestimate variability as the chains will not have sufficiently traversed the sample space. Thus, as the Markov chain approaches the target distribution, we expect $\hat{R}$ to decrease toward 1.

There are many other diagnostics that are useful. For example, Zellner and Min (1995) propose comparing the computable ratio

$$\frac{p(x_1)}{p(x_2)} = \frac{f(x_1)}{f(x_2)}$$

with estimates of this quantity to determine if convergence has been achieved. The comparison can be made using multiple values of (x_1, x_2). If the Rao–Blackwellization technique is to be used to estimate the joint density, then multiple chains are required. Alternatively, kernel density estimates can be used to estimate the ratio. In Brooks (1998) an interesting convergence diagnostic is described for use with log-concave densities that allows for a precise measure of how well the sampler has explored the state space. Johnson (1997, 1998) has employed coupling techniques to be used in the diagnosis of convergence. Cowles, Roberts and Rosenthal (1999) sound a cautionary note concerning diagnostics pointing out that even in ideal circumstances the naive use of diagnostics can lead to serious biases in estimates. These authors recommend that practitioners determine an appropriate number of iterations r for the burn-in period by running multiple pilot chains and then discarding the first r iterations of a chain that will be used for estimation; i.e. r is not determined based on the output from the chain used to construct estimates.

7.5.2 ESTIMATING THE ASYMPTOTIC VARIANCE

We assume now that the process has achieved stationarity and consider the estimation of σ_m^2. There are a number of different techniques that can be used and we refer the reader to Brockwell and Davis (1991) for an extensive discussion of many of these. For discussions related to simulation problems see Ripley (1987) and Geyer (1992).

We discuss only a relatively simple approach to this problem; namely the method of *batching*. For this we divide the sequence $m(x_1), \ldots, m(x_n)$ into nonoverlapping sequential batches of size l, calculate the mean in each batch obtaining $\bar{m}_1, \ldots, \bar{m}_{\lfloor n/l \rfloor}$ and then estimate σ_m^2/l using the sample variance obtained from the batch means; i.e.

$$s_l^2 = \frac{1}{\lfloor n/l \rfloor - 1} \sum_{i=1}^{\lfloor n/l \rfloor} (\bar{m}_i - \bar{m})^2$$

where

$$\bar{m} = \frac{1}{\lfloor n/l \rfloor} \sum_{i=1}^{\lfloor n/l \rfloor} \bar{m}_i .$$

The batch size l must be chosen large enough so that the batch means have low serial correlations but not so large as to leave few degrees of freedom for the estimation of σ_m^2.

We consider a specific example.

Example 7.12 *Application of Gibbs sampling, independence and random walk Metropolis chains*
We consider the integrals discussed in Examples 4.9, 6.4 and 6.15. This provides for an interesting comparison amongst the methods.

For Gibbs sampling the generating problem was solved, as in Verdinelli and Wasserman (1991), by adding the independent latent variables

$$u_{ij} \sim N(0, 1)$$
$$v_{ij} \sim \text{gamma}(\lambda/2, 1) \text{ for } i = 1, \ldots, 9 \text{ and } j = 1, \ldots, 5.$$

Table 7.2. Gibbs sampling estimates and estimated absolute coefficients of variation for Example 7.12 where R denotes the exact value and $\hat{R}_{\text{Gibbs}}$ denotes the Gibbs sampling estimate.

i	$R(\theta_i)$	$\hat{R}_{\text{Gibbs}}(\theta_i)$	$R(\theta_i^2)$	$\hat{R}_{\text{Gibbs}}(\theta_i^2)$
1	2.043	2.042 (0.001)	4.263	4.260 (0.002)
2	0.095	0.094 (0.015)	0.081	0.081 (0.008)
4	0.018	0.015 (0.088)	0.069	0.069 (0.008)
10	−0.073	−0.073 (0.019)	0.033	0.033 (0.010)

This gives that

$$z_{ij} = \sqrt{\frac{\lambda-2}{2}} \frac{u_{ij}}{\sqrt{v_{ij}}} \sim \sqrt{\frac{\lambda-2}{\lambda}} \text{Student}(\lambda).$$

The density $f(\beta_1, \ldots, \beta_9, \sigma, v_{11}, \ldots, v_{95})$ is then proportional to

$$\sigma^{-nk-1} \prod_{i=1}^{9} \prod_{j=1}^{5} \exp\left\{-\frac{1}{2}\left(\frac{2}{\lambda-2}\right) v_{ij} \left(\frac{y_{ij}-\beta_i}{\sigma}\right)^2\right\} v_{ij}^{(\lambda+1)/2-1} e^{-v_{ij}}.$$

Then denoting, for example, the conditional distribution of β_1 given the remaining variables by $\beta_1|\cdot$, we have that

$$\beta_i|\cdot \sim N\left(\frac{\sum_{j=1}^5 v_{ij} y_{ij}}{\sum_{j=1}^5 v_{ij}}, \frac{\lambda-2}{2}\left(\sum_{j=1}^{5} v_{ij}\right)^{-1} \sigma^2\right)$$

$$\frac{1}{\sigma^2}|\cdot \sim \left(\frac{1}{\lambda-2}\sum_{i=1}^{9}\sum_{j=1}^{5} v_{ij}(y_{ij}-\beta_i)^2\right)^{-1} \text{gamma}(45/2, 1)$$

$$v_{ij}|\cdot \sim \left[\frac{1}{\lambda-2}\left(\frac{y_{ij}-\beta_i}{\sigma}\right)^2 + 1\right]^{-1} \text{gamma}\left(\frac{\lambda+1}{2}, 1\right).$$

We ran the Gibbs sampler for approximately 50 seconds so that it used an equivalent amount of time as the other Monte Carlo algorithms. This required $n = 50\,000$ steps in the chain. The standard errors were estimated using batch size $l = 100$. We also tried $l = 1000$ and there was very little change in the estimates. Recall that $\theta_i = \beta_i$ for $i = 1, \ldots, 9$ and $\theta_{10} = \ln \sigma$. Table 7.2 gives the exact values and estimates together with their estimated absolute coefficients of variation for some selected integrals.

We also ran the Metropolis independence chain, using the importance sampler of Example 6.4. To provide a fair comparison with the other methods we ran this for

$n = 10^5$ steps. The standard errors were estimated using batch size $l = 100$ and the results were similar for $l = 1000$. This algorithm was substantially less accurate than Gibbs sampling. For example, the estimates and estimated absolute coefficients of variation, $\hat{R}_{\text{Indep}}(\theta_4) = 0.015(0.206)$ and $\hat{R}_{\text{Indep}}(\theta_{10}) = -0.071(0.027)$ were obtained. There was evidence of substantial serial correlation as the estimates of the standard errors for $l = 20$ were very different. This was not the case with the Gibbs sampling algorithm. The acceptance rate for this algorithm, i.e. the proportion of generated values that resulted in a move for the chain, was 0.28. We also ran the random walk Metropolis chain based on a multivariate Student proposal distribution with mean 0 and variance matrix equal to the inverse Hessian matrix of $-\log f$ evaluated at the mode of f. Following the advice given in Gelman, Roberts and Gilks (1996) we rescaled the variance matrix by $2.4/\sqrt{10}$. Again we ran this for $N = 10^5$ steps and estimated the standard errors using batch size $l = 100$. The random walk chain gave less accuracy than the independence chain. The acceptance rate for this algorithm was 0.33. Varying the degrees of freedom for the Student produced only minor differences.

We have seen that Gibbs sampling using data augmentation gave answers of similar accuracy and similar efficiency as importance sampling and adaptive importance sampling. Given the ease with which a program for Gibbs sampling can be written and debugged this is definitely the MCMC algorithm of choice in this example.

Mykland, Tierney and Yu (1995) discuss the use of *regeneration* techniques in MCMC. Effectively these allow a user to avoid the problems associated with assessing convergence to stationarity and estimating error based on dependent observations. These methods seem difficult to implement in general problems, however.

7.6 Web resources and software

There are a number of web sites that are useful in keeping up with the rapid developments in this area. In particular the **MCMC Preprint Service** at

http://www.stats.bris.ac.uk/MCMC/

contains an extensive set of downloadable preprints on the topics. It also provides links to other sites containing useful information and software for the implementation of MCMC methods. The BUGS (Bayesian Inference Using Gibbs Sampling) software is available at

http://www.mrc-bsu.cam.ac.uk/bugs/

and this offers the opportunity of using some of these methods while avoiding extensive coding.

7.7 Exercises

(1) Prove (7.3).

(2) Establish that the random walk in Example 7.1 is a Markov chain.

(3) Establish that if the initial distribution P_0 of a Markov chain is stationary then $P(X_n \in B) = P_0(B)$.

(4) Establish the exact form for $\text{Var}_P[\bar{m}_n]$ analogous to (7.7).

(5) Establish $\alpha(x, y)q(x, y)p(x) = \alpha(y, x)q(y, x)p(y)$ in (7.13).

(6) Consider the rejection algorithm where we generate pairs (Y, U) with $Y \sim g$ statistically independent of $U \sim U(0, 1)$ until $f(Y) \leq cUg(Y)$. Show that the final Y is distributed according to the density proportional to $\min\{f, cg\}$.

(7) Establish the form for the acceptance probability in Example 7.4.

(8) Determine the form of the transition kernels for the symmetric, independence, rejection and random walk Metropolis chains.

(9) Establish directly that $\prod_{j=1}^{k} p_j(y_j | y_1, \ldots, y_{j-1}, x_{j+1}, \ldots, x_k)$ is a density function as a function of $(y_1, \ldots, y_k)$ in the description of the Gibbs sampling algorithm.

(10) Verify that

$$\int_{R^k} K(x, B)p(x)\,dx = \int_B p(x)\,dx$$

in the Gibbs sampling chain.

(11) Suppose that we want to evaluate integrals of the form

$$\int_{R^k} m(x)p(x)\,dx$$

where p is a $N_k(\mu, \Sigma)$ density truncated to the rectangle $\prod_{i=1}^{k}(a_i, b_i)$. Develop a Gibbs sampling algorithm for this problem.

(12) Consider the sample $X_1, \ldots, X_n$ arising from the mixture distribution $X \mid \lambda_1, \ldots, \lambda_m, p_1, \ldots, p_m \sim \sum_{i=1}^{m} p_i \text{Poisson}(\lambda_i)$ where the p_i are assumed to be fixed known probabilities satisfying $\sum_{i=1}^{m} p_i = 1$ and $\lambda_i \sim \text{gamma}(\upsilon_i, \gamma)$ where γ and υ_i for $i = 1, \ldots, m$ are known constants. Obtain the form of the conditional distribution of $(\lambda_1, \ldots, \lambda_m)$ given $X_1, \ldots, X_n$. Derive a Gibbs sampling algorithm for this conditional distribution by introducing appropriate latent variables.

(13) Verify the conditional distributions obtained for Gibbs sampling in Example 7.8.

(14) Suppose the stochastic process $\{X_i : i = 1, 2, \ldots\}$ is second order stationary. Suppose that nN states are observed. Let $\bar{X}$ be the average of all nN

observations and let $\bar{X}_n$ be the average of every N-th observation. Prove that

$$\mathrm{Var}[\bar{X}] \leq \mathrm{Var}[\bar{X}_n].$$

(15) Suppose that the transition kernels $K_1, K_2, \ldots, K_m$ are repeated in fixed order and π is stationary for each component chain. Prove that π is stationary for the cyclic chain.

(16) Implement the chain discussed in Example 7.9 with $n_1 = n_2 = 10$, $y_1 = 4$, $y_2 = 8$, $\nu_1 = (1, 1, 1, 1)'$, $\nu_2 = (1, 1, -1, -1)'$ and $a_0 = 3$, $b_0 = 1$.

(17) Establish the form of the transition kernel for the hit-and-run algorithm.

(18) Prove that p is stationary for the hit-and-run algorithm.

(19) Prove for the slice sampling algorithm described in Section 7.3.4 that $x|\eta$ has density proportional to

$$f_0(x) I_{\{z: f_i(z) \geq \eta_i, i=1,\ldots,l\}}(x) = f_0(x) \prod_{i=1}^{l} I_{f_i^{-1}[\eta_i,\infty)}(x).$$

(20) Prove that for the monotone Monte Carlo context described in Section 7.3.6 we need only consider the behavior of the least and maximum elements in the state space in order to determine coalescence.

(21) Establish that the sampler described in Section 7.4.4 for simulated tempering has p as its stationary distribution.

(22) Establish that the chain described for tempered transitions in Section 7.4.4 satisfies detailed balance.

References

Achcar, J.A. and Smith, A.F.M. (1990). Aspects of reparameterization in approximate Bayesian inference. In *Bayesian and likelihood methods in statistics and econometrics* (ed. S. Geisser, J.S. Hodges, S.J. Press and A. Zellner) pp. 439–452. Elsevier North-Holland, Amsterdam.

Adler, S.L. (1981). Over-relaxation method for the Monte Carlo evaluation of the partition function for multiquadratic actions. *Phys. Rev. D*, **23**, 2901–2904.

Ahlfors, L.V. (1966). *Complex Analysis*. McGraw-Hill, New York.

Amit, Y. and Grenander, U. (1991). Comparing sweep strategies for stochastic relaxation. *J. Mult. Anal.*, **37**, 197–222.

Anderson, T.W., Olkin, I. and Underhill, L.G. (1987). Generation of a random orthogonal matrix. *SIAM J. Sci. Stat. Comput.*, **8**, 625–629.

Barndorff-Nielsen, O.E. and Cox, D.R. (1989). *Asymptotic techniques for use in statistics*. Chapman and Hall, London.

Barndorff-Nielsen, O.E. and Cox, D.R. (1994). *Inference and asymptotics*. Chapman and Hall, London.

Belisle, C. (1998). Slow convergence of the Gibbs sampler. *Canad. J. Statist.*, **26**, 629–641.

Belisle, C. (1999). Slow hit and run sampling. To appear in *Stat. and Prob. Letters*.

Belisle, C., Romeijin, H.E. and Smith, R.L. (1993). Hit-and-run algorithms for generating multivariate distributions. *Math. Oper. Res.*, **18**, 255–266.

Bernsten, J. (1989). Practical error estimation in adaptive multidimensional quadrature routines. *J. of Comput. and Appl. Math.*, **25**, 327–340.

Bernsten, J., Espelid, T.O. and Genz, A. (1991). An adaptive algorithm for the approximate calculation of multiple integrals. *ACM Trans. on Math. Software*, **17**, 437–451.

Besag, J. and Green, P.J. (1993). Spatial statistics and Bayesian computation. *J. Roy. Statist. Soc. B*, **55**, 25–37.

Bissell, A.F. (1986). Ordered random selection without replacement. *Appl. Stat.*, **35**, 73–75.

Bleistein, N. and Handelsman, R.A. (1986). *Asymptotic expansions of integrals*. Dover, New York.

Bogues, K., Morrow, C.R. and Patterson, T.N.L. (1981). An implementation of the method of Ermakov and Zolotukhin for multidimensional integration and interpolation. *Numer. Math.*, **37**, 49–60.

Bratley P., Fox, B.L. and Schrage, L.E. (1983). *A guide to simulation*. Springer-Verlag, New York.

Brockwell, P. and Davis, R. (1991). *Time series: theory and methods*. (2nd edn). Springer-Verlag, New York.

Bronstein, M. (1997). *Symbolic integration I: transcendental functions*. Springer-Verlag, Berlin.

Brooks, S. (1998). MCMC convergence diagnosis via multivariate bounds on log-concave densities. *Ann. Statist.*, **26**, 398–433.

Brooks, S.P. and Giudici, P. (1999). Convergence assessment for reversible jump MCMC simulations. *Bayesian statistics 6* (ed. J. Bernardo, J.O. Berger, A.P. Dawid and A.F.M. Smith) pp. 723–731. Oxford University Press.

Brooks, S.P. and Roberts, G.O. (1999). On quantile estimation and MCMC convergence, *Biometrika*, **86**, 710–717.

Bucklew, J.A. (1990). *Large deviation techniques in decision, simulation, and estimation.* Wiley, New York.

Casella, G. and Berger, R.L. (1990). *Statistical inference.* Duxbury Press, Belmont, California.

Chaitin, G.J. (1966). On the length of programs for computing finite binary sequences. *J. Assoc. Comput. Mach.*, **13**, 547–569.

Chan, K.S. (1993). Asymptotic behavior of the Gibbs sampler. *J. Amer. Statist. Assoc.*, **88**, 320–326.

Chan, K.S. and Geyer, C. (1994). Discussion of "Markov chains for exploring posterior distributions". *Ann. Statist.*, **22**, 1747–1758.

Char, B.W., Geddes, K.O., Gonnet, G.H., Leong, B.L., Monagan, M.B. and Heidelberg, S. (1992). *First leaves: a tutorial introduction to Maple V.* Springer-Verlag, Berlin.

Chen, C.-F. (1985). On the asymptotic normality of limiting density functions with Bayesian implications. *J. Roy. Statist. Soc. B*, **47**, 540–546.

Chen, M.-H. (1994). Importance-weighted marginal Bayesian posterior density estimation. *J. Amer. Stat. Assoc.*, **89**, 818–824.

Chen, M.-H. and Schmeiser, B.W. (1993). Performance of the Gibbs, hit-and-run and Metropolis samplers. *J. Comput. Graph. Statist.*, **2**, 251–272.

Chen, M.-H. and Schmeiser, B.W. (1996). General hit-and-run Monte Carlo sampling for evaluating multidimensional integrals. *Oper. Res. Lett.*, **19**, 161–169.

Chen, M.-H. and Shao, Q.-M. (1997). On Monte Carlo methods for estimating ratios of normalizing constants. *Ann. Statist.*, **25**, 1563–1594.

Chib, S. (1995). Marginal likelihood from the Gibbs output. *J. Amer. Statist. Assoc.*, **90**, 1313–1321.

Cools, R. and Rabinowitz, P. (1993). Monomial cubature rules since "Stroud": a compilation. *J. Comput. Appl. Math.*, **48**, 309–326.

Corcoran, J.N. and Tweedie, R.L. (1998a). Perfect sampling from independent Metropolis chains. Preprint. Colorado State University.

Corcoran, J.N. and Tweedie, R.L. (1998b). Perfect sampling of Harris recurrent Markov chains. Preprint. Colorado State University.

Cowles, M.K. and Carlin, B.P. (1996). Markov chain Monte Carlo convergence diagnostics: a comparative review. *J. Amer. Statist. Assoc.*, **91**, 883–904.

Cowles, M.K. Roberts, G.O. and Rosenthal, J.S. (1999). Possible biases induced by MCMC convergence diagnostics. *J. Stat. Comp. and Sim.*, **64**, 87–104.

Cranley, R. and Patterson, T.N.L. (1976). Randomization of number theoretic methods for multiple integration. *SIAM J. Numer. Anal.*, **13**, 904–914.

Dahlquist, G. and Björck, A. (1974). *Numerical Methods*, (trans. N. Anderson). Prentice-Hall, Englewood Cliffs, New Jersey.

Damien, P., Wakefield, J. and Walker, S. (1999). Gibbs sampling for Bayesian nonconjugate and hierarchical models using auxiliary variables. *JRSSB*, **61**, 2, 331–344.

Davis, P.J. (1975). *Interpolation and approximation*. Ginn (Blaisdell), Boston, Massachusetts.

Davis, P.J. and Rabinowitz, P. (1984). *Methods of numerical integration*, (2nd edn). Academic Press, Orlando, Florida.

de Bruijn, N.G. (1981). *Asymptotic methods in analysis*. Dover, New York.

Dellaportas, P. and Forster, J.J. (1999). Markov chain Monte Carlo model determination for hierarchical and graphical log-linear models. *Biometrika*, **86**, 615–633.

Dellaportas, P. and Wright, D. (1991). Positive embedded integration in Bayesian analysis. *Stat. and Compt.*, **1**, 1–12.

Devroye, L. (1986). *Non-uniform random variate generation*. Springer-Verlag, New York.

Devroye, L. (1987). *A first course in density estimation*. Birkhäuser, Boston.

Diaconis, P. (1988). Bayesian numerical analysis. In *Statistical decision theory and related topics IV*, 1 (ed. S.S. Gupta and J. Berger) pp. 163–175. Springer-Verlag, New York.

DiCiccio, T.J., Kass, R.E., Raftery, A. and Wasserman, L. (1997). Computing Bayes factors by combining simulation and asymptotic approximations. *J. Amer. Statist. Assoc.*, **92**, 903–915.

Durrett, R. (1991). *Probability: theory and examples*. Wadsworth and Brooks/Cole, Pacific Grove.

Eberlein, W.F. (1962). An integral over function space. *Canad. J. Math.*, **14**, 379–384.

Edwards, R.G. and Sokal, A.D. (1988). Generalization of the Fortuin–Kasteleyn–Swendsen–Wang representation in Monte Carlo. *Phys. Rev.*, **38**, 2009–2012.

Efron, B. and Stein, C. (1981). The jacknife estimate of variance. *Ann. Statist.*, **9**, 586–596.

Engels, H. (1980). *Numerical quadrature and cubature*. Academic Press, New York.

Ermakov, S.M. and Zolotukhin, V.G. (1960). Polynomial approximations and the Monte Carlo method. *Teor. Veroyatnost. i Primenen*, **5**, 473–476; translated as *Theor. Prob. and Appl.*, **5**, 428–431.

Escobar, M. (1994). Estimating normal means with a Dirichlet process prior. *J. Amer. Statist. Assoc.*, **89**, 268–277.

Evans, M. (1988). Monte Carlo computation of marginal posterior quantiles. Technical Report No. 11, Department of Statistics, University of Toronto.

Evans, M. (1991a). Adaptive importance sampling and chaining. In *Statistical multiple integration* (ed. N. Flournoy and R.K. Tsutakawa) pp. 137–143. *Contemporary Mathematics*, **115**, American Mathematical Society.

Evans, M. (1991b). Chaining via annealing. *Ann. Statist.*, **19**, 382–393.

Evans, M., Guttman, I., Haitovsky, Y. and Swartz, T.B. (1996). Bayesian analysis of binary data subject to misclassification. In *Bayesian statistics and econometrics: Essays in Honor of Arnold Zellner* (ed. D. Berry, J. Geweke, and K.M. Chaloner) pp. 67–77. Wiley, New York.

Evans, M., Guttman, I. and Olkin, I. (1993). Numerical aspects in estimating the parameters of a mixture of normal distributions. *J. Comp. Graph. Statist.*, **1**, 351–365.

Evans, M. and Swartz, T. (1988a). Monte Carlo computation of some multivariate normal probabilities. *J. Statist. Comput. Simul.*, **30**, 117–128.

Evans, M. and Swartz, T. (1988b). Sampling from Gauss rules. *SIAM J. Sci. Stat. Comput.*, **9**, 950–961.

Evans, M. and Swartz, T. (1995a). Methods for approximating integrals in statistics with special emphasis on Bayesian integration problems. *Stat. Sci.*, **10**, 254–272.

Evans, M. and Swartz, T. (1995b). Bayesian integration using multivariate Student importance sampling. In *Computing science and statistics*, **27** (ed. M.M. Meyer and J.L. Rosenberger) pp. 456–461. Interface Foundation of North America, Inc., Fairfax Station, VA.

Evans, M. and Swartz, T.B. (1997). An algorithm for the approximation of integrals with exact error bounds. Technical Report No. 19, Dept. of Statistics, University of Toronto, Toronto, Canada.

Evans, M. and Swartz, T.B. (1998a). Random variable generation using concavity properties of transformed densities. *J. Comp. Graph. Statist.*, **7**, 514–528.

Evans, M. and Swartz, T.B. (1998b). Higher order envelope random variate generators. Technical Report No. 2, Dept. of Statistics, University of Toronto, Toronto, Canada.

Fang, K.-T. and Wang, Y. (1994). *Number-theoretic methods in statistics*. Chapman and Hall, London.

Fang, K.-T., Wang, Y. and Bentler, P.M. (1994). Some applications of number-theoretic methods in statistics. *Stat. Sci.*, **9**, 416–428.

Feller, W. (1971). *An introduction to probability theory and its applications*. Wiley, New York.

Fishman, G.S. (1996). *Monte Carlo: concepts, algorithms and applications*. Springer, New York.

Fishman, G.S. (1999). An analysis of Swendsen–Wang and related sampling methods. *J. Roy. Statist. Soc. B*, **61**, 623–642.

Fishman, G.S. and Huang, B.D. (1983). Antithetic variates revisited. *Commun. ACM*, **26**, 964–971.

Flournoy, N. and Tsutakawa, R.K. (ed.) (1991). *Statistical multiple integration. Contemporary Mathematics*, **115**, American Mathematical Society.

Fox, B.L. (1986). Implementation and relative efficiency of quasirandom sequence generators. *ACM Trans. Math. Software*, **9**, 362–376.

Friedman, J.H. and Wright, M.H. (1981). A nested partitioning procedure for numerical multiple integration. *ACM Trans. Math. Software*, **7**, 76–92.

Gamerman, D. (1997). *Markov chain Monte Carlo: stochastic simulation for Bayesian inference*. Chapman and Hall, London.

Gautschi, W. (1997). *Numerical analysis, an introduction*. Birkhäuser, Boston.

Gelfand, A.E. and Dey, D.K. (1994). Bayesian model choice: asymptotics and exact calculations. *J. Roy. Statist. Soc. B*, **56**, 501–514.

Gelfand, A.E. and Smith, A.F.M. (1990). Sampling based approaches to calculating marginal densities. *J. Amer. Statist. Assoc.*, **85**, 398–409.

Gelman, A. and Meng, X.-L. (1998). Simulating normalizing constants: from importance sampling to bridge sampling to path sampling. *Stat. Sci.*, **13**, 163–185.

Gelman, A., Roberts, G.O. and Gilks, W.R. (1996). Efficient Metropolis jumping rules. In *Bayesian statistics 5* (ed. J.M. Bernardo, J.O. Berger, A.P. Dawid and A.F.M. Smith) pp. 599–607. Oxford University Press.

Gelman, A. and Rubin, D.B. (1992). Inference from iterative simulation. *Stat. Sci.*, **7**, 473–511 (with discussion).

Geman, S. and Geman, D. (1984). Stochastic relaxation, Gibbs distributions and the Bayesian restoration of images. *IEEE Trans. on Pattern Analysis and Machine Intelligence*, **6**, 721–741.

Gentle, J.E. (1998). *Random number generation and Monte Carlo methods*. Springer-Verlag, New York.

Genz, A. (1991). Subregion adaptive algorithms for multiple integrals. In *Statistical multiple integration* (ed. N. Flournoy and R.K. Tsutakawa) pp. 23–31. *Contemporary Mathematics*, **115**, American Mathematical Society.

Genz, A. (1992a). Numerical computation of multivariate normal probabilities. *J. Comput. Graph. Statist.*, **1**, 141–150.

Genz, A. (1992b). Statistics applications of subregion adaptive multiple numerical integration. In *Numerical integration: recent developments, software and applications* (ed. T.O. Espelid and A. Genz) pp. 267–280. Kluwer Academic Publishers, Dordrecht.

Genz, A. and Kass, R. (1991). An application of subregion adaptive numerical integration to a Bayesian inference problem. In *Computing science and statistics* 23 (ed. E. Keramidas) pp. 441–444. Interface Foundation of North America, Inc., Fairfax Station, VA.

Genz, A. and Kass, R. (1997). Subregion adaptive integration of functions having a dominant peak. *J. Comp. Graph. Statist.*, **6**, 92–111.

Genz, A. and Malik, A. (1980) Algorithm 019. Remarks on algorithm 006: An adaptive algorithm for numerical integration over an n-dimensional rectangular region. *J. Comp. Appl. Math.*, **6**, 295–302.

Genz, A. and Malik, A. (1983). An imbedded family of fully symmetric numerical integration rules. *SIAM J. of Numer. Anal.*, **20**, 580–587.

Genz, A. and Monahan, J. (1998a). Stochastic integration rules for infinite regions. *SIAM J. Sci. Comp.*, **19**, 426–439.

Genz, A. and Monahan, J. (1998b). Stochastic methods for multiple integrals over unbounded regions. *Math. and Comp. in Simul.*, **47**, 287–298.

Genz, A. and Monahan, J. (1999). A stochastic algorithm for high dimensional multiple integrals over unbounded regions with Gaussian weight. To appear in *J. Comp. Appl. Math.*

Genz, A., Monahan J., Schervish, M.J., Shaw, J.E.H., Wasserman, L. and Wolfinger, R. (1996). Comments on "Methods for approximating integrals in statistics with special emphasis on Bayesian integration problems" by Michael Evans and Tim Swartz. *Stat. Sci.*, **11**, 54–64.

Geweke, J. (1988). Antithetic acceleration of Monte Carlo integration in Bayesian inference. *J. of Econometrics*, **38**, 73–89.

Geweke, J. (1989). Bayesian inference in econometric models using Monte Carlo integration. *Econometrica*, **57**, 1317–1339.

Geweke, J. (1991). Evaluating the accuracy of sampling-based approaches to the calculation of posterior moments. In *Bayesian statistics 4* (ed. J.M. Bernardo, J.O. Berger, A.P. Dawid and A.F.M. Smith) pp. 169–193 (with discussion). Oxford University Press.

Geweke, J. (1996). Monte Carlo simulation and numerical integration. In *Handbook of computational economics* (ed. H. Amman, D. Kendrick and J. Rust) pp. 731–800. North-Holland, Amsterdam.

Geweke, J. (1997). Posterior simulators in econometrics. In *Advances in economics and econometrics: theory and applications*, III (ed. D. Kreps and K.F. Wallis) pp. 128–165. Cambridge University Press.

Geweke, J. (1999). Using simulation methods for Bayesian econometric models: inference, development and communication (with discussion and rejoinder). *Econometric Reviews*, **18**, 1–126.

Geyer, C.J. (1991). Monte Carlo maximum likelihood for dependent data. In *Computing science and statistics: 23rd Symposium on the Interface* (ed. E.M. Keramidas) pp. 156–163. Interface Foundation of North America, Fairfax Station, VA.

Geyer, C.J. (1992). Practical Markov chain Monte Carlo. *Stat. Sci.*, **7**, 473–511 (with discussion).

Geyer, C.J. and Thompson, E.A. (1992). Constrained Monte Carlo maximum likelihood for dependent data. *J. Roy. Statist. Soc. B*, **54**, 657–699 (with discussion).

Geyer, C.J. and Thompson, E.A. (1995). Annealing Markov chain Monte Carlo with applications to ancestral inference. *J. Amer. Statist. Assoc.*, **90**, 909–920.

Ghosh, J.K. (1991). *Higher order asymptotics*. Vol. 4, IMS lecture Notes: NSF-CBMS Regional Conference Series in Probability and Statistics. IMS, Hayward CA.

Gilks, W.R., Richardson, S. and Spiegelhalter, D.J. (ed.) (1996). *Markov chain Monte Carlo in practice*. Chapman and Hall, London.

Gilks, W.R. and Roberts, G.O. (1996). Strategies for improving MCMC. In *Markov chain Monte Carlo in practice* (ed. W.R. Gilks, S. Richardson, and D.J. Spiegelhalter) pp. 89–114. Chapman and Hall, London.

Gilks, W.R., Roberts, G.O. and George, E.I. (1994). Adaptive direction sampling. *The Statistician*, **43**, 179–189.

Gilks, W.R. and Wild, P. (1992). Adaptive rejection sampling for Gibbs sampling. *Appl. Statist.*, **41**, 337–348.

Giudici, P. and Green, P.J. (1998). Decomposable graphical gaussian model determination. Technical Report, University of Pavia.

Giudici, P. and Roberts, G. (1999). On the automatic choice of reversible jumps. In *Bayesian statistics 6* (ed. J.M. Bernardo, J.O. Berger, A.P. Dawid and A.F.M. Smith). Oxford University Press.

Givens, G.H. and Raftery, A.E. (1996). Local adaptive importance sampling for multivariate densities with strong nonlinear relationships. *J. Amer. Statist. Assoc.*, **91**, 132–141.

Glynn, P.W. and Iglehart, D.L. (1989). Importance sampling for stochastic simulations. *Mgmnt. Sc.*, **35**, 1367–1392.

Granovsky, B.L. (1981). Optimal formulae of the conditional Monte Carlo. *SIAM J. Alg. Disc. Meth.*, **2**, 289–294.

Green, P.J. (1995). Reversible jump Markov chain Monte Carlo computation and Bayesian model determination. *Biometrika*, **82**, 711–732.

Green, P.J. and Han, X. (1992). Metropolis methods, gaussian proposals and antithetic variables. In *Stochastic models, statistical methods and algorithms in image analysis* (ed. P. Barone, A. Frigessi and M. Piccioni) pp. 142–164. Lecture Notes in Statistics, Springer-Verlag, Berlin.

Green, P.J. and Murdoch, D.J. (1999). Exact sampling for Bayesian inference: towards general purpose algorithms. In *Bayesian statistics 6* (ed. J.M. Bernardo, J.O. Berger, A.P. Dawid and A.F.M. Smith) pp. 301–321 (with discussion). Oxford University Press.

Greenwood, P.E., McKeague, I.W. and Wefelmeyer, W. (1996). Outperforming the Gibbs sampler empirical estimator for nearest neighbour random fields. *Ann. Statist.*, **24**, 1433–1456.

Haber, S. (1966). A modified Monte Carlo quadrature. *Math. Comp.*, **20**, 361–368.

Haber, S. (1968). A combination of Monte Carlo and classical methods for evaluating multiple integrals. *Bull. Amer. Math. Soc.*, **74**, 683–686.

Haber, S. (1969). Stochastic quadrature formulas. *Math. Comp.*, **23**, 751–764.

Haber, S. (1970). Numerical evaluation of multiple integrals. *SIAM Review*, **12**, 481–526.

Hall, P. and Heyde, C.C. (1980). *Martingale limit theory and its application*. Academic Press, San Diego.

Hammersley, J.M. (1960). Monte Carlo methods for solving multivariable problems. *Ann. New York Acad. Sci.*, **86**, 844–874.

Hammersley, J.M. and Handscomb, D.C. (1964). *Monte Carlo methods*. Methuen, London.

Hammersley, J.M. and Morton, K.W. (1956). A new Monte Carlo technique: antithetic variates. *Proc. Cambridge Philos. Soc.*, **52**, 449–475.

Hastings, W.K. (1970). Monte Carlo sampling methods using Markov chains and their applications. *Biometrika*, **57**, 97–109.

Heard, A.H. and Swartz, T.B. (1999). Extended voting measures. *Canadian Journal of Statistics*, **27**, 177–186.

Heller, B. (1991) *MacSyma for statisticians*. Wiley, New York.

Hesterberg, T. (1991). Importance sampling for Bayesian estimation. In *Computing and graphics in statistics*, Vol. 36 of the Institute for Mathematics and Its Applications Volumes in Mathematics and Its Applications (ed. A. Buja and P.A. Tukey) pp. 63–75. Springer-Verlag, New York.

Hesterberg, T. (1995). Weighted average importance sampling and defensive mixture distributions. *Technometrics*, **37**, 185–194.

Higdon, D.M. (1996). Auxiliary variable methods for Markov chain Monte Carlo with applications. *J. Amer. Statist. Assoc.*, **93**, 585–595.

Hills, S.E. and Smith, A.F.M. (1992). Parameterization issues in Bayesian inference. In *Bayesian statistics 4* (ed. J.M Bernardo, J.O. Berger, A.P. Dawid and A.F.M. Smith) pp. 227–246 (with discussion). Oxford University Press.

Hlawka, E. (1961). Funktionen von beschränkter Variation in der Theorie der Gleichverteilung. *Ann. Math. Pura Appl.*, **54**, 325–333.

Hochstadt, H. (1986). *The functions of mathematical physics*. Dover, New York.

Hörmann, W. (1995). A rejection technique for sampling from T-concave distributions. *ACM Transactions on Mathematical Software*, **21**, 182–193.

Hull, T.E. and Dobell, A.R. (1962). Random number generators. *SIAM Review*, **4**, 230–254.

IMSL MATH/LIBRARY, STAT/LIBRARY, SFUN/LIBRARY: User's Manual—Version 3.0 (1994). Visual Numerics Inc., Houston, Texas.

Jackson, D. (1941). Fourier series and orthogonal polynomials. Carus Monograph Series, No. 6. Mathematical Association of America.

Jensen, J.L. (1995). *Saddle point approximations*. Clarendon Press, Oxford.

Johnson, V.E. (1992). A technique for estimating marginal posterior densities in hierarchical models using mixtures of conditional densities. *J. Amer. Statist. Assoc.*, **87**, 852–860.

Johnson, V.E. (1997). Studying convergence of Markov chain Monte Carlo algorithms using sample paths. *JASA*, **91**, 154–166.

Johnson, V.E. (1998). A coupling regeneration scheme for diagnosing convergence in Markov chain Monte Carlo algorithms. *J. Amer. Statist. Assoc.*, **93**, 238–248.

Kass, R.E., Tierney, L. and Kadane, J.B. (1988). Asymptotics in Bayesian computation. In *Bayesian statistics 3* (ed. J. Bernardo, M. DeGroot, D.V. Lindley and A.F.M. Smith) pp. 263–267 (with discussion). Oxford University Press.

Kass, R.E., Tierney, L. and Kadane, J.B. (1990). The validity of posterior expansions based on Laplace's method. In *Bayesian and likelihood methods in statistics and econometrics* (ed. S. Geisser, J.S. Hodges, S.J. Press and A. Zellner) pp. 473–488. Elsevier Science North Holland, Amsterdam.

Keng, H.L. and Yuan, W. (1981). *Applications of number theory to numerical analysis.* Springer-Verlag, Berlin.

Kennedy, W.J. and Gentle, J.E. (1980). *Statistical computing*. Marcel Dekker, New York.

Kimeldorf, G.S. and Wahba, G. (1970). A correspondence between Bayesian estimation on stochastic processes and smoothing by splines. *Ann. Math. Stat.*, **41**, 495–502.

Kinderman, A.J. and Monahan, J. (1977). Computer generation of random variables using the ratio of uniform deviates. *ACM Transactions on Mathematical Software*, **3**, 257–260.

Kinderman, A.J. and Ramage, J.G. (1976). Computer generation of normal random variables. *J. Amer. Statist. Assoc.*, **71**, 893–896.

Kipnis, C. and Varadhan, S.R.S. (1986). Central limit theorem for additive functionals of reversible Markov chains ans applications to simple exclusions. *Commun. Math. Phys.*, **104**, 1–19.

Kirkpatrick, S., Gelatt, C.D. Jr. and Vecchi, M.P. (1983). Optimization by simulated annealing. *Science*, **220**, 671–680.

Kloek, T. and van Dijk, H.K. (1978). Bayesian estimates of equation system parameters: an application of integration by Monte Carlo. *Econometrica*, **46**, 1–19.

Knuth, D.E. (1981). *The art of computer programming*, Vol. 2. *Seminumerical algorithms* Addison-Wesley, Reading, Mass.

Kolmogorov, A.N. (1963). On tables of random numbers *Sankhya, The Indian J. of Stat., Ser. A*, **25**, 369–376.

Krommer, A.R. and Ueberhuber, C.W. (1998). *Computational integration.* SIAM, Philadelphia.

Kronmal, R.A. and Peterson, A.V. (1981). A variant of the acceptance–rejection method for computer generation of random variables. *J. Amer. Statist. Assoc.*, **76**, 446–451.

Kuipers, L. and Niederreiter, H. (1974). *Uniform distribution of sequences.* Wiley Interscience, New York.

Lamperti, J. (1977). *Stochastic processes. A survey of the mathematical theory.* Springer-Verlag, New York.

L'Ecuyer, P. (1988). Efficient and portable combined random number generators. *Comm. of the ACM*, **31**, 742–749.

Lepage, G.P. (1978). A new algorithm for adaptive multidimensional integration. *J. Comput. Phys.*, **27**, 192–203.

Lewis, P.A.W. and Orav, E.J. (1989). *Simulation methodology for statisticians, operations analysts and engineers.* Vol. I. The Wadsworth and Brooks/Cole Statistics/Probability Series. Wadsworth and Brooks/Cole, Pacific Grove.

Lewis, T.G. and Payne, W.H. (1973). Generalized feedabck shift register pseudorandom number algorithm. *J. Assoc. Comput. Mach.*, **20**, 456–468.

Li, M. and Vitányi, P. (1993). *An introduction to Kolmogorov complexity and its applications*. Texts and Monographs in Computer Science. Springer-Verlag, New York.

Lindley, D.V. (1961). The use of prior probability distributions in statistical inference and decisions. Proc. 4th Berkeley Symp. **1**, 453–468.

Lindley, D.V. (1980). Approximate Bayesian methods. In *Bayesian statistics* (ed. J.M. Bernardo, M.H. De Groot, D.V. Lindley and A.F.M. Smith) pp. 223–245 (with discussion). University of Valencia Press, Valencia, Spain.

Liu, J.S. (1996). Metropolized independent sampling with comparisons to rejection sampling and importance sampling. *Stat. and Comput.*, **6**, 113–119.

Liu, J.S., Wong, W.H. and Kong, A. (1995). Covariance structure and convergence rate of the Gibbs sampler with various scans. *J. Roy. Statist. Soc. B*, **57**, 157–169.

Lyness, J.N. (1983). When not to use an automatic quadrature routine. *SIAM Rev.*, **25**, 63–87.

Ma, J., Rokhlin, V. and Wandzura, S. (1996). Generalized Gaussian quadrature rules for systems of arbitrary functions. *SIAM J. Numer. Anal.*, **33**, 971–996.

MacEachern, S.N. and Berliner, L.M. (1994). Subsampling the Gibbs sampler. *Amer. Statist.*, **48**, 188–190.

MacEachern, S.N. and Peruggia, M. (1999). Subsampling the Gibbs Sampler: variance reduction. To appear in *Stat. and Prob. Lett.*

MacLaren, M.D. and Marsaglia, G. (1965). Uniform random number generators. *Journal of the ACM*, **12**, 83–89.

Madras, N. and Slade, G. (1993). *The self-avoiding walk*. Birkhäuser, Boston.

Madras, N. and Piccioni, M. (1999). Importance sampling for families of distributions. To appear in *Ann. Appl. Prob.*

Marinari, E. and Parisi, G. (1992). Simulated tempering: a new Monte Carlo scheme. *Europhys. Lett.* **19**, 451–458.

Marsaglia, G. (1985). A current view of random number generator. In *Computer Science and Statistics: 16th Symposium on the Interface* (ed. L. Billard) pp. 3–10. North Holland, Amsterdam.

Marsaglia, G. (1995). The Marsaglia Random Number CD-ROM including the DIEHARD Battery of Tests of Randomness. Department of Statistics and Super Supercomputer Research Institute, Florida State University.

Marsaglia, G., MacLaren, M.D. and Bray, T.A. (1964). A fast procedure for generating normal random variables. *Communications of the ACM*, **7**, 4–10.

Matthews, P. (1991). A slowly mixing markov chain with implications for Gibbs sampling. *Stat. and Prob. Lett.*, **17**, 231–236.

McCullagh, P. (1987). *Tensor methods in statistics*. Chapman and Hall, London.

McKay, M.D., Beckman, R.J. and Conover, W.J. (1979). A comparison of three methods for selecting values of input variables in the analysis of output from a computer code. *Technometrics*, **21**, 239–245.

Meng, X.L. and Wong, W.H. (1996). Simulating ratios of normalizing constants via a simple identity: a theoretical exploration. *Statist. Sinica*, **6**, 831–860.

Metropolis, N., Rosenbluth, A.W., Teller, A.H. and Teller, E. (1953). Equations of state calculations by fast computing machines. *J. Chemical Physics*, **21**, 1087–1091.

Metropolis, N. and Ulam, S. (1949). The Monte Carlo method. *J. Amer. Statist. Assoc.*, **44**, 335–341.

Meyn, S.P. and Tweedie, R.L. (1993). *Markov chains and stochastic stability*. Springer-Verlag, London.

Mira, A. (1998). Ordering, slicing and splitting Monte Carlo Markov chains. Thesis, Department of Statistics, University of Minnesota.

Mira, A. and Tierney, L. (1999). On the use of auxiliary variables in Markov chain Monte Carlo sampling. Preprint. School of Statistics, University of Minnesota, Minneapolis.

Moller, J. (1999). Perfect simulation of conditionally specified models. *J. Roy. Statist. Soc. B*, **61**, 251–264.

Monahan, J. and Genz, A. (1997). Spherical-radial integration rules for Bayesian computation. *J. Amer. Statist. Assoc.*, **92**, 664–674.

Moore, R.E. (1979). *Methods and applications of interval analysis*, SIAM, Philadelphia.

Müller, P. (1991). A generic approach to posterior integration and Gibbs sampling. Technical Report #91-09, Department of Statistics, Purdue University.

Murdoch, D.J. and Green, P.J. (1998). Exact sampling for a continuous state space. Scand. *J. Statist.*, **50**, 483–502.

Murray, J.D. (1984). *Asymptotic analysis*. Springer-Verlag, New York.

Mykland, P., Tierney, L. and Yu, B. (1995). Regeneration in Markov chain samplers. *J. Amer. Statist. Assoc.*, **90**, 233–241.

NAG Fortran Library Manual—Mark 17 (1996). NAG Ltd., Oxford.

Naylor, J.C. and Smith, A.F.M. (1982). Applications of a method for the efficient computation of posterior distributions. *Appl. Stat.*, **31**, 214–225.

Naylor, J.C. and Smith, A.F.M. (1988). Econometric illustrations of novel integration strategies for Bayesian inference. *J. of Econometrics*, **38**, 103–125.

Neal, R. (1993). Probabilistic inference using Markov chain Monte Carlo methods. Technical Report CRG-TR-93-1, Department of Computer Science, University of Toronto.

Neal, R. (1996). Sampling from multimodal distributions using tempered transitions. *Stat. and Comput.*, **6**, 353–366.

Neal, R. (1997). Markov chain Monte Carlo methods based on "slicing" the density function. Technical Report No. 9722, Department of Statistics, University of Toronto.

Neal, R. (1998a). Suppressing random walks in Markov chain Monte Carlo using ordered overrelaxation. In *Learning in graphical models* (ed. M.I. Jordan) pp. 205–225. Kluwer Academic, Dordrecht.

Neal, R. (1998b). Annealed importance sampling. Technical Report No. 9805, Department of Statistics, University of Toronto.

Newton, M.A. and Raftery, A.E. (1994). Approximate Bayesian inference with the weighted likelihood bootstrap. *J. Roy. Statist. Soc. B*, **56**, 141–168 (with discussion).

Niederreiter, H. (1992). *Random number generation and quasi-Monte Carlo methods*. CBMS-NSF Regional Conference Series in Applied Mathematics. SIAM.

Nivens, I. and Zuckerman, H.S. (1972). *An introduction to the theory of numbers*. Wiley, New York.

Nummelin, E. (1984). *General irreducible Markov chains and non-negative operators*. Cambridge University Press.

Ogata, Y. (1989) A Monte Carlo method for high-dimensional integration. *Numer. Math.*, **55**, 137–157.

Ogata, Y. (1990). A Monte Carlo method for an objective Bayesian procedure. *Ann. Inst. Statist. Math.*, **42**, 403–433.

Oh, M.-S. (1991). Monte Carlo integration via importance sampling: dimensionality effect and an adaptive algorithm. In *Statistical multiple integration* (ed. N. Flournoy and R.K. Tsutakawa) pp. 165–187. *Contemporary Mathematics*, **115**, American Mathematical Society.

Oh, M.-S. and Berger, J. (1992). Adaptive importance sampling in Monte Carlo integration. *J. Stat. Comp. and Simul.*, **41**, 143–168.

Oh, M.-S. and Berger, J.O. (1993). Integration of multimodal functions by Monte Carlo importance sampling. *J. Amer. Statist. Assoc.*, **88**, 450–456.

O'Hagan, A. (1991). Bayes–Hermite integration. *J. Statist. Plan. Inf.*, **29**, 245–260.

O'Hagan, A. (1992). Some Bayesian numerical analysis. In *Bayesian statistics 4* (ed. J.M. Bernardo, J.O. Berger, A.P. Dawid and A.F.M Smith) pp. 345–363 (with discussion). Oxford University Press.

O'Hagan, A. (1996). Iterative rescaling for Bayesian quadrature. In *Bayesian statistics 5* (ed. J.M. Bernardo, J.O. Berger, A.P. Dawid and A.F.M. Smith) pp. 639–645. Oxford University Press.

Owen, A.B. (1992). A central limit theorem for latin hypercube sampling. *J.R. Statist. Soc. B*, **54**, 541–551.

Owen, A.B. (1997a). Monte Carlo variance of scrambled equidustribution quadrature. *SIAM J. Numer. Anal.*, **34**, 1884–1910.

Owen, A.B. (1997b). Scrambled net variance for integrals of smooth functions. *Ann. Statist.*, **25**, 1541–1562.

Patterson, T.N.L. (1987). On the construction of a practical Ermakov-Zolotukhin multiple integrator. In *Numerical integration* (ed. P. Keast and G. Fairweather) pp. 269–290. D. Reidel Publishing Co., Dordrecht.

Peskun, P.H. (1973). Optimum Monte Carlo sampling using Markov chains. *Biometrika*, **60**, 607–612.

Piessens, R., de Doncker-Kapenga, E., Überhuber, C.W. and Kahaner, D.K. (1983). *Quadpack a subroutine package for automatic integration*. Springer-Verlag, Berlin.

Pinkham, R.S. (1987). An efficient algorithm for drawing a simple random sample. *Appl. Stat.*, **36**, 370–372.

Polson, N. (1996). Convergence of Markov chain Monte Carlo algorithms. In *Bayesian statistics 5* (ed. J.M. Bernardo, J. Berger, A.P. Dawid and A.F.M. Smith) pp. 297–321 (with discussion). Oxford University Press.

Popper, K.R. (1959). *The logic of scientific discovery*. Harper and Row, New York.

Powell, M.J.D. and Swann, J. (1966). Weighted uniform sampling—a Monte Carlo technique for reducing variance. *J. Inst. Math. and Appl.*, **2**, 228–236.

Propp, J.G. and Wilson, D.B. (1996). Exact sampling with coupled Markov chains and applications to statistical mechanics. *Random Structures and Algorithms*, **9**, 223–252.

Reid, N. (1988). Saddlepoint methods and statistical inference. *Stat. Sci.*, **3**, 213–238 (with discussion).

Richardson, S. and Green, P.J. (1997). On Bayesian analysis of mixtures with an unknown number of components. *J. Roy. Statist. Soc. B*, **59**, 731–792 (with discussion).

Ripley, B.D. (1987). *Stochastic simulation*. Wiley, New York.

Robert, C.P. (1998). *Discretization and MCMC convergence assessment*. Lecture Notes in Statistics. Vol. 135, Springer-Verlag, New York.

Robert, C.P. and Casella, G. (1999). *Monte Carlo statistical methods*. Springer-Verlag, New York.

Roberts, G.O. and Polson, N.G. (1994). On the geometric convergence of the Gibbs sampler. *J. Roy. Statist. Soc. B*, **56**, 377–384.

Roberts, G.O. and Rosenthal, J.S. (1997). Geometric ergodicity and hybrid Markov chains. *Elect. Comm. in Probab.*, **2**, 13–25.

Roberts, G.O. and Rosenthal, J.S. (1998). Markov chain Monte Carlo: some practical implications of theoretical results. *Can. J. Stat.*, **26**, 5–31 (with discussion).

Roberts, G.O. and Rosenthal, J.S. (1999). Convergence of slice sampler Markov chains. *J. Roy. Statist. Soc. B*, **61**, 643–660.

Roberts, G.O. and Sahu, S.K. (1997). Updating schemes, correlation structure, blocking and parametrization for the Gibbs sampler. *J. Roy. Statist. Soc. B*, **59**, 291–317.

Roberts, G.O. and Smith, A.F.M. (1994). Simple conditions for the convergence of the Gibbs sampler and Metropolis-Hastings algorithms. *Stochastic Process. Appl.*, **49**, 207–216.

Roberts, G.O. and Tweedie, R.L. (1996). Geometric convergence and central limit theorems for multidimensional Hastings and Metropolis algorithms. *Biometrika*, **83**, 96–110.

Rosenthal, J. (1999). A review of asymptotic convergence for general state space Markov chains. Technical Report No. 9903, Department of Statistics, University of Toronto.

Ross, S.M. (1996). Bayesians should not resample a prior sample to learn about the posterior. *The American Statistician*, **50**, 116.

Royden, H.L. (1968). *Real analysis*. Macmillan, New York.

Rubin, D.B. (1988). Using the SIR algorithm to simulate posterior distributions. In *Bayesian statistics 3* (ed. J.M. Bernardo, D.V. Lindley and A.F.M. Smith) pp. 395–402. Oxford University Press.

Rubinstein, R.Y. (1981). *Simulation and the Monte Carlo method*. Wiley, New York.

Rubinstein, R.Y. (1986). *Monte Carlo optimization, simulation and sensitivity of queueing networks*. Wiley, New York.

Sacks, J. and Ylvisaker, D. (1970). Statistical design and integral approximation, 115–135, Proc. 12th Bien. Sem. Canad. Math. Cong., Canad. Math. Soc.

Sacks, J. and Ylvisaker, D. (1985). Model robust design in regression: Bayes theory. Proceedings of the Berkeley Conference in Honor of Jerzy Neyman and Jack Kiefer, 2 (ed. L. LeCam and R. Olshen) pp. 667–679, Wadsworth.

Schervish, M. and Carlin, B.P. (1992). On the convergence of successive substitution sampling. *J. Comp. Graph. Stat.*, **1**, 111–127.

Serfling, R.J. (1980). *Approximation theorems of mathematical statistics*. Wiley, New York.

Shaw, J.E.H. (1988). A quasirandom approach to integration in Bayesian statistics. *Ann. Statist.*, **16**, 895–914.

Shaw, J.E.H. (1999). Some numerical integration diagnostics in Bayesian analysis. In *Bayesian statistics 6* (ed. J.M Bernardo, J.O. Berger, A.P. Dawid and A.F.M. Smith). Oxford University Press.

Siegel, A.F. and O'Brien, F. (1985). Unbiased Monte Carlo integration methods with exactness for low order polynomials. *SIAM J. Sci. Stat. Comput.*, **6**, 169–181.

Siegel, C.L. (1980). *Lectures on the geometry of numbers*. Springer-Verlag, Berlin.

Silverman, R.A. (1974). *Complex analysis with applications*. Prentice-Hall, Englewood Cliffs, New Jersey.

Sinclair, A. (1993) *Algorithms for random generation and counting: a Markov chain approach*. Birkhäuser, Boston.

Sloan, I.H. and Joe, S. (1994). *Lattice methods for multiple integration*. Oxford University Press.

Smith, A.F.M. and Roberts, G.O. (1993). Bayesian computation via the Gibbs sampler and related Markov chain Monte Carlo methods. *J. R. Statist. Soc. B*, **55**, 3–23.

Smith, A.F.M., Skene, A.M., Shaw, J.E.H. and Naylor, J.C. (1987). Progress with numerical and graphical methods for practical Bayesian statistics. *The Statistician*, **36**, 75–82.

Smith, A.F.M., Skene, A.M., Shaw, J.E.H., Naylor and Dransfield, M. (1985). The implementation of the Bayesian paradigm. *Commun. Statist.-Theor. Meth.*, **14**, 1079–1102.

Smith, R.L. (1984). Efficient Monte Carlo procedures for generating points uniformly distributed over bounded regions. *Oper. Res.*, **32**, 1298–1308.

Spanier, J. (1979). A new family of estimators for random walk problems. *J. Inst. Math. Appl.*, **23**, 1–31.

Spanier, J. and Maize, E.H. (1994). Quasi-random methods for estimating integrals using relatively small samples. *SIAM Review*, **36**, 18–44.

Stein, E. and Weiss, G. (1971). *Introduction to Fourier analysis on Euclidean spaces*. Princeton University Press, NJ.

Stein, M. (1987). Large sample properties of simulations using latin hypercube sampling. *Technometrics*, **29**, 143–151.

Stewart, G.W. (1980). The efficient generation of random orthogonal matrices with an application to condition estimators. *SIAM J. Numer. Anal.*, **17**, 403–409.

Stroud, A.H. (1971). *Approximate calculation of multiple integrals*. Prentice-Hall, Englewood Cliffs, New Jersey.

Stroud, A.H. and Secrest, D.H. (1966). *Gaussian quadrature formulas*. Prentice-Hall, Englewood Cliffs, New Jersey.

Swendsen, R.H. and Wang, J.-S. (1987). Nonuniversal critical dynamics for the Ising model. *Phys. Rev. Lett.*, **58**, 86–88.

Szegö, G. (1985). Orthogonal polynomials. American Mathematical Society Colloquium Publications, Volume XXIII, American Mathematical Society.

Tang, B. (1993). Orthogonal array-based latin hypercubes. *J. Amer. Statist. Assoc.*, **88**, 1392–1397.

Tanner, M. (1993). *Tools for statistical inference. Observed data and data augmentation methods* (2nd edn). Springer-Verlag, New York.

Tanner, M.A. and Wong, W.H. (1987). The calculation of posterior distributions by data augmentation. *J. Amer. Statist. Assoc.*, **82**, 528–540.

Tchakaloff, V. (1957). Formules de cubature mécaniques à coefficients non négatifs. *Bull. Sci. Math.*, (2) **81**, 123–134.

Thisted, R.A. (1988). *Elements of statistical computing*. Chapman and Hall, London.

Tierney, L. (1994). Markov chains for exploring posterior distributions. *Ann. Statist.*, **22**, 1701–1728.

Tierney, L. (1995). A note on Metropolis–Hastings kernels for general state spaces. Technical Report No. 606, School of Statistics, University of Minnesota.

Tierney, L. and Kadane, J.B. (1986). Accurate approximations for posterior moments and marginal densities. *J. Amer. Statist. Assoc.*, **81**, 82–86.

Tierney, L., Kass, R. and Kadane, J.B. (1989). Fully exponential Laplace approximations to expectations and variances of nonpositive functions. *J. Amer. Statist. Assoc.*, **84**, 710–716.

Torrie, G.M. and Valleau, J.P. (1977). Nonphysical sampling distributions in Monte Carlo free energy estimation: umbrella sampling. *Jour. Comp. Phys.*, **23**, 187–199.

Traub, J.F. and Wozniakowski, H. (1991). Information-based complexity: new questions for mathematicians. *Math. Int.*, **3**, 34–43.

van Dooren, P. and de Ridder, L. (1976). Algorithm 6. An adaptive algorithm for numerical integration over an n-dimensional cube. *J. Comput. Appl. Math.*, **2**, 207–217.

Verdinelli, I. and Wasserman, L. (1991). Bayesian analysis of outlier problems using the Gibbs sampler. *Statistics and Computing*, **1**, 105–117.

Wakefield, J. (1992). Comment on Hills and Smith (1992). In *Bayesian statistics 4* (ed. J.M. Bernardo, J.O. Berger, A.P. Dawid and A.F.M. Smith) pp. 243–244. Oxford University Press.

Walker, A.J. (1974). New fast method for generating discrete random numbers with arbitrary frequency distributions. *Electronics Lett.*, **10**, 127–128.

Walker, A.J. (1977). An efficient method for generating discrete random variables with general distributions. *ACM Transaction Mathematical Software*, **3**, 253–256.

Wang, M.C. and Kennedy, W.J. (1994). Self validating computations of probabilities for selected central and non-central univariate probability functions. *J. Amer. Statist. Assoc.*, **89**, 878–887.

West, M. (1992). Modelling with mixtures. In *Bayesian statistics 4* (ed. J.M. Bernardo, J.O. Berger, A.P. Dawid and A.F.M. Smith) pp. 503–524 (with discussion). Oxford University Press.

West, M. (1993). Approximating posterior distributions by mixtures. *J. Roy. Statist. Soc. B*, **55**, 409–422.

Williamson, R.E., Crowell, R.H. and Trotter, H.F. (1968). *Calculus of vector functions*. Prentice-Hall, Englewood Cliffs, New Jersey.

Wolfram, S. (1996) *The Mathematica Book*. Cambridge University Press.

Wolpert, R.L. (1990). Monte Carlo integration in Bayesian statistical analysis. In *Statistical multiple integration* (ed. N. Flournoy and R.K. Tsutakawa) pp. 101–116. *Contemporary Mathematics*, **115**, Amer. Math. Soc.

Wong, R. (1989). *Asymptotic approximations of integrals*. Academic Press, San Diego.

Wozniakowski, H. (1991). Average case complexity of multivariate integration. *Bull. AMS*, **24**, 185–194.

Yakowitz, S., Krimmel, J.E. and Szidarovsky, F. (1978). Weighted Monte Carlo integration. *SIAM J. Numer. Anal.*, **15**, 1289–1300.

Zellner, A. and Min, C. (1995). Gibbs sampler convergence criteria. *J. Amer. Statist. Assoc.*, **90**, 921–927.

Zhang, P. (1996). Nonparametric importance sampling. *J. Amer. Statist. Assoc.*, **91**, 1245–1253.

Author index

Subject index